METHODS IN CELL PHYSIOLOGY

VOLUME II

Methods in
Cell Physiology

Edited by

DAVID M. PRESCOTT

DEPARTMENT OF ANATOMY
UNIVERSITY OF COLORADO MEDICAL CENTER
DENVER, COLORADO

VOLUME II

1966

ACADEMIC PRESS · New York and London

ACADEMIC PRESS INC.
111 Fifth Avenue, New York, New York 10003

United Kingdom Edition published by
ACADEMIC PRESS INC. (LONDON) LTD.
Berkeley Square House, London W.1

LIBRARY OF CONGRESS CATALOG CARD NUMBER: 64–14220

PRINTED IN THE UNITED STATES OF AMERICA

LIST OF CONTRIBUTORS

Numbers in parentheses indicate the pages on which the authors' contributions begin.

S. Aaronson, Haskins Laboratories, New York, New York (217)

Herman Baker, Haskins Laboratories, New York, New York (217)

D. P. Evenson, Department of Anatomy, University of Colorado Medical Center, Denver, Colorado (131)

Charles J. Flickinger, Department of Anatomy, University of Colorado Medical Center, Denver, Colorado (311)

Oscar Frank, The New Jersey College of Medicine and Dentistry, Jersey City, New Jersey (217)

Morton E. Freiman, Department of Microbiology and Immunology, Albert Einstein College of Medicine, Bronx, New York (93)

Joseph G. Gall, Department of Biology, Yale University, New Haven, Connecticut (37)

Seymour Gelfant, Department of Zoology, Syracuse University, Syracuse, New York (359)

S. H. Hutner, Haskins Laboratories, New York, New York (217)

Thomas J. King, The Institute for Cancer Research, Philadelphia, Pennsylvania (1)

H. Kroeger, Zoologisches Institut, Eidgenössische Technische Hochschule, Zürich, Switzerland (61)

Joseph J. Maio, Department of Cell Biology and the Unit for Research in Aging, Albert Einstein College of Medicine, Bronx, New York (113)

Philip I. Marcus, Department of Microbiology and Immunology, Albert Einstein College of Medicine, Bronx, New York (93)

D. M. Prescott,[1] Department of Anatomy, University of Colorado Medical Center, Denver, Colorado (131)

M. V. N. Rao,[2] Department of Anatomy, University of Colorado Medical Center, Denver, Colorado (131)

Miriam M. Salpeter, Laboratory of Electron Microscopy, Department of Engineering Physics, Cornell University, Ithaca, New York (229)

Carl L. Schildkraut, Department of Cell Biology and the Unit for Research in Aging, Albert Einstein College of Medicine, Bronx, New York (113)

E. J. Stadelmann, Department of Plant Pathology and Physiology, University of Minnesota, St. Paul, Minnesota (143)

---

[1] Present address: Institute of Developmental Biology, University of Colorado, Boulder, Colorado.

[2] Permanent address: Department of Zoology, Andhra University, Waltair, India.

A. R. STEVENS, Department of Anatomy, University of Colorado Medical Center, Denver, Colorado (255)

G. E. STONE, Department of Anatomy, University of Colorado Medical Center, Denver, Colorado (131)

MAURICE SUSSMAN, Department of Biology, Brandeis University, Waltham, Massachusetts (397)

J. D. THRASHER,[3] Department of Anatomy, University of Colorado Medical Center, Denver, Colorado (131, 323)

A. C. ZAHALSKY, Haskins Laboratories, New York, New York (217)

---

[3] Present address: Department of Anatomy, The Center for the Health Sciences, University of California, Los Angeles, California.

PREFACE

Much of the information on experimental techniques in modern cell biology is scattered in a fragmentary fashion throughout the research literature. In addition, the general practice of condensing to the most abbreviated form materials and methods sections of journal articles has led to descriptions that are frequently inadequate guides to techniques. The aim of this volume is to bring together into one compilation complete and detailed treatment of a number of widely useful techniques which have not been published in full detail elsewhere in the literature.

In the absence of firsthand personal instruction, researchers are often reluctant to adopt new techniques. This hesitancy probably stems chiefly from the fact that descriptions in the literature do not contain sufficient detail concerning methodology; in addition the information given may not be sufficient to estimate the difficulties or practicality of the technique or to judge whether the method can actually provide a suitable solution to the problem under consideration. The presentations in this volume are designed to overcome these drawbacks. They are comprehensive to the extent that they may serve not only as a practical introduction to experimental procedures but also to provide, to some extent, an evaluation of the limitations, potentialities, and current applications of the methods. Only those theoretical considerations needed for proper use of the method are included.

The book may be particularly useful to those working with intact cells; several chapters deal with culturing and experimental manipulation of a variety of cell types. There are numerous descriptions of techniques designed for use with single cells and of procedures for application to cellular activities in relation to the cell life cycle. Two chapters are devoted to technical advances in high-resolution autoradiography, particularly at the electron microscope level.

Finally, special emphasis has been placed on inclusion of much reference material in order to guide readers to early and current pertinent literature.

June, 1966   DAVID M. PRESCOTT

CONTENTS

8. Culture Media for Euglena gracilis

S. H. Hutner, A. C. Zahalsky, S. Aaronson, Herman Baker, and Oscar Frank

9. General Area of Autoradiography at the Electron Microscope Level

Miriam M. Salpeter

10. High Resolution Autoradiography

A. R. Stevens

CONTENTS OF PREVIOUS VOLUME

METHODS IN CELL PHYSIOLOGY

VOLUME II

Chapter 1

Nuclear Transplantation in Amphibia

THOMAS J. KING

The Institute for Cancer Research, Philadelphia, Pennsylvania

I. Introduction

It has been realized for some time that the most direct way to obtain evidence concerning the role of the nucleus and cytoplasm in embryonic differentiation would be to transfer these components from one cell type to another (Ephrussi, 1951; Hämmerling, 1934; Lopashov, 1945; Lorch and Danielli, 1950; Rauber, 1886; Rostand, 1943; Schultz, 1952; Spemann, 1938). However, this approach has been technically difficult to realize because of the fragility of living nuclei, which are usually damaged if merely touched with operating instruments and frequently killed when exposed to media other than that of the intact cytoplasm. Nevertheless, nuclei have been successfully transferred from cell to cell in the amoeba (Comandon and de Fonbrune, 1939) and amphibia (Briggs and

1

King, 1952), as well as in the ciliate *Stentor* (Tartar, 1953, 1961), the unicellular alga *Acetabularium* (Zetsche, 1962), the honeybee (DuPraw, 1963), and the mold *Neurospora crassa* (Wilson, 1963). Thus the success of nuclear transplantation in both unicellular and multicellular organisms indicates that, with improved methodology, it should become feasible to make nuclear transfers in other forms.

The first demonstration that nuclear transplantation could be performed successfully was reported by Comandon and de Fonbrune (1939) with *Amoeba sphaeronucleus*. When host and donor amoebae are held in apposition, in a hanging drop chamber, the nucleus of one cell can be pushed into the cytoplasm of another by means of a blunt microprobe. The subsequent utilization of this technique, in numerous studies of nucleo-cytoplasmic interactions in this form, has recently been reviewed by Danielli (1960) and Goldstein (1963, 1964). In 1952 a different method of nuclear transplantation was devised for the frog *Rana pipiens* (Briggs and King, 1952). In this form, the method consists of transplanting nuclei of different cell types of the developing embryo back into an egg, the nucleus of which has been removed. This has the theoretical advantage, over the procedures used in unicellular organisms, of testing the properties of nuclei by combining them with cytoplasm primed to develop into a complete organism with a diversity of specialized cell types.

In this chapter, the details of the methods of nuclear transplantation devised for amphibia will be described, and the types of problem to which they have been applied will be discussed.

II. Nuclear Transplantation in Anurans

A. *Rana*

The nuclear transplantation procedure developed for the leopard frog *Rana pipiens* is carried out according to the following plan. First an adult female is injected with 2 or 3 macerated frog pituitaries, and then left for $1\frac{1}{2}$–2 days at 18°C (see Hamburger, 1960). When the eggs have been ovulated and are packed in the uteri, a few are stripped into a sperm suspension made up in 10% amphibian Ringer's solution. After 10 minutes the eggs are flooded with the same medium used for the sperm suspension, and allowed to develop at 18°–20°C. If, as is usually the case, 90% or more of the eggs develop normally, they are used at the appropriate developmental stage to provide nuclei for transplantation. Eggs of a given female can be used for several days, provided she is kept at 4°–5°C after ovulation is completed and allowed to come to room temperature (18°C) before stripping. When donor embryos have reached the desired stage, the nuclear transplantation procedure is carried out in two separate operations (Fig. 1) (Briggs and King, 1953).

Nuclear Transplantation in *Rana*

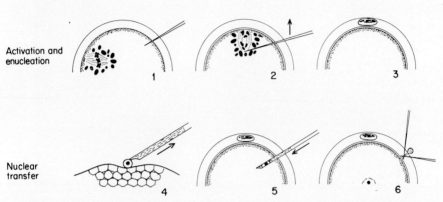

Activation and enucleation

Nuclear transfer

FIG. 1.   Diagram illustrating the steps in the procedure for transplanting embryonic nuclei into enucleated eggs of the frog *Rana pipiens*. In actual dimensions the maturation spindle of the egg and the micropipette are much smaller than indicated in this diagram. The internal diameter of the transfer pipette varies from 10μ to 50μ, while the diameter of the recipient egg is approximately 1800μ. See text (Section II,A, 1 and 2) for description.

1. REMOVAL OF THE EGG NUCLEUS

In order to assess the results of amphibian nuclear transplantation, it was first necessary to develop a completely dependable technique for removing the egg nucleus. This was accomplished by using a modification of Porter's technique for producing androgenetic haploids (Porter, 1939). Eggs from an ovulating female first are stripped into rows on a clean, dry microscope slide. The slide is then submerged in a 90-mm Petri dish containing 10% amphibian Ringer's solution. If the eggs do not remain firmly attached to the slide after immersion, an alternate method is to moisten a clean, flat Syracuse dish with just enough 10% Ringer's solution to cover the bottom of the dish. Eggs from an ovulating female are stripped into the dish, and carefully flooded with 10% Ringer's about 10 minutes *after* activation (Fig. 2A). In either case the eggs are immediately stimulated parthenogenetically by pricking, at some point in the animal hemisphere other than the animal pole area, with a clean glass needle. The majority of eggs so treated will be activated, and within a few minutes will begin to rotate so that the animal pole is uppermost

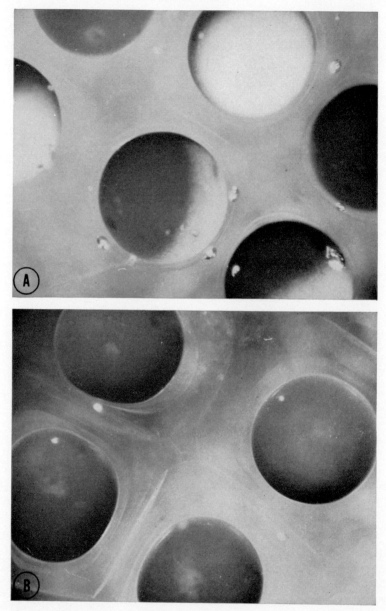

Fig. 2. A: Photograph showing appearance of eggs of *Rana pipiens* immediately after being stripped from an ovulating female. B: Photograph showing animal hemisphere of rotated eggs about 10 minutes after activation with a clean glass needle.

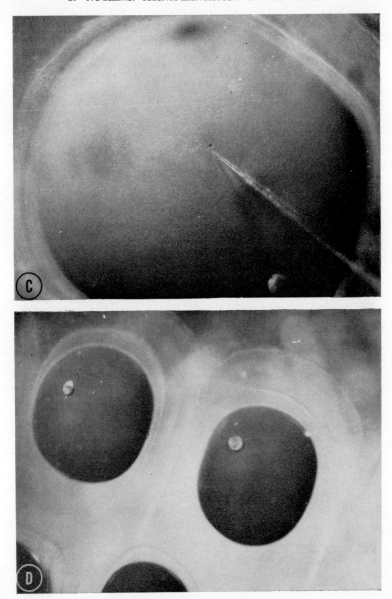

FIG. 2. (*Cont'd.*) C: Photograph showing animal hemisphere of egg about 15 minutes after activation. The small "black dot," which localizes the position of the egg nucleus, is seen immediately in front of the tip of the needle used for enucleation. D: Photograph showing appearance of an enucleated egg approximately 30 minutes after activation and enucleation. The exovate which contains the egg nucleus is trapped in the inner jelly coat surrounding the egg.

(Fig. 2B). Within 10–15 minutes after activation, at 18°C, the cortical granules have migrated away from the area over the nucleus, at the center of the animal pole; this allows the operator to locate the exact position of the egg nucleus, which appears as a small but distinct *black dot* (Fig. 2C). This appearance is due to the absence of light-reflecting pigment granules over the second maturation spindle.

The second part of the procedure, the enucleation operation, requires a magnification of about 60 diameters and a bright light to reveal clearly the "black dot." Eggs with distinct black dots conveniently located are enucleated by inserting a second glass needle diagonally through the jelly membranes, vitelline membrane, and surface coat of the egg, slightly to one side of and directly beneath the dot. By moving the needle straight up through the far side of the dot and the overlying surface coat of the egg, an exovate is produced which always contains the egg nucleus (Fig. 2D). Exovates formed in this manner are trapped in the inner jelly layer that surrounds the egg. The black dot is visible for about 10–20 minutes after activation, at 18°C, whereupon it disappears, signaling the completion of the second maturation division. When the enucleation operation is done carefully, 20 or more eggs can be prepared out of each group of 100–150 eggs activated.

This part of the method presents no problems. With practice, one can be sure that virtually all of the eggs operated on will be enucleated. After enucleating the desired number of eggs, the outer jelly layers are removed with watchmaker's forceps and fine scissors. The activated and enucleated eggs are then ready to receive transplanted nuclei.

2. TRANSPLANTATION OF NUCLEI INTO ENUCLEATED EGGS

While the enucleated eggs are being prepared, the donor tissue whose nuclei are to be tested is dissociated into individual cells. Blastula or early gastrula tissue (St. 8 to St. 10, Shumway, 1940), placed in amphibian Ringer's or Niu-Twitty solution (Niu and Twitty, 1953) (see Table I), can be dissociated into individual cells mechanically with fine glass needles. A less damaging isolation procedure is to expose the tissue to media lacking the divalent ions calcium and magnesium, e.g., modified Niu-Twitty or modified Steinberg's solution (modified from Steinberg, 1957) (see Table I). Within 10–20 minutes after exposure to calcium- and magnesium-free media, individual cells round up and separate completely from their neighbors.

In post-gastrula embryos it is more difficult to mechanically isolate individual cells of known type without injury. This difficulty has been overcome by the use of the proteolytic enzyme, trypsin, and the chelating agent, Versene (King and Briggs, 1955). For example, the neural plate

of a mid-neurula embryo (St. 14, Shumway, 1940) is first teased away
from the underlying mesoderm and endoderm in sterile 0.5% trypsin
(1-300, Nutritional Biochemicals Corp., Cleveland, Ohio), made up in
modified Niu-Twitty or modified Steinberg's solution. In this solution the
different tissues of the embryo can be cleanly separated from each other,
apparently as a result of a preferential effect of the enzyme on the
material cementing the layers together (Moscona, 1962). The enzyme
in this concentration does not readily dissociate individual cells of amphibian tissues. To accomplish this, a portion of the isolated tissue is
rinsed in modified Niu-Twitty or modified Steinberg's solution, and then
placed in sterile $1 \times 10^{-3} M$ Versene (ethylenedinitrilotetraacetic acid
disodium salt, Eastman Kodak Co., Rochester, New York) made up in
modified Niu-Twitty or modified Steinberg's medium. Within 5 minutes
individual cells begin to round up. After a 10–20-minute exposure to
Versene, dissociated cells are rinsed in Niu-Twitty or Steinberg's solution,
and placed on the pigmented surface of the animal hemisphere portion
of a late blastula in a wax-bottomed operating dish (60-mm Syracuse
dish) containing regular Niu-Twitty or Steinberg's solution. The operating dish is prepared by mixing lampblack with melted paraffin and a
small amount of beeswax. The melted mixture is poured into a Syracuse
dish, and anchored to the bottom of the dish by means of two 3-mm
glass rods of about 20-mm length. The soft surface coat of the ruptured
blastula serves admirably as an operating platform on which to keep the
donor cells isolated, and permits manipulations to be carried out without
damaging the cells or the tip of the injection pipette. A previously prepared enucleated host egg, rinsed in Niu-Twitty or Steinberg's solution,
is placed adjacent to the blastula platform, in a hemispherical depression
of the operating dish. An alternative method is to use a siliconed Romi-
cron slide[1] as an operating dish. Hemispherical depressions to hold the
host eggs are drilled into the glass surface of the well-like depression of
the slide. Then the entire surface of the well is siliconed. In this case,
donor cells are placed directly on the siliconed surface adjacent to the
hemispherical depressions of the well.

There is no indication that the trypsin-Versene treatment damages the
cells. When returned to regular media, the cells begin to reaggregate
within 30 minutes and continue their normal differentiation as explants.
Furthermore, control transfers of blastula nuclei can be successfully performed directly in $10^{-3}–10^{-4} M$ Versene without apparent damage to the
nuclei. The great advantage of this procedure is that it makes possible an
accurate separation of the closely knit tissues of advanced embryos, thus

[1] Paul Rosenthal, 505 Fifth Avenue, New York, New York.

allowing isolated cells of known origin to be obtained without cellular or nuclear damage. This greatly facilitates the transplantation of nuclei from advanced cell types (King and Briggs, 1955).

The second part of the operation, the transplantation of the donor cell nucleus into the enucleated egg, is done under a magnification of 90–216 diameters, using a dissecting microscope fitted with a foot-focusing attachment (Fig. 3A and B).

The donor cell is carefully drawn up into the mouth of a thin-walled glass micropipette, the lumen of which is somewhat smaller than the diameter of the cell. The micropipette is held in a Leitz microinjection

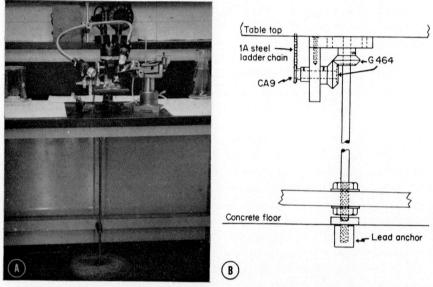

FIG. 3. A: Photograph of the apparatus used in carrying out nuclear transplantation. The transformer and variac mounted on the shelf behind the microscope control the two lights required to illuminate the operating field. The chain on the left focusing knob of the microscope is attached to the foot-focusing apparatus below the table. B: Assembly of foot-focusing apparatus. The ladder chain on the left focusing knob of the microscope (Fig. 3A) is mounted on a 2-inch steel sprocket (CA 34), and passes through a slit in the table top to the foot-focusing device below. The lower end of the chain is attached to a ½-inch diameter sprocket (CA 9), which in turn is attached to a brass bevel gear (G 464). A second bevel gear engages the first and in turn is attached to a rod leading to a horizontal wheel mounted close to the floor. The microscope is focused by moving the wheel with either foot. The numbers shown are the catalogue numbers of parts obtained from the Boston Gear Works (Boston, Mass.). This is a variation of a foot-focusing apparatus used in the Biology Department of Swarthmore College (Swarthmore, Pa.). The current model was designed and built by Mr. William Hafner, Machinist, The Institute for Cancer Research (Philadelphia, Pa.).

apparatus, supported in either an Emerson or Leitz micromanipulator, and connected by means of $\frac{1}{8}$-inch bore plastic tubing to a 2-ml hypodermic syringe. The system may be filled with air or water. If air is used, all of the system except the tip of the injection pipette is filled with air. If water is used, the water column in the plastic tubing should not be allowed to enter the hollow metal holder that carries the pipette. The tip of the pipette contains a small column of the injection medium that is drawn up with the donor cell. Provided the pipette is really clean, the movements of the fluid column can be controlled accurately. A pipette is cleaned by placing it in a Leitz micropipette apparatus to which is attached a $3\frac{1}{4}$-inch rubber bulb. The hand-held rubber bulb is used to draw up and expel hot acid (a mixture of 95% sulfuric acid and 5% nitric acid) from the tip of the micropipette. Following acid cleaning, the pipette is rinsed several times in hot distilled water and tested for neutrality by means of the indicator, bromthymol blue.

A single cell, gently drawn into the tip of a micropipette, is compressed and distorted in such a way as to break the cell surface without dispersing the cytoplasm in the immediate vicinity of the nucleus. This part of the procedure is quite difficult to do without diluting the cytoplasm with the medium, thereby damaging or killing the nucleus. However, when this is properly done the translucent nucleus can be seen near the center of the broken cell. The pipette, carrying the broken cell, is then inserted into the center of the previously prepared enucleated egg at a point along the animal-vegetal axis roughly one third of the distance from the animal pole. The donor nucleus is liberated into the egg cytoplasm by carefully injecting the broken cell with a minimal amount of fluid.

The injection is controlled by watching the meniscus of the fluid column within the pipette. The broken cell is kept near the tip of the pipette while the meniscus of the column is higher but still within the field of the microscope. Following the injection the pipette is slowly withdrawn from the egg. As it is withdrawn, it usually pulls the surface coat up against the vitelline membrane so that a small canal is formed through which the egg substance may leak. This can be prevented by severing the connection between the egg surface and the vitelline membrane with glass needles. The egg is then removed from the operating dish and placed in a small Stender dish containing spring water.

The principal criterion of the success of this method is that it should give normal development in experiments in which nuclei of undetermined cells are transferred into enucleated eggs. To meet this requirement, transplantation tests were first carried out with nuclei of late blastulae (Briggs and King, 1952). When such nuclei are transplanted, one third or more of the recipient eggs cleave completely (Fig. 4A and

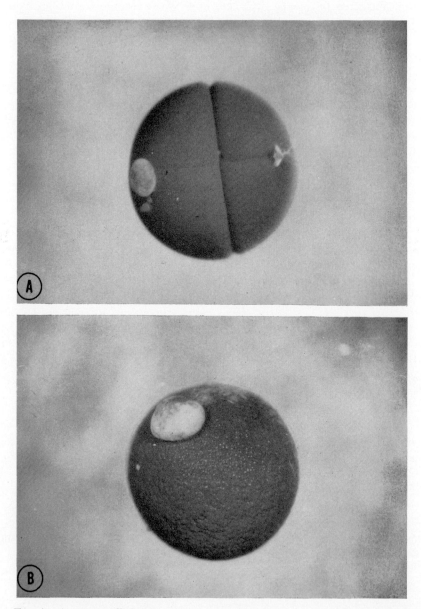

Fig. 4. A: Four-celled stage of an enucleated *Rana pipiens* egg injected with a late blastula nucleus. B: Normal blastula of *Rana pipiens* resulting from the transplantation of a late blastula nucleus.

B) and continue to develop normally through embryonic and larval life (Fig. 4C). If one wishes to maintain such nuclear-transplant larvae, many of them will metamorphose into young frogs (Fig. 4D). Early gastrula nuclei give essentially the same results (King and Briggs, 1955); however, the possibility that some of these nuclei may possess a limited ability to promote the differentiation of growing oocytes in the meta-morphosing frog has not been excluded (Briggs and King, 1960).

FIG. 4. (*Cont'd.*) C: Normal nuclear-transplant larva. Donor nucleus from the animal pole region of a late blastula. D: Nuclear-transplant frogs (*Rana pipiens*). Each individual developed from an enucleated egg which received a transplanted blastula nucleus. Trunk length equals 30 mm.

These experiments demonstrated two things: (1) that it is technically feasible to transplant living nuclei of amphibian cells in an undamaged condition; (2) that blastula and early gastrula nuclei, capable of promoting completely normal development, are undifferentiated and as such are equivalent to the zygote nucleus at the beginning of development. These results have been thoroughly confirmed in *Rana pipiens* by Subtelny (1958), Moore (1958b), McKinnell (1960), Grant (1961), Hennen (1963), and Markert and Ursprung (1963). Comparable results have been obtained in several other *Rana* species: *Rana nigromaculata* (Sambuichi, 1957, 1961), *Rana temporaria* and *Rana arvalis* (Stroeva and Nikitina, 1960; Nikitina, 1964).

3. Controls for the Nuclear Transplantation Procedure

When frogs' eggs are simply pricked with a clean glass needle they rotate, form the ephemeral black dot which localizes the second maturation spindle, and, within a few hours at 18°C, show puckering of the surface coat or abortive and irregular cleavage furrows. Within 10 hours after activation the abortive furrows fade away and the eggs begin to cytolyze. There are no signs of normal cleavage or blastula formation. By contrast, eggs which are pricked and then enucleated fail to show any signs of cleavage. When observed at 5–10 hours after activation these eggs, at 18°C, show none of the surface activity present in practically all of the pricked eggs at this time. This difference in behavior of the two types of eggs provides a convenient criterion for determining how reliable the enucleation operation is, and in our hands has indicated that at least 99% of the attempted enucleations are in fact successful.

Another estimate of the success of the enucleation operation is obtained by inseminating eggs and then enucleating them. This is done at the end of each transfer experiment. From such eggs one obtains androgenetic haploids if the operation is successful, and diploids when it is not. Invariably these eggs develop into androgenetic haploids in nearly 100% of the cases. This again demonstrates the enucleation procedure to be completely dependable.

4. The Transplantation Medium

As has been stated previously (Section II,A,2), the method involves transplanting the cytoplasm as well as the nucleus into the enucleated egg. This does not, however, bring about a significant dilution of the egg cytoplasm, since the volume of cytoplasm that must be injected along with the nucleus is minute compared with the volume of the recipient egg (1:40,000 or less). Nevertheless, the necessity of including it introduces into the method a complication which must be considered. This complication is not, however, a serious one, since enucleated eggs into which endoderm cytoplasm of blastula or gastrula cells alone is injected fail to cleave, while the injection of donor cell cytoplasm (without a nucleus) into fertilized eggs has no effect on development (Briggs and King, 1957; Gurdon, 1960a). Therefore, there is no evidence that the donor cell cytoplasm by itself can either elicit cleavage in enucleated eggs, or alter the cleavage or differentiation of normally nucleated eggs.

A more important consideration is the protection that the donor cell cytoplasm affords the nucleus until the moment it is liberated into the enucleated egg. In the case of animal pole cells of St. 9 blastulae, the volume of cytoplasm is still relatively large $(7 \times 10^{-5}$ mm$^3)$ and apparently affords adequate protection to the nucleus in the course of the transplantation. Gastrula cells from comparable areas of the embryo have less than half this amount of cytoplasmic protection (Briggs and King, 1953). Except for endoderm (Briggs and King, 1957), the amount of cytoplasm surrounding the nucleus within the transfer pipette is considerably smaller in donor cells of older embryos; therefore, there is an increased likelihood that these nuclei might be damaged in the transfer procedure. Theoretically, the likelihood of damage to the smaller cells of advanced embryos could be reduced, either by improving the medium or by arranging the mechanics of the transfer procedure so that the surface of the donor cell is ruptured only after the cell is within the cytoplasma of the recipient egg. In attempts to accomplish the latter, various modifications of the construction of the transfer pipettes have been tried but so far none have been successful. The injection of unbroken cells into enucleated eggs, in the hope that the outer membranes of the cell might later break down and the nucleus be liberated into the egg cytoplasm, was also ineffective (Briggs and King, 1953). Successful nuclear transfers require that the donor cell surface be broken, and this is done while the cell is within the transfer pipette. Thus there is bound to be some dilution of the donor cell cytoplasm by the surrounding medium, and probably the only way to reduce the damage that this might cause would be to improve the medium.

When the method was first worked out, the Niu-Twitty modification of Holtfreter's solution was arbitrarily selected as a medium in which to carry out the transfer operation. This is a simple solution of the salts of Na, K, Ca, and Mg, plus bicarbonate and phosphate buffer (Table I). Numerous experiments have since been carried out in an attempt to find an improved medium. These include experiments with (1) various alterations of the salt mixture; (2) additions to the Niu-Twitty medium of ATP, NAD, glutathione, plasma extenders (dextran, oxypolygelatin, polyvinylpyrrolidone), bovine serum albumin, frog body cavity fluid, and coconut milk; (3) various sucrose-salt mixtures (Briggs and King, 1953); (4) a modification of Barth's medium, used by Moore (1960), which differs from Niu-Twitty saline in that it contains 1.5 times as much

TABLE I

MEDIA USED IN THE NUCLEAR TRANSPLANTATION PROCEDURES OF AMPHIBIA[a]

	Niu-Twitty	Modified Niu-Twitty
Solution A (500 ml)		
NaCl	3400 mg	2943 mg
KCl	50 mg	50 mg
$Ca(NO_3)_2 \cdot 4H_2O$	80 mg	—
$MgSO_4 \cdot 7H_2O$	197 mg	—
Solution B (250 ml)		
Na_2HPO_4	110 mg	1300 mg
KH_2PO_4	20 mg	116 mg
Solution C (250 ml)		
$NaHCO_3$	200 mg	200 mg

	Steinberg	Modified Steinberg
NaCl (17.0%)	20 ml	20 ml
KCl (0.5%)	10 ml	10 ml
$Ca(NO_3)_2 \cdot 4H_2O$ (0.8%)	10 ml	—
$MgSO_4 \cdot 7H_2O$ (2.05%)	10 ml	—
HCl (1.00 N)	4 ml	4 ml
Glass-distilled water	946 ml	966 ml
Tris(hydroxymethyl) aminomethane	560 mg	691 mg
Streptomycin sulfate	50 mg	—
Penicillin-G, sodium	50,000 units	—

[a] Solutions A, B, and C of the Niu-Twitty media, made up in glass-distilled water, are heat-sterilized separately and mixed when cool. Steinberg solutions are made up in one flask and autoclaved. The addition of antibiotics is optional; if they are added, a portion of the water is reserved for putting them into solution, just before use. The solution of antibiotics is sterilized by Seitz filtration and then added to the major portion of the medium.

NaCl; (5) modified Stearn's medium, used by Grant (1961); and (6) Steinberg's solution, which substitutes Tris-HCl for the phosphates and bicarbonate of the Niu-Twitty medium (Steinberg, 1957).

Transfers of nuclei from undifferentiated embryonic cells were performed in the usual way in each medium tested. The effect of each medium on the nuclei was then determined by observing the cleavage and development of the recipient eggs. Tests of media in categories (1),

(2), and (3) above were conducted on nuclei from the animal pole region of late blastulae. Although not exhaustive, these tests were adequate to show that none of these media gives results as good as those obtained with the Niu-Twitty salt solution. Transplantation tests involving the modifications of Barth's and Stearn's media, as well as Steinberg's solution, were done on the smaller undifferentiated cells of the animal pole area of early gastrulae (Table II) (St. 10, Shumway, 1940). In this series, each medium was tested on 75–150 recipient eggs (King and DiBerardino, unpublished work, 1966). Of these three media, the only one permitting a high proportion of the successful nuclear-transplant embryos to develop normally through larval stages was that suggested by Steinberg. In view of these results we use Steinberg's solution routinely as an operating medium (Table I).

5. CONSTRUCTION OF MICRONEEDLES AND MICROPIPETTES

The glass needle used for the enucleation procedure must have a tip fine enough to penetrate the jelly coats, vitelline membrane, and egg surface without appreciably deforming the egg, and yet must be stiff enough near the tip so that the exovate can be properly formed in the manner described (Section II,A,1). Appropriate microneedles can be conveniently and reproducibly fashioned from 1-mm standard flint glass rod[2] on the Livingston Needle Puller.[3]

As has been indicated, the most important single step in the transplantation operation is that in which the donor cell is pulled into the tip of the micropipette. If this is done carefully, and the pipette is of the right size and construction, the cell surface will be broken, leaving the contents otherwise undisturbed. Uniform bore glass capillary tubing of about 1-mm O.D.[2] can be drawn out by the Livingston machine into long thin-walled pipettes of 1–50μ (I.D.) as required. The drawn-out pipette is mounted in a metal sleeve and inserted into the spring clamp of a de Fonbrune microforge. The pipette holder is rotated until the tip of the micropipette is positioned vertically near the platinum-iridium filament of the microforge. Then the heating element is adjusted until the filament can be seen from the side in the microscope field, and the filament and micropipette are centered in the field of the horizontally mounted binocular microscope. A small glass bead, previously placed on the tip of the filament, is heated to a dull-red glow. The region of uniform bore and desired internal diameter is brought into contact with the glass bead, with just enough pressure to fuse the pipette to the

[2] Drummond Scientific Company, 524 N. 61st Street, Philadelphia, Pennsylvania.
[3] Designed by Dr. L. G. Livingston and built by Mr. Otto Hebel, Biology Department, Swarthmore College, Swarthmore, Pennsylvania.

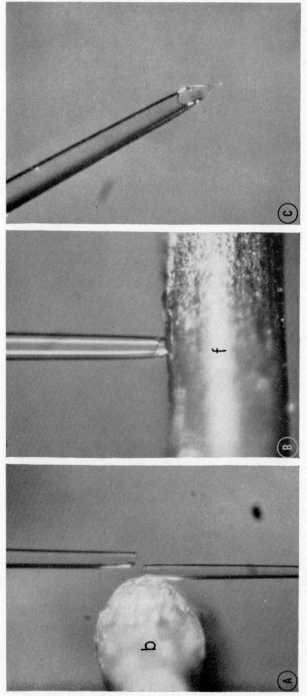

FIG. 5. A: A small glass bead (b), placed on the tip of the platinum-iridium filament of a de Fonbrune microforge, is heated to a dull-red glow. The tapered end of a micropipette, brought into contact with the molten bead, fuses with the bead. When the current is cut off quickly, the retraction of the rapidly cooled filament sections the pipette at the point of contact. B: To fashion the tip of a micropipette, one side of the sectioned pipette (Fig. 5A) is brought into contact with a flat, uncoated area of the filament (f). The final tip is formed by pulling the heated element away from the pipette to produce a sharp tip (Fig. 5C). C: A fine-pointed micropipette with an internal diameter at the orifice of 35μ.

molten bead without constricting the lumen of the pipette. The current is cut off quickly and the retraction of the rapidly cooled filament sections the pipette cleanly at the point of contact with the glass bead (Fig. 5A). One edge of the open tip of the pipette is then brought into contact with a flat, uncoated area of the platinum filament, which is heated to a somewhat higher temperature than that used for sectioning (Fig. 5B). The final tip is fashioned by pulling the heated filament away from the pipette to produce a sharp tip, whose sides are beveled like a hypodermic needle (Fig. 5C).

To eliminate most of the trial and error in selecting the appropriate size of pipette for nuclear transplantations, a series of measurements have

TABLE II

DIAMETERS OF ISOLATED DIPLOID EMBRYONIC CELLS (*Rana pipiens*)[a]

Stage of embryonic development	Cell type	Mean cell diameter (μ)
8 (mid-blastula)	Animal pole region	65
10 (early gastrula)	Animal pole region	42
12 (late gastrula)	Floor of archenteron	68
14 (mid-neurula)	Medullary plate	26
	Floor of mid-gut	49
17 (tail bud)	Floor of mid-gut	43

[a] Diploid embryonic cells of the frog *Rana pipiens* were isolated via the trypsin-versene technique described in the text (Section II,A,2). The mean diameters of approximately 100 free cells, from three or more specimens of each developmental stage, were measured with a Filar micrometer at a magnification of × 100. Micropipettes with an internal diameter of $\frac{1}{2}$–$\frac{3}{4}$ the diameter of the cell are used for nuclear transfers.

been made of the diameters of various diploid embryonic cells of the frog *Rana pipiens* (Table II). Cells were isolated by the trypsin-versene procedure (Section II,A,2) and were approximately spherical when measured.

B. Xenopus

Fischberg and his collaborators have applied the principle of this technique to the South African clawed toad *Xenopus* (Elsdale *et al.,* 1960; Gurdon, 1960a). The main modification of the technique as applied to *Xenopus* concerns the method of preparing the recipient eggs. First, the activation of the recipient egg is not carried out as a separate procedure. Second, the greater strength and elasticity of the vitelline membrane of *Xenopus* eggs preclude the use of mechanical enucleation as a reliable method of removing the egg pronucleus (Gurdon, 1960b). Therefore, the egg pronucleus is not removed but instead inactivated by irradiation. Nuclei of freshly laid unfertilized eggs are killed by a short exposure to ultraviolet irradiation.[4] This exposure also activates the egg. After irradiation the killed nucleus is allowed to degenerate in the egg cytoplasm. The cytoplasmic organization of the egg appears not to be damaged by the irradiation treatment. A beneficial secondary effect of the ultraviolet irradiation is that it weakens the external membranes of the egg, rendering them more easily penetrable by the injection pipette.

In *Xenopus*, as in *Rana*, the successful transplantation of blastula nuclei promotes unrestricted normal development of the recipient eggs (Fischberg *et al.*, 1958a). To prove that the nuclei of transplant-embryos are derived from the transplanted nucleus, and not from the recipient egg nucleus, a nuclear marker was used (Elsdale *et al.*, 1958). The nuclear marker consists of a spontaneous mutation, which presumably suppresses the nucleolar organizer region of the chromosome that forms the nucleolus in each haploid set of chromosomes (Kahn, 1962; Barr and Esper, 1963). It has been found that this mutation reduces the concentration of cytoplasmic ribonucleic acid (RNA) (Esper and Barr, 1964) by preventing the synthesis of ribosomal RNA (Brown and Gurdon, 1964). Individuals heterozygous for this mutation possess only one nucleolus per diploid cell. When donor nuclei from such individuals are transplanted into eggs from wild-type toads, nearly all the diploid transplant-embryos contain only one nucleolus in their nuclei. The nuclear marker is recognized in squash preparations of interphase nuclei of post-blastula transplant embryos by the use of phase-contrast microscopy. The availability of such a marker makes *Xenopus* a valuable form in which to combine the method of nuclear transplantation with genetic analyses of somatic cells. Furthermore, the ease with which this amphibian is raised makes it possible to rear adult, sexually mature toads from eggs receiving transplanted nuclei in less than 1 year under laboratory conditions (Gurdon *et al.*, 1958; Gurdon, 1962a,b).

III. Nuclear Transplantation in Urodeles

Early attempts to extend the method of nuclear transplantation to Urodele eggs met with limited success. Lopashov (1945) and later Waddington and Pantelouris (1953) transplanted diploid blastula nuclei into

---

[4] A 110-W medium-pressure Hanovia mercury arc lamp with a quartz condenser is used.

nonnucleated fragments of *Triton* eggs. A reasonable percentage of the injected blastomeres cleaved, but none developed beyond the blastula stage. Lehman (1955, 1957) repeated these experiments on whole eggs of *Triton palmatus* and *Triton alpestris*, and found that the method of simply pricking the egg with a glass needle, which works so well on the frogs' egg, fails to activate the *Triton* egg. This difficulty was partially circumvented by fertilizing eggs with sperm, the nucleus of which had been inactivated by X-rays. A second characteristic of Urodele eggs is that, like those of *Xenopus*, they cannot be enucleated reliably by

Nuclear Transplantation in Axolotl

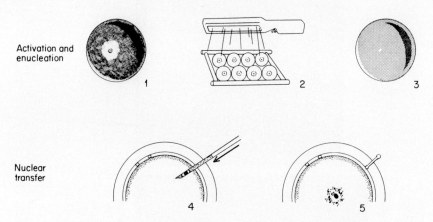

FIG. 6. Diagram showing steps in the procedure for transplanting embryonic nuclei into eggs of the Mexican axolotl (*Ambystoma mexicanum*). See text for details (Section III).

mechanical means (Gallien, 1960; Lehman, 1957; Pantelouris and Jacob, 1958). Lehman (1955) attacked this difficulty by making a small hole through the vitelline membrane and sucking out the egg nucleus by means of a micropipette. This procedure was found to be successful in only 50% of the cases and, as in previous attempts, the subsequent nuclear transfers never promoted the development of the test eggs beyond the late blastula stage.

A more effective method of excluding the nucleus of Urodele eggs has been worked out by Signoret *et al.* (1962), using eggs of the Mexican axolotl (*Ambystoma mexicanum*). Recipient eggs were prepared by exposing unfertilized eggs to a mild heat shock. It was subsequently found that the application of a mild electrical shock is the preferred method

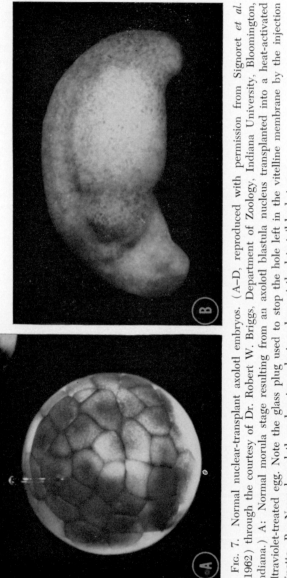

FIG. 7. Normal nuclear-transplant axolotl embryos. (A–D, reproduced with permission from Signoret *et al.* (1962) through the courtesy of Dr. Robert W. Briggs, Department of Zoology, Indiana University, Bloomington, Indiana.) A: Normal morula stage resulting from an axolotl blastula nucleus transplanted into a heat-activated ultraviolet-treated egg. Note the glass plug used to stop the hole left in the vitelline membrane by the injection pipette. B: Normal axolotl nuclear-transplant embryo at the late tail bud stage.

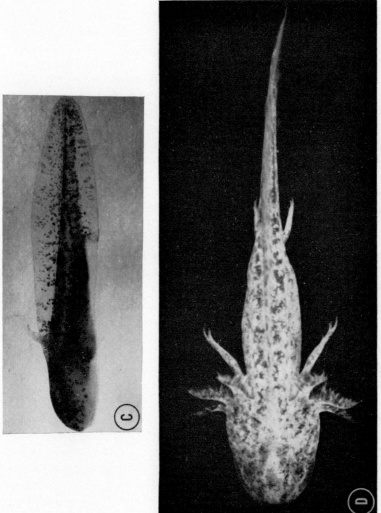

FIG. 7. (Cont'd.) C: Axolotl larva derived from transplanting a DD (or Dd) blastula nucleus into an egg from a dd female. The extensive distribution of black pigment cells is characteristic of the nuclear donor. D: Nuclear-transplant axolotl larva which developed from an egg from a dd female injected with a DD (or Dd) blastula nucleus. Age 51 days.

for activating the axolotl egg (Briggs *et al.*, 1964). After the egg is activated, the maternal chromosomes are inactivated by the ultraviolet irradiation technique known to be effective in *Xenopus* eggs (Section II,B).

The method by which successful nuclear transfers have been achieved in axolotl is illustrated in Fig. 6. Unfertilized eggs from an FSH-injected female are collected within 30–60-minute intervals after being spawned. Eggs, freed of their capsules, are pipetted into a finger bowl of charcoal-filtered tap water and activated by electrical shock. After activation the eggs are rinsed in 10% Steinberg's solution, and transferred in this solution into an irradiation chamber consisting of glass rods partially embedded in a wax-bottomed dish. The rods are arranged in the form of a rectangle of a size that will fit well within the beam of an ultraviolet lamp. To ensure equal exposure of the egg nuclei, which at this time are located directly beneath the egg surface at the animal pole, the fluid level of the chamber is lowered until the animal poles of the eggs are just exposed. The eggs are then subjected to the ultraviolet beam of a Mineralite lamp (Model SL-2537) for 4 minutes at a distance of 4 cm. Following irradiation, the eggs are transferred back into filtered water and, after 1½–2 hours, examined for activation under strong tangential illumination. Eggs showing a smooth surface, even pigment distribution over the animal hemisphere, and second polar body emission (Fig. 6-3) are judged to be activated and are set aside for nuclear transplantation.

Control experiments in which normally fertilized eggs are treated in the same manner develop into haploids, showing that this method of preparation effectively inactivates the chromosomes of the recipient eggs.

The validity of this method was tested by transplanting blastula nuclei into enucleated eggs between 2 and 4 hours after activation. In general, the transfer procedure is that described for *Rana* (Section II,A,2), with minor modifications. The principal modification concerns the handling of the axolotl egg immediately after injecting the donor nucleus. When the transfer pipette is withdrawn, the cytoplasm of the axolotl egg has a propensity to leak through the small hole left in the vitelline membrane by the micropipette. The most satisfactory method of controlling this is simply to plug the hole in the vitelline membrane by inserting a piece of glass thread, with rounded ends, about 25μ in diameter and 0.8 mm long (Fig. 7A). Stopping the hole in this manner prevents the leakage from continuing and appears not to interfere with the normal development of the test eggs. All operations are carried out under aseptic conditions.

After receiving a transplanted nucleus, the recipient egg is left overnight in the operating dish in Steinberg's solution. On the next day, while it is still in the blastula stage, it is transferred to sterile filtered water containing antibiotics (0.01% each of penicillin, streptomycin, and sulfadiazine).

Blastula nuclei carrying the dominant gene D (for black pigmentation pattern) were transplanted into enucleated eggs from a recessive white (dd) female. The majority of the recipient eggs cleaved (Fig. 7A) and developed into complete blastulae of normal appearance. Most of these blastulae completed gastrulation (Fig. 7B) and developed to larval stages (Fig. 7C). A total of 37 larvae were obtained. All showed the black phenotype (Fig. 7C and D), proving that the nuclear transplantations were successful.

As was noted for Xenopus (Section II,B), the axolotl too has several distinct advantages over Rana pipiens in nuclear transplantation studies. When reared in the laboratory, it reaches sexual maturity in 10–12 months. Furthermore, it is the best known genetically of all the amphibia. In addition to the "white" mutant, used to test the reliability of the nuclear transplantation method, seven other mutant genes are known to affect development at times varying from cleavage to early larval stages (Humphrey, 1960; Signoret et al., 1962). These genes should prove extremely useful as markers in nuclear transplantation studies of nuclear differentiation.

The method of axolotl nuclear transplantation has been applied to the newt Pleurodeles waltlii (Signoret and Picheral, 1962; Picheral, 1962; Signoret and Fagnier, 1962). As has been shown in Rana, Xenopus, and axolotl, the successful transplantation of blastula nuclei in Pleurodeles also promotes the complete differentiation of the test eggs. Like axolotl and Xenopus, Pleurodeles flourishes under laboratory conditions and has a sufficiently short generation time (18 months) to make it feasible to use this species in genetic analyses, even though there are as yet no mutants available.

IV. Differences in the Methods of Amphibian Nuclear Transplantation

Activation of the recipient egg is the most important single factor determining success of nuclear transplantation in amphibia. Nonactivated eggs never cleave following nuclear transfer. However, when activated eggs receive transplanted blastula nuclei, the cleavage of the recipient eggs approaches 100% in some experiments. This striking difference emphasizes the fact that the condition of the recipient egg cytoplasm is of prime importance in determining whether nuclear transfers succeed or fail. In the frogs' egg, successful activation is achieved by simply pricking the egg with a clean glass needle (Briggs and King, 1952, 1953). *Triton* (Lehman, 1955) and axolotl eggs fail to respond to this method of activation. Exposure to a temperature of 35°C for 5 minutes can successfully activate the axolotl egg but, as was noted above, the application of a mild electric shock is the preferred method for activating eggs of this form (Briggs *et al.*, 1964). In *Xenopus*, ultraviolet irradiation appears to be the most reliable way to both activate the egg and eliminate the egg nucleus (Elsdale *et al.*, 1960; Gurdon, 1960b, 1964).

The optimal time for removing or inactivating the egg nucleus also varies between eggs of different amphibian species. In the frogs' egg, the black dot indicating the position of the egg nucleus is visible for about 10–20 minutes after activation (Briggs and King, 1953). In order to be assured that the egg nucleus of the axolotl is inactivated while still at the egg surface, the ultraviolet irradiation must be performed within 5 minutes of the time when the eggs are activated (Signoret *et al.*, 1962). In *Xenopus*, if fertilized eggs are irradiated with ultraviolet more than 20 minutes after fertilization, the pronucleus of most eggs will have sunk below the surface and not be affected by the ultraviolet (Gurdon, 1960b).

Cytological observations of the behavior of blastula nuclei introduced into activated, nonenucleated *Rana pipiens* eggs show that after about 30 minutes a small yolk-free area, which later becomes the center of the astral system, develops around the transplanted nucleus (Subtelny and Bradt, 1963). During this time the introduced nucleus swells in size and begins to descend into the interior of the egg. About 120 minutes after activation, the yolk-free area surrounding the transplanted nucleus fades away and the asters of the first mitotic spindle appear on opposite sides of the nucleus. These data suggest that nuclear transfers into nonenucleated frogs' eggs should be performed within 90 minutes after activation at 18°C. By contrast, the *Xenopus* egg cleaves more rapidly than that of *Rana*. Since nuclei transplanted into host eggs of this species have been found to be at the anaphase of the first mitosis within 60 minutes after injection (Gurdon, 1960b), nuclear transfers should be performed well in advance of this time. The optimal time for nuclear transplantation in the axolotl egg is of longer duration between 2 and 4 hours following activation (Signoret *et al.*, 1962).

Thus the eggs of different amphibian species appear to require distinctly different methods of activation, different times for eliminating the recipient egg nucleus, and specific optimal times for receiving transplanted nuclei. Since these factors have been found to be crucial, it is evident that the precise conditions must be found before nuclear transplantation procedures can be expected to succeed in other forms.

V. Some Applications of the Methods of Amphibian Nuclear Transplantation

A. The Role of the Nucleus in Embryonic Differentiation

The techniques of nuclear transplantation in amphibia have been used principally to investigate the question of whether or not nuclei of somatic cells undergo stabilized genetic changes in the course of embryonic cellular differentiation. Earlier studies by Spemann (1914) and Seidel (1932) on newt and dragonfly eggs, respectively, indicated that nuclei remain identical and totipotent at least during the early cleavage stages (see review by Briggs and King, 1959). As we have seen in this chapter, the results of nuclear transplantation in several species of amphibia have extended this conclusion to include blastula nuclei. However, transplantation tests of nuclei from late gastrula and older embryonic stages of *Rana, Bufo, Xenopus, Pleurodeles,* and axolotl have shown that, as development proceeds, fewer and fewer of the nuclei promote the normal development of test eggs.

Most of the transfers with nuclei of later stages have been done on nuclei of endoderm cells. In *Rana pipiens,* an extensive series of tests carried out on endoderm nuclei of late gastrula, mid-neurula, and post-neurula stages (all from the anterior mid-gut region) indicated that a progressive nuclear change occurs during embryonic differentiation (Briggs and King, 1957). These changes were best expressed in embryos that gastrulated normally, but then showed deficiencies, especially in ectodermal derivatives, at later stages. Serial transplantation of late gastrula endoderm nuclei demonstrated that the nuclear condition responsible for these developmental deficiencies is highly stabilized and heritable (King and Briggs, 1956). Theoretically, we might expect that a nucleus which has undergone a specific differentiation would, when submitted to the nuclear transplantation test, promote normal differentiation of donor-type cells, and that other cell types would be absent or abnormal. Although most of the "endoderm" embryos that develop to post-neurula stages display a pattern of differentiation expected on the basis of specific nuclear differentiation, others do not. In order to provide such embryos with an opportunity to express their full range of potentialities for differentiation, representative embryos of both types were placed in parabiosis with normal embryos at the early gastrula stage (Briggs *et al.*, 1961). These experiments showed that the development of nuclear-transplant embryos is unaffected by parabiosis with normal embryos. Embryos which would have developed into abnormal gastrulae or post-gastrulae, if left undisturbed, displayed the same deficiencies when united with normal embryos; experimental embryos that would have developed normally, if permitted to do so, developed as well as the normal parabionts to which they were joined. However, a definite correlation was found between chromosome constitution and the pattern of deficiencies seen in histological sections. Abnormal nuclear-transplant parabionts not conforming to the "endoderm" pattern of deficiencies were found to have abnormal chromosome constitutions of varying degree and in histological sections exhibited deficiencies in endoderm as well as ectoderm and mesoderm derivatives, whereas abnormal embryos exhibiting a pattern of deficiencies consistent with the endodermal origin of their nuclei were found to have a euploid or predominantly euploid chromosome complement. Thus nuclear-transplant embryos with gross chromosome abnormalities do not provide a valid test of the properties of the nuclei as they exist in the donor cell prior to transplantation. The valid cases for the significance of the nuclear changes that may accompany embryonic development, are those in which the chromosome complement remains euploid following nuclear transfer, and then either develop normally, indicating that some late gastrula nuclei have undergone no stable differentiation, or become abnormal and exhibit a pattern of deficiencies consistent with the endodermal origin of their nuclei.

Endoderm nuclei taken from progressively older embryos of *Xenopus* also show a decrease in their ability to promote normal development after transplantation (Fischberg *et al.*, 1958b; Fischberg and Blackler, 1963; Gurdon, 1960c). There is, however, a difference between the results in *Rana* and *Xenopus*, in that the nuclear change in *Xenopus* is not evident until the neurula stage. This, along with the fact that a limited number of macroscopically normal embryos have been obtained from nuclei of visibly differentiated intestinal epithelium cells of *Xenopus* tadpoles, has led Gurdon (1962c,d, 1963, 1964) to question whether the restrictive changes observed in both *Rana* and *Xenopus* reflect the normal state of the nucleus in differentiated somatic cells, or are a consequence of replicating in the cytoplasm of the test egg.

Thus the transplantation studies of endoderm nuclei have left unresolved two important questions: (1) the specificity of the nuclear restrictions encountered in the majority of endoderm nuclear-transplant embryos, (2) the nature of these restrictions. The restrictive developmental capacity noted for endoderm nuclei is also seen in nuclear-transplant embryos derived from nuclei of other cell types. Transplantation tests of nuclei from ectodermal and mesodermal cells of neurula and later stages of *Rana* (DiBerardino and King, 1966), *Xenopus* (Gurdon, 1962c), *Bufo* (Nikitina, 1964), and axolotl (Briggs *et al.*, 1964) have shown that the majority of these nuclei fail to promote normal development. However, before the question of specificity can be settled, histological and chromosome examination of normal and abnormal nuclear-transplant embryos from donor cells of different embryonic origin will have to be performed to see if they display a pattern of differentiation similar to, or distinct from, that found in embryos resulting from endoderm nuclear transfers.

There is also very little definite information concerning the precise nature of the nuclear restrictions encountered in transplant embryos derived from nuclei of endoderm or other cell types. A cytological examination of nuclear-transplant embryos from presumptive neural cells of *Rana* revealed that the high frequency of developmental arrest was due to the presence of pronounced chromosome abnormalities (DiBerardino and King, 1966). This inability to go through mitosis normally, following transfer into egg cytoplasm, appears to be a common characteristic of nuclei from differentiated amphibian cells. Abnormal mitoses have been found in *Rana* nuclear-transplant embryos derived from endoderm (Briggs *et al.*, 1961; Hennen and Briggs, personal communication, 1964; Subtelny, 1965), as well as in embryos derived from nuclei of neural plate (DiBerardino and King, 1966) and normal kidney cells (DiBerardino and King, 1965). Similar abnormalities have been found in endoderm nuclear-transplant embryos of *Pleurodeles* (Gallien *et al.*, 1963) and in endoderm and notochord nuclear-transplants of axolotl (Briggs *et al.*, 1964). The evidence at hand indicates that these chromosome abnormalities are not the result of technical damage sustained during the transfer procedure, but rather represent a genuine restriction in the capacity of somatic nuclei from advanced cell types to function normally following transfer into egg cytoplasm. Whereas many of the transplant embryos obtained from nuclei of endoderm cells of late gastrula and older embryos possess a high frequency of abnormal chromosome constitutions of varying degree (Briggs *et al.*, 1961), control nuclear transplants from blastula and early gastrula cells only occasionally have been observed to contain abnormal chromosome complements. Second, in axolotl, donor nuclei from the more recently differentiated cells of the posterior end of the notochord and endoderm displayed a greater capacity to promote development of the recipient eggs than did nuclei from the previously determined anterior regions of these organs. This difference can be related to the "stage" of differentiation of the donor cells, but not to any known technical factors in the experiments (Briggs *et al.*, 1964). Third, despite the technical difficulties involved in disaggregating and manipulating primordial germ cells in the transfer pipette, nuclear transfers of such cells from early larvae gave normal development in 40% of the completely cleaved test eggs (Smith, 1965). By contrast, somatic nuclei from endoderm cells of younger donors (tail bud stage), although easier to handle in the transfer procedure, are more restricted in their ability to promote the normal development of test eggs (Briggs and King, 1957; Subtelny, 1965). These three facts suggest that the abnormal mitotic behavior of nuclei of somatic cells is somehow an expression of the differentiated state of the donor nuclei, affecting their capacity to replicate normally following transfer into egg cytoplasm. The precise time and exact manner in which these abnormalities originate are unknown.

B. Interspecific Nuclear Transfers

In the past, one of the principal attacks on the problem of nucleo-cytoplasmic interaction in development has been made through studies of hybrids. In this field of investigation the method of nuclear transplantation has permitted certain types of experiment to be carried out that could not be done in any other way. One such experiment was designed to test the effects of a nucleus of one species replicating in the cytoplasm of a distantly related one (Moore, 1958a,b). *Rana sylvatica* eggs first were fertilized with *Rana pipiens* sperm and then enucleated. The androgenetic hybrids were allowed to develop to the mid-blastula stage, at which time their nuclei were transferred back into enucleated *pipiens* eggs. Although the nuclei were then in the egg cytoplasm of their own species, they were unable to support normal development, indicating that the *pipiens* nuclei had been altered as a consequence of replicating in *sylvatica* cytoplasm.

Fischberg *et al.* (1958b) performed experiments similar to those of Moore by transferring diploid nuclei between *Rana temporaria* and *Xenopus laevis*. Here, as in Moore's system, nuclei exposed to a foreign cytoplasm, and then returned to the egg cytoplasm of their own species, were unable to sustain normal development. However, if the *temporaria* nuclei were repeatedly exposed to their own type of cytoplasm, via the serial transplantation technique (King and Briggs, 1956), the resulting embryos developed further than those of the original back-transfer generation. Five successive back-transfers to *pipiens* egg cytoplasm, in the *R. sylvatica–R. pipiens* system, failed to confirm this recovery (Moore, 1960, 1962). Successive back-transfers to *Xenopus laevis* egg cytoplasm, in the *Xenopus laevis–Xenopus tropicalis* system (Gurdon, 1962d), also failed to produce any improvement.

To determine whether or not the restrictions in the capacity of the nuclei to promote development after repeated division in foreign cytoplasm are accompanied by detectable alterations in their chromosome complement, Hennen (1963) analyzed the chromosomes of abnormal embryos resulting from diploid *R. pipiens* nuclei transplanted into *R. sylvatica* cytoplasm and then back-transferred into enucleated *pipiens* eggs. All but one of the embryos derived from these eggs developed abnormally and were found to possess abnormal chromosome constitutions. Such karyotypic abnormalities can account both for the changes in the developmental capacity of *pipiens* nuclei as a consequence of replicating in *sylvatica* cytoplasm, and for the results, obtained by Moore (1960), showing that these changes are irreversible. However, the mechanism by which these chromosomal aberrations are brought about and the precise time at which they occur remain unknown.

C. Effects of Changes in Ploidy on Development

Nuclear transplantation affords a convenient means of obtaining embryos of different ploidy, since a delay in cleavage following transplantation allows for complete duplication of the chromosomes of the donor nucleus before the egg cytoplasm is prepared to cleave.

If a delay of one cleavage interval occurs following the successful transplantation of a diploid nucleus, a tetraploid embryo results (Briggs and King, 1952, 1953). Delayed cleavages on subsequent retransfers can result in higher polyploids, e.g., retransplantation of tetraploid blastulae, obtained from presumptive medullary plate nuclei of *Rana pipiens*, has given rise to octoploids (King and DiBerardino, 1966). In *Rana,* an increased proportion of polyploid nuclear-transplant embryos was found to be associated with nuclei of later stages (Briggs and King, 1957) whereas, in *Xenopus,* the highest frequency of tetraploids came from donor nuclei of both early and later stages (blastula and tadpole stages). Gurdon (1959) has studied the development of tetraploid embryos resulting from *Xenopus* nuclear transfers. Although these embryos were slightly smaller in size than their diploid controls, their rate of growth was not found to be significantly different. In fact they developed normally in every respect except for gametogenesis. Both oogenesis and spermatogenesis were abnormal and, so far, tetraploid adult nuclear-transplant individuals have been found to be entirely sterile.

Embryos with chromosome constitutions different from that of the donor nucleus also have been obtained by transplanting blastula nuclei into body cavity and uterine eggs of *Rana pipiens*, the nucleus of which was not removed (Subtelny and Bradt, 1960, 1961). Following the transplantation of androgenetic haploid nuclei, host eggs that cleave at the normal time after activation give rise to diploid embryos. By contrast, the majority of host eggs that cleave on time following diploid nuclear transfers give rise to triploid embryos, as a result of a fusion of the transplanted diploid nucleus with the resident haploid egg pronucleus prior to the initial cleavage of the recipient egg (Subtelny and Bradt, 1963). In the latter case, a delay of one cleavage interval results in hexaploidy. Thus, the transplantation of nuclei into activated, nonenucleated eggs can provide a whole range of amphibian polyploidy in which the developmental consequences of increased ploidy can be explored.

Nuclear transplantation studies of haploids also have been performed. Amphibian haploid embryos develop normally to the beginning of gastrulation, but thereafter become retarded and display a combination of abnormalities (deficiencies in the gut, central nervous system, sense organs, pronephros, and cardiovascular system) that has come to be known as the "haploid syndrome." To determine whether the poor viability of haploids is attributable to lethal factors associated with the haploid chromosome complement, Subtelny (1958) transplanted androgenetic haploid nuclei into enucleated frogs' eggs. Recipient eggs which cleaved on time developed into typical haploids and displayed the characteristic deficiencies described above. Test eggs which showed a delay of approximately one cleavage interval following haploid nuclear transfers gave rise to homozygous diploids, as a result of the duplication of the haploid set of chromosomes. The normal development of homozygous diploids through early post-gastrula stages indicates that the deficiencies of haploid embryos at these stages are not due to gene lethality per se, but rather to the presence of but a single set of chromosomes. Whether gene lethality or some other nuclear condition is responsible for the later abnormal development of homozygous diploids, with a duplicate set of haploid chromosomes, is not yet known.

D. Biochemical Studies

Grant (1961) has used the nuclear transplantation procedure to analyze the effects of nitrogen mustard on nucleo-cytoplasmic interactions during early amphibian development. When blastula nuclei from nitrogen mustard-treated eggs were transplanted into untreated cytoplasm, development was arrested at the blastula stage. However, continued exposure to untreated cytoplasm, via serial transplantations, brought about some improvement in the developmental capabilities of the descendants of the original treated nuclei. This is explained by the fact that two kinds of nuclear lesion were found to result from exposure to nitrogen mustard: (1) direct damage produced as a result of an intimate combination between the nucleus and free nitrogen mustard, and (2) an indirect cytoplasmically mediated damage from which recovery may occur after repeated duplication in normal cytoplasm.

To test the hypothesis that cellular differentiation is based upon complexes formed between chromosomes and specific macromolecules, Markert and Ursprung (1963) injected various liver fractions from mature *Rana pipiens* into fertilized eggs of the same species. The albumin and, to a lesser extent, the histone fraction were the only ones that affected development in a reproducible way; both arrested development at the late blastula stage. To test the persistence of this effect, nuclei from arrested blastulae were retransplanted into a new series of host eggs. This was repeated for seven generations without the treated nuclei ever regaining the capacity to promote development beyond the blastula stage. Chromosome analyses of arrested blastulae revealed that the arrest of development is due to the formation of chromosomal aberrations (Ursprung and Markert, 1963). These results have been interpreted as indicating that an irreversible change in the zygote chromosomes has been brought about either directly by an enzymatic fragmentation of the chromosomes (or by an abnormal association of the deoxyribonucleic acid (DNA) of the zygote with the injected macromolecules in such a way as to lead to breaks during replication), or indirectly through a cytoplasmic intermediate. Assays of the adult liver extract suggest that the active material is a protein (Melton, 1963). Since comparable extracts from various tissues of other animals cause less frequent blastula arrest (Kimmel, 1963a,b), it is suggested that adult liver contains species-specific proteins which are important in regulating gene activity, and that premature exposure of the zygote chromosomes to such molecules interferes with the normal function of developmental genes, thereby arresting development.

The entire process of embryogenesis in amphibia takes place in a "closed system" where all the newly synthesized macromolecules are formed entirely from precursor substances which were present in the unfertilized egg. This fact has permitted the characterization of RNA and protein synthesis during development, uncomplicated by the needs of a growing organism. In *Rana pipiens*, no synthesis of ribosomal RNA from P^{32}-labeled phosphate could be detected until the neural tube stage (Brown and Caston, 1962a). The most active ribosome synthesis was found to begin after the embryo had hatched from its protective membranes (Brown and Caston, 1962b). Concomitant with the increase in RNA synthesis, new ribosomes appear in the cytoplasm and a marked increase occurs in the rate of protein synthesis. In the case of *Xenopus laevis*, the embryo does not need new ribosomal RNA until after the early swimming stage. This is evidenced by the fact that *Xenopus* embryos, homozygous for the anucleolate mutant (Elsdale *et al.*, 1958), are incapable of synthesizing new ribosomal RNA, yet these embryos develop normally until the early swimming stages (Brown and Gurdon, 1964). These studies have been extended to a study of the pattern of RNA metabolism in nuclear-transplant embryos (Gurdon and Brown, 1965). Endoderm nuclei transplanted into P^{32}-labeled host eggs were found to revert to the type of RNA metabolism characteristic of the zygote nucleus following fertilization. At the beginning of gastrulation the nucleolus reappears and, as development continues, the transplanted nucleus takes on the pattern of RNA synthesis found in normal fertilized controls (Brown and Littna, 1964a,b).

E. Transplantation of Tumor Nuclei

Transplantation of nuclei from spontaneously occurring frog renal adenocarcinomata (Lucké, 1934), or from renal tumors cultured in the anterior eye chambers of adult frogs, can give rise to normally cleaved blastulae when transplanted into enucleated egg cytoplasm (King and McKinnell, 1961). A limited number of such blastulae developed into abnormal embryos with primitive but recognizable embryonic organ systems. Subsequent transfers, from both primary and intraocularly grown tumors, gave rise to a few advanced larvae (McKinnell, 1962).

Because of the technical difficulties involved in handling the small dissociated tumor cells in the transfer pipette, most of these advanced tumor nuclear-transplant embryos were obtained from the transfer of several ruptured cells into a single enucleated host egg. Since more than one nucleus may have participated in the development of the test eggs, the developmental capacity of nuclei from individual donor cells of the tumor remained unclear. Improved optical conditions and greater facility in handling dissociated tumor cells have permitted the successful transplantation of nuclei from *single* cells of primary, intraocular, and short-term cultured frog renal adenocarcinomata (King and DiBerardino, 1965). Despite their normal appearance, none of the complete blastulae derived from nuclei of these three sources of tumor developed into normal embryos. The most advanced tumor nuclear-transplant embryos arrested as abnormal larvae. A histological examination of a number of these embryos revealed cellular deficiencies in all of the three primary germ layer derivatives. Although these deficiencies presented no pattern that could be associated with the tumor origin of their donor nuclei, the degree of organ differentiation and the extent of development of such embryos were found to be related to their chromosome constitution (Di-Berardino and King, 1965). Nuclear-transplant embryos that arrested as blastulae or gastrulae, or donors which gave rise to nuclear clones that arrested at these stages, were severely aneuploid and contained gross chromosome abnormalities (rings, fragments, and anaphase bridges). Those embryos that developed to post-neurula stages were also aneuploid, but contained less severe chromosome aberrations. Nuclear-transplant embryos that developed to larval stages, although mainly euploid, were found to possess abnormal and mosaic karyotypes. In some cases it is known that the metaphase karyotype of the tumor nuclear donor was normal, and that the chromosome abnormalities arose during replication in the cytoplasm of the test egg (DiBerardino and King, 1965). The precise time and exact manner in which these chromosome abnormalities originate are unknown. Neither is it known whether their origin is the same or different in transplant embryos from nuclei of tumor versus normal cells (see Section V,A). Another difficulty in interpreting the results of tumor nuclear transplantation experiments is the presence of nontumor cells in the donor material. Although histologically observed to be only 1 in 12 (DiBerardino et al., 1963), the chance transplantation of a nucleus from a "nontransformed" stroma cell would influence the interpretation of these results.

VI. Concluding Remarks

The advantage of being able to study nuclear function in development with the amphibian embryo is considerable. It is one of the classical objects of experimental embryology and thereby provides us with the opportunity of correlating the properties of the nuclei, as they may be revealed by nuclear transplantation, with the known properties of tissues that have been demonstrated by numerous transplantation and explantation experiments, as well as with the biochemical properties that are now beginning to unfold. To be sure, these opportunities are just beginning to be realized and the material holds much more promise. Eventually the nuclear transplantation technique may be refined to detect how genes exert their control over embryonic differentiation. Another possibility is the development of an optimal nuclear medium—a matter that should have real significance for future studies of nuclear biochemistry.

34   THOMAS J. KING

Acknowledgments

The author is indebted to Dr. Robert Briggs, Department of Zoology, Indiana University, and to Dr. Marie DiBerardino and Dr. Jack Schultz, The Institute for Cancer Research, Philadelphia, for their critical reading of this manuscript.

It is a pleasure also to acknowledge the assistance of Mrs. Christiane Fiedler in preparing the line drawings, Mr. Lawrence Anderson for the photography, and Miss Nancy J. Evans for calculating the cell diameters recorded in Table II.

The investigations of the author have been supported by research grants from the U.S. Public Health Service, National Institutes of Health (C-913 and CA-05755), and in part by the American Cancer Society (E-71 and E-194).

References

Barr, H. J., and Esper, H. (1963). *Exptl. Cell Res.* **31**, 211.
Briggs, R., and King, T. J. (1952). *Proc. Natl. Acad. Sci. U.S.* **38**, 455.
Briggs, R., and King, T. J. (1953). *J. Exptl. Zool.* **122**, 485.
Briggs, R., and King, T. J. (1957). *J. Morphol.* **100**, 269.
Briggs, R., and King, T. J. (1959). In "The Cell" (J. Brachet and A. E. Mirsky, eds.), Vol. I, p. 537. Academic Press, New York.
Briggs, R., and King, T. J. (1960). *Develop. Biol.* **2**, 252.
Briggs, R., King, T. J., and DiBerardino, M. A. (1961). In "Symposium on Germ Cells and Development" (Inst. Intern. Embryol., S. Ranzi, ed.), p. 441. Fondazione A. Baselli, Milan.
Briggs, R., Signoret, J., and Humphrey, R. R. (1964). *Develop. Biol.* **10**, 233.
Brown, D. D., and Caston, J. D. (1962a). *Develop. Biol.* **5**, 435.
Brown, D. D., and Caston, J. D. (1962b). *Develop. Biol.* **5**, 412.
Brown, D. D., and Gurdon, J. B. (1964). *Proc. Natl. Acad. Sci. U.S.* **51**, 139.
Brown, D. D., and Littna, E. (1964a). *J. Mol. Biol.* **8**, 669.
Brown, D. D., and Littna, E. (1964b). *J. Mol. Biol.* **8**, 688.
Comandon, J., and de Fonbrune, P. (1939). *Compt. Rend. Soc. Biol.* **130**, 744.
Danielli, J. F. (1960). In "New Approaches in Cell Biology" (P. M. B. Walker, ed.), p. 15. Academic Press, New York.
DiBerardino, M. A., and King, T. J. (1965). *Develop. Biol.* **11**, 217.
DiBerardino, M. A., and King, T. J. (1966). In preparation.
DiBerardino, M. A., King, T. J., and McKinnell, R. G. (1963). *J. Natl. Cancer Inst.* **31**, 769.
DuPraw, E. J. (1963). *Proc. 16R Intern. Congr. Zool., Washington, D.C., 1963* Vol. 2, p. 238.
Elsdale, T. R., Fischberg, M., and Smith, S. (1958). *Exptl. Cell Res.* **14**, 642.
Elsdale, T. R., Gurdon, J. B., and Fischberg, M. (1960). *J. Embryol. Exptl. Morphol.* **8**, 437.
Ephrussi, B. (1951). In "Genetics in the 20th Century" (L. C. Dunn, ed.), p. 241. Macmillan, New York.
Esper, H., and Barr, H. J. (1964). *Develop. Biol.* **10**, 105.
Fischberg, M., and Blackler, A. W. (1963). In "Biological Organization at the Cellular and Supercellular Level" (R. J. C. Harris, ed.), p. 111. Academic Press, New York.
Fischberg, M., Gurdon, J. B., and Elsdale, T. R. (1958a). *Nature* **181**, 424.
Fischberg, M., Gurdon, J. B., and Elsdale, T. R. (1958b). *Exptl. Cell Res.* Suppl. **6**, 161.

Gallien, L. (1960). *Compt. Rend.* **250**, 4038.
Gallien, L., Picheral, B., and Lacroix, J. (1963). *Compt. Rend.* **257**, 1721.
Goldstein, L. (1963). *In* "Cytology and Cell Physiology," 3rd ed. (G. H. Bourne ed.), p. 559. Academic Press, New York.
Goldstein, L. (1964). *In* "Methods in Cell Physiology" (D. M. Prescott, ed.), Vol. 1, p. 97. Academic Press, New York.
Grant, P. (1961). *In* "Symposium on Germ Cells and Development" (Inst. Intern. Embryol., S. Ranzi, ed.), p. 483. Fondazione A. Baselli, Milan.
Gurdon, J. B. (1959). *J. Exptl. Zool.* **141**, 519.
Gurdon, J. B. (1960a). *J. Embryol. Exptl. Morphol.* **8**, 327.
Gurdon, J. B. (1960b). *Quart. J. Microscop. Sci.* **101**, 299.
Gurdon, J. B. (1960c). *J. Embryol. Exptl. Morphol.* **8**, 505.
Gurdon, J. B. (1962a). *J. Heredity* **53**, 5.
Gurdon, J. B. (1962b). *Develop. Biol.* **4**, 256.
Gurdon, J. B. (1962c). *J. Embryol. Exptl. Morphol.* **10**, 622.
Gurdon, J. B. (1962d). *Develop. Biol.* **5**, 68.
Gurdon, J. B. (1963). *Quart. Rev. Biol.* **38**, 54.
Gurdon, J. B. (1964). *Advan. Morphogenesis* **4**, 1.
Gurdon, J. B., and Brown, D. D. (1965). *J. Mol. Biol.* **12**, 27.
Gurdon, J. B., Elsdale, T. R., and Fischberg, M. (1958). *Nature* **182**, 64.
Hamburger, V. (1960). "A Manual of Experimental Embryology," p. 28. Univ. of Chicago Press, Chicago, Illinois.
Hämmerling, J. (1934). *Arch. Entwicklungsmech. Organ.* **132**, 424.
Hennen, S. (1963). *Develop. Biol.* **6**, 133.
Humphrey, R. R. (1960). *Develop. Biol.* **2**, 105.
Kahn, J. (1962). *Quart. J. Microscop. Sci.* **103**, 407.
Kimmel, D. L., Jr. (1963a). *Genetics* **48**, 896.
Kimmel, D. L., Jr. (1963b). *Am. Zoologist* **3**, 486.
King, T. J., and Briggs, R. (1955). *Proc. Natl. Acad. Sci. U.S.* **41**, 321.
King, T. J., and Briggs, R. (1956). *Cold Spring Harbor Symp. Quant. Biol.* **21**, 271.
King, T. J., and DiBerardino, M. A. (1965). *Ann. N.Y. Acad. Sci.* **126**, 115.
King, T. J., and DiBerardino, M. A. (1966). In preparation.
King, T. J., and McKinnell, R. G. (1961). *In* "Cell Phiysiology of Neoplasia" (R. W. Cumley, M. Abbott, and J. McCay, eds.), 14th Symp. Fundamental Cancer Res., 1960, p. 591. Univ. of Texas Press, Austin, Texas.
Lehman, H. E. (1955). *Biol. Bull.* **108**, 138.
Lehman, H. E. (1957). *In* "Beginnings of Embryonic Development," Publ. No. 48, p. 201. Am. Assoc. Advance. Sci., Washington, D.C.
Lopashov, G. V. (1945). *Ref. rab. biol. Otd. Akad. nauk. SSSR* p. 88. (Abstr. Pub. Biol. Div. Acad. Sci., U.S.S.R., 1945.)
Lorch, I. J., and Danielli, J. F. (1950). *Nature* **166**, 329.
Lucké, B. (1934). *Am. J. Cancer* **20**, 352.
McKinnell, R. G. (1960). *Am. Naturalist* **94**, 187.
McKinnell, R. G. (1962). *Am. Zoologist* **2**, 430.
Markert, C. L., and Ursprung, H. (1963). *Develop. Biol.* **7**, 560.
Melton, C. G. (1963). *Genetics* **48**, 901.
Moore, J. A. (1958a). *Exptl. Cell Res.* **14**, 532.
Moore, J. A. (1958b). *Exptl. Cell Res. Suppl.* **6**, 179.
Moore, J. A. (1960). *Develop. Biol.* **2**, 535.

Moore, J. A. (1962). *J. Cellular Comp. Physiol.* **60**, Suppl. 1, 19.

Moscona, A. (1962). *J. Cellular Comp. Physiol.* **60**, Suppl. 1, 65.

Nikitina, L. A. (1964). *Dokl. Akad. Nauk SSSR* **156**, 1461.

Niu, M. C., and Twitty, V. C. (1953). *Proc. Natl. Acad. Sci. U.S.* **39**, 985.

Pantelouris, E. M., and Jacob, J. (1958). *Experientia* **14**, 99.

Picheral, B. (1962). *Compt. Rend.* **255**, 2509.

Porter, K. R. (1939). *Biol. Bull.* **77**, 233.

Rauber, A. (1886). *Zool. Anz.* **9**, 166.

Rostand, J. (1943). *Rev. Sci. (Paris)* **81**, 454.

Sambuichi, H. (1957). *J. Sci. Hiroshima Univ.* **B17**, 33.

Sambuichi, H. (1961). *J. Sci. Hiroshima Univ.* **B20**, 1.

Schultz, J. (1952). *Exptl. Cell Res. Suppl.* **2**, 17.

Seidel, F. (1932). *Arch. Entwicklungsmech. Organ.* **126**, 213.

Shumway, W. (1940). *Anat. Record* **78**, 139.

Signoret, J., and Fagnier, J. (1962). *Compt. Rend.* **254**, 4079.

Signoret, J., and Picheral, B. (1962). *Compt. Rend.* **254**, 1150.

Signoret, J., Briggs, R., and Humphrey, R. R. (1932). *Develop. Biol.* **4**, 134.

Smith, L. D. (1965). *Proc. Natl. Acad. Sci. U.S.* **38**, 455.

Spemann, H. (1914). *Verhandl. Deut. Zool. Ges.* p. 216.

Spemann, H. (1938). "Embryonic Development and Induction." Yale Univ. Press, New Haven, Connecticut.

Steinberg, M. (1957). *Carnegie Inst. Wash. Year Book* **56**, 347.

Stroeva, O. G., and Nikitina, L. A. (1960). *Zh. Obshch. Biol. (SSSR)* **21**, 335.

Subtelny, S. (1958). *J. Exptl. Zool.* **139**, 263.

Subtelny, S. (1965). *J. Exptl. Zool.* **159**, 59.

Subtelny, S., and Bradt, C. (1960). *Develop. Biol.* **2**, 393.

Subtelny, S., and Bradt, C. (1961). *Develop. Biol.* **3**, 96.

Subtelny, S., and Bradt, C. (1963). *J. Morphol.* **112**, 45.

Tartar, V. (1953). *J. Exptl. Zool.* **124**, 63.

Tartar, V. (1961). "The Biology of Stentor." Pergamon Press, Oxford.

Ursprung, H., and Markert, C. L. (1963). *Develop. Biol.* **8**, 309.

Waddington, C. H., and Pantelouris, E. M. (1953). *Nature* **172**, 1050.

Wilson, J. F. (1963). *Am. J. Botany* **50**, 780.

Zetsche, K. (1962). *Naturwiss.* **49**, 404.

Chapter 2

Techniques for the Study of Lampbrush Chromosomes

JOSEPH G. GALL

Department of Biology, Yale University, New Haven, Connecticut

I. Introduction

The largest known chromosomes are found in the developing oocytes of salamanders. These and the similar chromosomes from oocytes of other animals were long ago compared to an old-fashioned lampbrush (Rückert, 1892), and this epithet has been applied to them ever since. The lampbrush comparison refers to the hundreds of lateral projections—in reality, loops—which give these chromosomes a hairy or woolly appearance. The loops are the site of very active ribonucleoprotein synthesis, and as such are of great interest for studies of chromosome physiology.

Although best known from the amphibians, lampbrush chromosomes are found in oocytes of many animals, both vertebrate and invertebrate (Callan, 1957; Wilson, 1928). It is also possible that the well-known fuzziness of spermatocyte chromosomes is due to looplike projections (Ris, 1945). Although this fuzziness is usually indistinct, Hess and Meyer

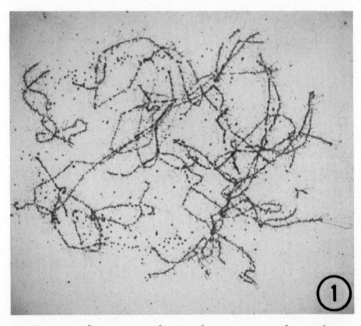

Fig. 1. Low magnification view of entire chromosome complement from an oocyte nucleus of the newt *Triturus viridescens*. Isolation of nucleus and chromosomes carried out in 0.1 *M* saline; fixation in the vapor of acid formaldehyde. The nucleus contains eleven meiotic bivalents. The small dark granules are some of the several hundred free nucleoli found in the oocyte nucleus. × 65.

(1963) have recently described typical lampbrush loops on the Y chromosome in developing spermatocytes of several *Drosophila* species. In oocytes the lampbrush condition persists for many weeks or months, but

Fig. 2. A single meiotic bivalent chromosome from an oocyte nucleus of the newt *Triturus viridescens*. The fuzzy appearance of the chromosome, which gave rise to the name "lampbrush," is due to the numerous laterally projecting loops. The two homologues are held together at four points, three of which are chiasmata; the fourth is a fusion of the centromere regions (*arrow*). Isolation of nucleus and chromosomes carried out in 0.1 *M* saline. Fixation for 3 minutes in vapor of buffered formaldehyde. Phase-contrast, electronic flash photograph. × 540.

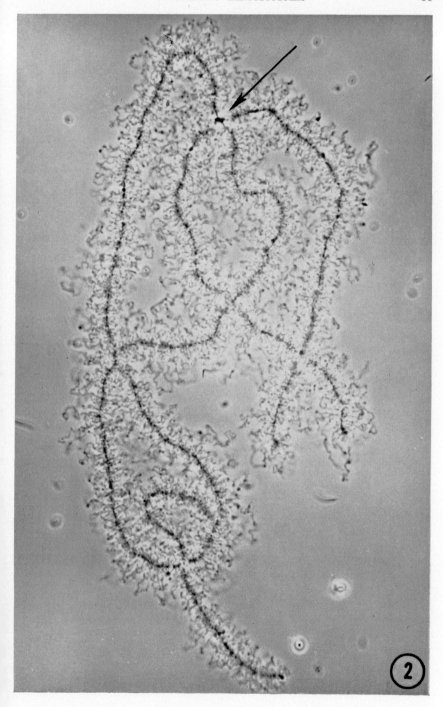

eventually the chromosomes contract to form typical rod-shaped elements
at the first meiotic metaphase. The lampbrush condition is not, therefore, an aberrancy limited to a few vertebrate species, but is a modification of normal chromosomes characteristic of a wide variety of animals.
Indeed, the structural modifications seen during the lampbrush stage may
very well characterize synthetically active chromosomes (Callan, 1963;
Callan and Lloyd, 1960a,b; Dodson, 1948; Gall, 1958, 1963b).

Lampbrush chromosomes are in many ways the most favorable of all
for experimental manipulation. Although techniques for their isolation
from living cells were worked out more than 25 years ago by Duryee
(1937, 1941, 1949, 1950), very few workers have availed themselves of
this remarkable material. The lack of interest may stem from the opinion
that these chromosomes are difficult to handle, or require elaborate
equipment. Quite the opposite is true as I will try to show in the following instructions.

It may be helpful to summarize briefly some of the major advantages
and disadvantages of the lampbrush chromosomes as experimental material.

(1) The large size and wealth of structural detail permit studies at
what is probably the gene level (Callan and Lloyd, 1956, 1960b; Gall,
1963b). Typical lampbrush chromosomes from several species of *Triturus*
are 400–800 μ in length (Figs. 1, 2). Several hundred pairs of loops
project from the main axis of each chromosome, and the loops themselves
average about 50 μ in length at maximal extension. Many of the loop
pairs are of distinctive morphology, permitting identification of particular

---

Fig. 3.   A pair of lateral loops in the "double bridge" configuration (Callan,
1963). Double bridges such as this occur when the chromosome is stretched. They
provide evidence that the loops are regions where the two sister chromatids are
bowed out laterally from the main chromosome axis. The characteristic tapering of
the loops is evident. This tapering defines a polarity which is constant for a given
loop pair. Phase-contrast, electronic flash. × 680.

Fig. 4.   A single lateral loop during digestion by DNase, photographed at 7,
18, 22, 27, and 29 minutes (A–E, respectively) after isolation in the enzyme. Note
that the fragmentation occurs without decrease in diameter of the pieces. The kinetics
of the fragmentation suggest that loop continuity is maintained by one double helix
of DNA (Gall, 1963a). Phase-contrast, electronic flash of unfixed chromosome. × 300.

Fig. 5.   Unfixed lampbrush chromosome 14 minutes after isolation in a saline solution containing DNase. Phase-contrast, electronic flash. × 200.

Fig. 6.   The same chromosome at 36 minutes. The enzyme has caused extensive
fragmentation of the loops, as well as breaks along the main chromosome axis. Phase-
contrast, electronic flash. × 200.

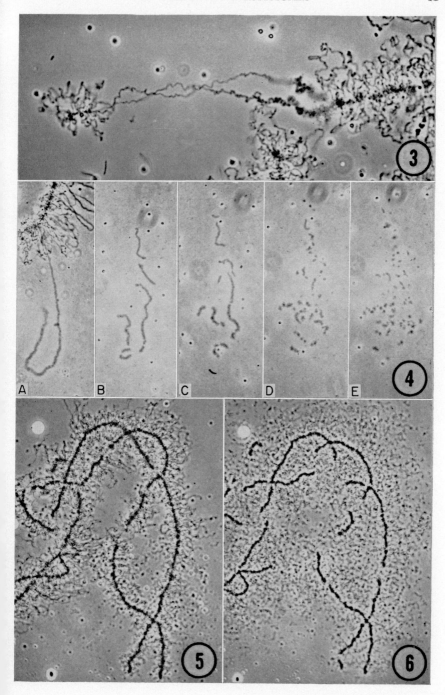

chromosome regions. Experimental studies may, therefore, be directed to specific loop pairs (Gall and Callan, 1962; Macgregor, 1963; Macgregor and Callan, 1962).

(2) The chromosomes are obtained unfixed within a minute or two after the oocyte is first broken. They may be subjected to a wide variety of chemical and physical treatments. The earlier studies gave information on the effects of pH, ions, enzymes, and radiation (Callan, 1952; Duryee, 1937, 1941, 1949, 1950; Gall, 1954, 1955, 1956; Gersch, 1940; Guyénot and Danon, 1953; Wischnitzer, 1957). Both Duryee (1941) and Callan (1956) have studied the behavior of the chromosomes when stretched between microneedles; Callan's experiments decisively tested Guyénot and Danon's suggestion (1953) that the loops are paired (cf. Fig. 3). Recently Macgregor and Callan (1962; Callan and Macgregor, 1958) have studied in detail the effects of enzymes on the chromosomes, and it has even been possible to obtain useful data on the kinetics of DNase action (Gall, 1963a) (cf. Figs. 4–6). Macgregor (1963) has examined the changes in lampbrush chromosomes resulting from the experimental manipulation of hormone levels in the female newt.

(3) Certain chemical analyses are possible, using either whole nuclei or isolated chromosomes. Several subfractions of the nucleus (chromosomes, nucleoli, sap) can be obtained with a high degree of morphological purity. G. L. Brown et al. (1950) published data on the amino acid composition of the nuclear sap proteins. Edström and Gall (1963) have recently obtained base ratios for chromosomal, nucleolar, and nuclear sap ribonucleic acid (RNA). The analyses were performed on manually separated parts of individual nuclei. The nuclear sap has been separated into several distinct components by acrylamide gel electrophoresis (Rogers, 1965).

(4) The lampbrush chromosomes readily incorporate amino acids and various RNA precursors (Ficq et al., 1958; Gall and Callan, 1962; Izawa et al., 1962; Pantelouris, 1958). Different loops show varying patterns and rates of incorporation, which may be reflections of differing gene activity (Gall and Callan, 1962). Since the oocyte is in a modified diplotene condition, the chromosomes have already completed their replication, and there is no thymidine incorporation into deoxyribonucleic acid (DNA) (Izawa et al., 1962; also unpublished observations, 1961, of Gall and Callan); RNA synthesis can, therefore, be studied without the complication of concomitant DNA synthesis. The oocytes can be easily subjected to experimental treatment during incorporation studies. Izawa et al. (1962), for instance, have recently investigated the effect of actinomycin D on RNA metabolism in lampbrush chromosomes.

Some important limitations of oocyte chromosomes should also be mentioned.

(1) The life span of salamanders is long, generally 1–2 years from the egg to sexual maturity. Cytogenetic analysis, therefore, requires not only careful planning but extreme patience. Callan and Lloyd (1960b) have studied lampbrush chromosomes from four races of *Triturus cristatus*, and have shown that they differ in a number of evident details. They have found (unpublished observations) that the parental chromosomes are recognizable in the F_1 interracial hybrids, and are currently studying backcrosses to the parental strains. Callan and Spurway (1951) previously described various abnormalities in the male meiosis of these same interracial hybrids. An advantage of the newt material is that chromosome studies can be made on the same individuals later used for breeding experiments. Fankhauser and Humphrey (1959) have bred many generations of the Mexican axolotl, and several mutant strains are now available. Their important studies on heteroploidy have included extensive analysis of the somatic chromosomes in tail tip preparations. Further study of the lampbrush chromosomes in their stocks should prove rewarding.

(2) Not only is the life span of the salamander long, but the chromosomes remain in the lampbrush condition for at least several months. It is, therefore, difficult to perform experiments extending over the entire meiotic cycle. The egg reaches the metaphase of the first meiotic division at the time of fertilization and laying. On the other hand, for certain purposes the slow development of the oocyte is an advantage. For instance, experiments may extend over a period of days or weeks with no significant change in the state of the treated oocytes (Gall and Callan, 1962).

(3) It is not easy to obtain large numbers of oocyte nuclei for chemical studies. Bulk isolation procedures are hampered by the large amount of yolk and fat in the cytoplasm and by the relative fragility of the nuclei. Nevertheless, some success has been obtained along these lines (England and Mayer, 1957). Macgregor (1962) has shown that an oocyte nucleus of *Triturus* has a dry mass of roughly 2–3 µg, most of which is "nuclear sap." Manual isolation of several hundred nuclei is not at all difficult (G. L. Brown *et al.*, 1950; Finamore and Volkin, 1958). This means that one can collect up to about a milligram dry weight without undue effort. Larger amounts present severe technical problems, which can be somewhat lessened by using the enormous nuclei of *Necturus*.

With these introductory comments on the uses of the material we may next turn to more explicit practical directions.

II. Animals and Their Care

Lampbrush chromosomes are known to occur in developing oocytes of most vertebrates and a number of invertebrates. The ease with which the chromosomes may be studied varies widely from group to group, and even between members of a single family. First attempts should be made with a tailed amphibian, preferably *Triturus viridescens** in the United States, or *Triturus cristatus** in Europe. The consistency of the nuclear sap is an important variable: the sap must be sufficiently gelatinous to permit certain manipulations, yet not so rigid that it fails to disperse after the nuclear membrane has been removed. These two species possess nearly ideal sap consistency, but the same cannot be said for many other species (many Plethodontid salamanders, for instance).

Either species of *Triturus* may be kept in pond water or boiled tap water in culture dishes.* A dozen or more of the small *viridescens* will do well in one 8-inch finger bowl, somewhat fewer of the larger *cristatus*. The animals may be kept at room temperature, although they will survive quite happily at temperatures as low as 5–10°C. They should be fed twice weekly on small bits of chopped liver, chopped earthworms, or living *Tubifex* worms. The bowl should be cleaned several hours after feeding in order to prevent fouling. Hand feeding is not necessary when *Tubifex* is used, as the wriggling of the worms attracts the newts. When several newts are kept together in a bowl, they may stimulate one another to feed, even if nonmotile food is given. However, valuable specimens should be fed individually, since some animals always starve when group feeding is practiced.

Humphrey (1962) has prepared detailed instructions on the care and feeding of the Mexican axolotl, *Ambystoma* (*Siredon*) *mexicanum*, which should be consulted by anyone working with these salamanders.

Female newts can be recognized at all seasons by examining the cloaca: the lips of the cloaca are much smaller in the female than in the male. In addition, males in the breeding condition exhibit striking secondary sexual characteristics. In *T. viridescens* these include enlargement of the cloaca and appearance of black patches on the hind legs. In *T. cristatus* a high dorsal crest develops over the body and tail of breeding males.

Ovaries in suitable condition may be found in some individuals during most of the year, especially if the animals have been maintained in the laboratory and have not laid eggs. However, it is best not to plan work with *Triturus* during the late summer when most of the oocytes are small.

* Starred items are discussed in Section V, where sources of supply are listed.

III. Routine Preparations for Visual Microscopy

A. Removal of the Ovary

It is so simple to obtain bits of ovary surgically that no animals need be killed. Chromosomes can be studied, for instance, from animals that are used later for breeding experiments (Callan and Lloyd, 1960b). Newts are anesthetized for 5–15 minutes in 0.1% MS222 solution* or until voluntary movements stop (anesthesia may be extended for many hours if desired). Alternatively, one may use light ether anesthesia, although here there is some danger of overdose. The animal should be exposed to the vapor only.

Place the anesthetized animal on its back and make a small incision in the ventral body wall on either the left or right side, depending upon

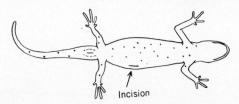

Incision

Fig. 7. Ventral view of *Triturus* female, showing correct location of incision for removing ovary.

which ovary is desired (Fig. 7). A 2–3-mm incision is adequate and will require only one stitch. A single snip will usually penetrate both the skin and the underlying musculature. With a pair of jeweler's tweezers* push the intestines aside until you can see either oocytes or the orange fat body attached to the ovary. Pull out as many oocytes as desired, or the whole ovary. Snip the mesentery while holding the ovary with fine scissors, trying not to damage any oocytes.

Transfer the oocytes immediately to a *dry* embryological watch glass ("solid" watch glass).* The coelomic fluid brought over to the watch glass provides adequate moisture for preservation of the oocytes. Smear the top of the watch glass with vaseline and seal with a flat glass cover. Place in the refrigerator at about 4°C. Oocytes thus preserved can be used for chromosome preparations for up to 2 days without evident signs of deterioration.

Most incisions can be closed with one or at most two stitches. Use a small, curved surgical needle and plain surgical suture (catgut), size 000.* Nylon thread is best avoided, as it may cut the delicate skin of the newt. It is not necessary to stitch the musculature and skin separately.

Leave at least a millimeter between the edge of the incision and the needle hole on each side, *and avoid tying the knot tightly*. Place operated individuals in separate aquaria and do not feed for several days. The major source of trouble comes from stitching too close to the incision, the thread then wearing through to the incision on one or both sides. If a gaping incision is detected soon enough, it may be repaired with a second stitching.

B. Isolation of the Nucleus

Place a small bit of ovary in "5 : 1" solution. This is a mixture of 5 parts 0.1 M KCl and 1 part 0.1 M NaCl, and contains the two cations in the proportion existing within the oocyte nucleus.[1] The solution may be buffered to pH 6.8–7.2 with 0.01 M phosphate, if desired. Do *not* use Ringer's solution as the calcium will clump the yolk and destroy the finer morphology of the chromosomes. Note that the smallest oocytes are nearly transparent. At about 0.5-mm diameter, evident yolk accumulation begins and the oocytes become cloudy. Those over about 0.8-mm diameter are completely opaque. In *cristatus* the larger oocytes are whitish or slightly greenish. In *viridescens* the medium-sized oocytes are yellowish but the larger ones are brown or brownish black due to melanin pigmentation.

If oocytes of 1.0–1.2-mm diameter are available (brown in *viridescens*, white in *cristatus*), choose one of these for your first attempt. In these the nuclei are of maximal size and are easy to handle.

Note that the ovary is a hollow sac with the oocytes attached on the inner surface. Tear open the sac. Each oocyte is covered by a thin layer of follicle cells and prominent blood vessels. It is not necessary to remove the oocyte from the follicular membrane. Grasp an oocyte with one pair of jeweler's tweezers at some point on its surface. Try to pick up a fold of the follicular membrane together with the oocyte surface without actually puncturing the cell. Grasp this same fold with a second pair of tweezers and pull laterally. The oocyte will break open and the yolky contents will spill out. Easily seen somewhere within the opaque yolk is the translucent nucleus, some 0.3–0.4 mm in diameter.

Alternatively one may puncture the cell with the forceps or a needle and squeeze gently. The nucleus appears as a clear "bubble" embedded

---

[1] The potassium and sodium determinations were made on nuclei of *Triturus cristatus carnifex* by W. T. W. Potts (personal communication, 1960, quoted in Callan and Lloyd, 1960b). Naora *et al.* (1962) state that the Na/K ratio is 1.11 in oocyte nuclei of *Rana pipiens*. For morphological studies on the lampbrush chromosomes the Na/K ratio is not critical.

in a ribbon of yolk flowing from the hole. With the tweezers gently push the nucleus away from the main bulk of yolk. In some cases the nucleus may be nearly clean of yolk at this stage; at other times it is fairly heavily encrusted. Using a pipette with a 0.75-mm bore* suck the nucleus in and out several times to remove the adherent yolk. The pipette should be filled with 5 : 1 solution before starting the cleaning procedure and care should be taken not to include an air bubble in the pipette. The nucleus is really quite sturdy and may be bounced off the bottom of the dish in order to remove bits of yolk. However, it should not be allowed to settle on the glass bottom, as it may adhere firmly and tear when sucked loose.

The nucleus will begin to swell immediately after isolation. The swelling cannot be prevented by increasing the ionic strength of the medium (Battin, 1959; Callan, 1952; Hunter and Hunter, 1961). In any event swelling serves the useful purpose of separating the nuclear membrane from the underlying chromosomes. Eventually, however, the nuclear sap will become fluid and the chromosomes will sink to one side of the nucleus. Therefore work as rapidly as possible once the oocyte has been opened.

C. Isolation of the Chromosomes

As soon as the yolk has been removed from the surface, transfer the nucleus to a flat-bottomed well slide,* previously filled with 5 : 1 solution. The well should be completely filled so that the liquid has a slightly convex surface (Fig. 8).

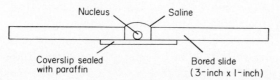

FIG. 8. Diagram of the well slide used in studies of lampbrush chromosomes. After the nuclear envelope is removed with forceps, the chromosomes will sink to the lower coverslip. There they can be observed by means of an inverted phase-contrast microscope.

The chromosome isolation can be done under a magnification of 20–40×, depending upon personal preference. Illumination is rather critical. I prefer a strong lateral illumination with no transmitted light. Generally I use a dissecting microscope with the substage removed. The specimen stage rests directly on the bench and a black background is used. Under these conditions the nucleus appears brilliant white against the black field.

Grasp the top of the nucleus with one pair of jeweler's tweezers, taking care to secure a good grip while not actually rupturing the membrane. Take hold with the second pair of tweezers very near the first. Now pull the two pairs of tweezers apart with a slight downward motion. If everything goes well the nuclear contents will pop out as a gelatinous mass completely separate from the membrane, which remains attached to one or both tweezers. Callan and Macgregor both prefer one pair of jeweler's tweezers and a very fine tungsten needle pointed in molten sodium nitrite.

If at any time the nuclear contents begin to extrude spontaneously through a small hole in the nucleus, one should immediately abandon the preparation and begin with a fresh nucleus. Only fragmented chromosomes will be found in such preparations. The ease of the isolation depends a great deal on the stiffness of the nuclear contents, the stiffer the nuclear sap the easier the dissection. Individuals which have received gonadotropic hormone injection, or been collected in the breeding season, have stiffer nuclear sap than those in poor breeding condition (Macgregor, 1963). Certain species have naturally stiff sap, in some cases so gelatinous that the nuclear membrane can be peeled off piece by piece without disrupting the contents (e.g., *Necturus, Pleurodeles*, many Plethodontids). Such very stiff nuclei, however, often yield poor preparations, since the chromosomes may never sink to the bottom coverslip.

Preparations may be observed unfixed for as long as a day, especially if kept cold between examinations. Ultimately, however, the loops become thin and contract as their ribonucleoprotein matrix dissolves. These changes may be entirely prevented, without altering the cytological appearance of the chromosomes, by a short fixation in the vapor of buffered formaldehyde. As soon as the nuclear membrane has been removed, transfer the slide to a formaldehyde chamber* for 3–5 minutes. Or place a drop of concentrated *neutral* formaldehyde solution in the bottom of a watch glass and invert over the preparation for a like time.

A top coverslip should now be added. Hold a coverslip at an angle above the depression, resting one margin of the coverslip on the slide. Then drop the coverslip into place. This technique is likely to trap air bubbles unless the meniscus has the proper convex curvature. The advantage lies in the fact that the material in the depression is not rolled around, as happens when a coverslip is slid laterally into place. The top coverslip should be sealed along its sides with melted vaseline.

Preparations of this sort may be kept several weeks with no trouble. Eventually air bubbles may creep in, if the seal is not good. These can be removed if a valuable preparation is involved. Remove part of the seal, push the top coverslip very carefully sideways until a corner of the depression is exposed, tilt the slide, tap gently until the bubbles come out, add a little 5 : 1 solution, and reseal.

D. Observation

The chromosomes will eventually spread out evenly on the bottom of the depression. Scattered among them will be several hundred nucleoli (Fig. 1). In addition there may be a very large number of smaller granules, particularly if the nucleus came from a large oocyte. Dispersal of the sap and settling of the chromosomes may take as little as 5–10 minutes or may extend over a period of several hours. As already mentioned, much depends on the species, the size of the oocyte, and the breeding condition of the individual.

Critical observations at high magnification require the use of an inverted microscope,* since the chromosomes lie on the bottom coverslip. Phase-contrast optics are essential. If an inverted microscope is not available, low power observations may be made through the liquid from above. It is possible (though we have not experimented along these lines) that one could use a very thin bored slide in constructing the observation chamber, thus accommodating a 25× or even a 40× objective from above. Under such conditions resolution would be slightly impaired.

The chromosomes can be attached firmly to the coverslip by lowering the pH, as described below. The preparation may then be inverted and observed with a conventional microscope. Such attached chromosomes provide relatively poor material for critical observations, since the loops contract and thereby obscure the chromomere axis. A chief advantage of fresh preparations, the Brownian motion, is also lost. The constant movement of parts relative to one another in fresh preparations often provides clarification of the finer points of structure.

IV. Additional Techniques

A. Other Isolation Media

Dilute media tend to dissolve the loop matrix and increase the rate of sap dispersal. They are particularly useful when one wants to study the chromomeres. They are sometimes helpful when the nuclear sap is so stiff that it will not disperse in 0.1 M saline (Callan and Lloyd, 1960b). One may merely dilute the usual 0.1 M 5 : 1 (buffered or unbuffered) with distilled water. In dilute media the nucleus swells rapidly and the sap becomes relatively fluid. Consequently one must work faster for good preparations. Chromosomes and nucleoli isolated in 0.05 M saline are rather similar to those in 0.1 M. Rapid dissolution of the loop matrix occurs in 0.025 M saline; partial dissolution of the nucleoli also takes place. More dilute salines cause even more extreme dissolution. However, the chromomere axis never dissolves, even in distilled water, although it becomes of exceedingly low optical contrast.

Calcium at a concentration of $10^{-3}\,M$, as in Ringer's solution, causes almost instantaneous liquefaction of the nuclear contents. At the same time the chromosomes contract and the ribonucleoprotein matrix of the loops dissolves. Hence calcium is generally excluded from the isolation medium (Duryee, 1941).

Concentrations of Ca^{++} between $10^{-5}\,M$ and $10^{-4}\,M$ (in the usual 0.1 M 5 : 1) cause fairly rapid dispersal of the nuclear sap without noticeable effects on the chromosomes. A trace of calcium in the isolation medium is, therefore, quite useful when one is working with particularly "stiff" nuclei. The nucleus should be isolated and cleaned in calcium-free 5 : 1, as even the slightest trace of calcium tends to clump the yolk and impede the cleaning procedure. The clean nucleus can then be transferred to 0.1 M 5 : 1 containing $0.5 \times 10^{-4}\,M$ $CaCl_2$.

B. Permanent Mounts

If chromosomes are properly attached to a slide, they may be carried through some very harsh treatments without damage (e.g., Feulgen reaction).

(1) Seal an ordinary 3-inch $\times$ 1-inch slide to a bored slide with paraffin wax. The chromosomes will eventually be attached to the ordinary slide. If one wants the chromosomes on a coverslip, then use the standard depression slide described previously.

(2) Isolate and clean a nucleus in the usual 5 : 1 medium without calcium. Transfer the nucleus quickly through a watch glass containing 0.1 M 5 : 1 plus $0.5 \times 10^{-4}\,M$ $CaCl_2$ and then into a depression slide containing the same calcium medium (or omit the calcium, as desired).

(3) Remove the nuclear envelope. Place the slide immediately into the formaldehyde chamber for 3–5 minutes; this step is not essential if dispersal is expected to be rapid. Do not add a top coverslip.

(4) Place the slide in a moist chamber until the chromosomes come to lie flat on the bottom of the depression. A convenient moist chamber consists of a 4-inch Petri plate with a piece of filter paper in the bottom; add 2 ml H_2O to saturate the filter paper. It is best to leave preparations 1–2 hours in the chamber, although shorter times are possible with fluid

nuclei. Use one slide as a control to check the rate of sap dispersal, but do not move the slides intended for permanent mounts.

(5) Make up a 1% solution of acetic acid in full strength commercial formaldehyde. Pour 1 ml of this onto the filter paper in the moist chamber, and replace the cover of the Petri plate. Leave the preparations undisturbed for 15 minutes, during which time the chromosomes will attach to the slide.

(6) Place the whole preparation in a staining jar filled with 2–4% unbuffered formaldehyde (5–10% "formalin"). One may buffer this solution to pH 4–5 if desired, but the solution should not be neutral. Leave for an hour or more. Subsequently the slides can be handled as casually as ordinary histological preparations.

(7) Separate the standard slide (carrying the chromosomes) from the bored slide by prying apart with a razor.

(8) Rinse the slide, dehydrate through an alcohol series, and remove the paraffin with xylene. Stain by whatever technique desired. Iron hematoxylin gives good staining for routine observation of loops and chromomeres. The Feulgen reaction will be faintly positive (in the chromomeres only). Tests for RNA, such as azure B at pH 4 (Flax and Himes, 1952) will be positive in the loops, although weak. For simple staining of the loops with basic dyes, use the stain at pH 7.

Comments: Until recently it was our practice to obtain attachment by leaving the slides in an unbuffered formaldehyde chamber for 1–3 hours. This was always a frustrating procedure since some slides showed good attachment while others were unsuccessful, even after prolonged fixation. We have finally realized that formaldehyde vapor fixation, by itself, does not alter the appearance of the chromosomes so long as the pH is maintained near neutrality. Attachment, with its consequent shrinkage and increase in optical contrast, occurs when the pH is lowered, and in our earlier work was probably dependent upon the traces of formic acid found in commercial formaldehyde. The chromosomes contract and attach to the slide when the pH of the fluid surrounding them is lowered to about 5.2–5.5.

Formaldehyde fixation does not prevent sap dispersal. On the other hand, osmium fixation does. If osmium-fixed chromosomes are desired, the sequence should be: chromosome isolation in saline; dispersal in moist chamber; osmium vapor fixation; attachment with vapor of 1% acetic acid (made up in water).

Chromosomes can also be attached to glass slides by means of ethanol vapor. Although we have not as yet had much experience with ethanol, it appears probable that the fixation is at least as good as with acid. In step (5) above, add 2 ml of 95% ethanol to the filter paper in the Petri plate. This will give a final concentration of about 50% ethanol in the chamber. The chromosomes should attach within 30–45 minutes.

C. Autoradiographs

Lampbrush chromosomes readily incorporate certain RNA and protein precursors. Doses will obviously depend on the precursor, the isotope used, and the specific activity available. Tritium-labeled compounds are desirable for best resolution. The following schedules are known to be adequate and can serve as a guide for further experiments (Gall and Callan, 1962; Izawa *et al.*, 1962).

(1) Inject 100–300 microcuries uridine-H^3 (specific activity 0.5–5.0 curies per millimole) into the coelomic cavity of a female *T. cristatus* or *T. viridescens*. Twelve to 24 hours later the loops will be well labeled, although the level will continue to go up for 3–4 days after the single injection. In order to obtain much higher levels of labeling and to study short-term incorporation, evaporate 100 microcuries of uridine-H^3 in a watch glass at room temperature. Remove an ovary from a female newt and place directly into the dried precursor. Swirl the ovary about to bring the uridine into solution. Seal the dish with vaseline and a glass cover. The incubation may be carried out in a small volume of Ringer's solution, but the chromosome isolation itself must be performed as usual in a calcium-free medium. Radioactivity is easily detectable in the chromosomes after an hour. By 6 hours the loops are intensely labeled.

(2) Prepare isolated chromosomes as described under *Permanent Mounts*. After fixation and washing, place the preparations for 5 minutes in 5% trichloroacetic acid at about 4°C to remove any free precursor, and finally wash thoroughly in water. Dip the slide in chrome alum "subbing" solution and dry in a vertical position.

(3) Cover with autoradiographic stripping film or liquid emulsion by standard procedures. For Kodak AR-10 stripping film or Kodak NTB-2 emulsion, exposure times of 1–4 weeks are adequate with the above isotope levels.

(4) Develop 3–4 minutes in full strength Kodak D-19. Rinse briefly in distilled water, and fix about 5 minutes in Kodak acid fixer. Wash one half hour in water, and stand slides in front of a small fan to dry.

(5) For observations place a drop of water on the film above the chromosome group and add a coverslip. The chromosomes will stand out beautifully with phase-contrast optics. We have not found a successful staining procedure for the chromosomes, nor do they show up well if preparations are dehydrated in an alcohol series and mounted in a medium of high refractive index.

For observing autoradiographs of all kinds, a most serviceable lens is the Zeiss 40× apochromatic phase (immersion). The chief advantages of this objective are (a) high numerical aperture, (b) relatively great depth of focus, permitting both the specimen and the silver grains to be in focus at the same time (excellent for photography), (c) less contrast than achromats or fluorites, an advantage since the silver grains then stand out sharply against the specimen, and (d) a built-in iris diaphragm which can be adjusted to remove the prominent halo around the silver grains. Photographs taken with this lens are shown in Figs. 9 and 10.

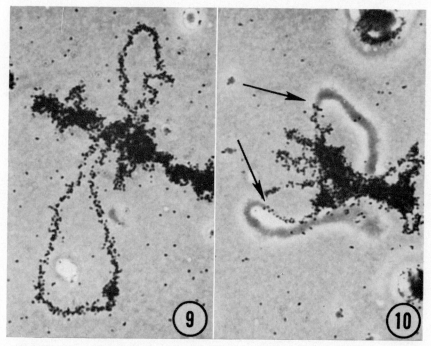

Fig. 9. Autoradiograph of a pair of loops on a lampbrush chromosome of *Triturus viridescens*. A piece of ovary was incubated in 50 microcuries uridine-H³ (1.0 curie per millimole) for 5 hours before the chromosomes were isolated. Most loops show this type of uniform labeling. Apparent labeling of the main chromomere axis is due to shorter loops which contract and tangle about this region during the fixation and drying which precede application of the autoradiographic emulsion. Exposure 117 days. Developed 4 minutes in Kodak D-19. Mounted in water and photographed with Zeiss 40× apochromatic phase objective. The silver grains completely obscure the underlying loops. × 1250.

Fig. 10. From the same preparation as Fig. 9. However, this is a pair of sequentially labeling loops, and radioactivity extends only a short distance up the thin arm of each loop (*arrows*). The bulk of the loop is not radioactive 5 hours after administration of the isotope. However, over a period of days, radioactivity would extend farther and farther toward the thick end of the loops, until after about 10 days the loops would be uniformly labeled (Gall and Callan, 1962). Photography, etc., as in Fig. 9. × 1250.

(6) The following procedure will yield permanent wet-mounts, if they are desired. Callan and I originally made such mounts (Gall and Callan, 1962), but they offer no particular advantage over dry storage. With a razor blade remove all of the photographic emulsion from the slide except for a square over the chromosomes. Breathe on the emulsion to facilitate the scraping procedure. Add a drop of water containing 1 : 10,000 Merthiolate (Thiomersal) as a preservative, and put on a coverslip. Let excess liquid evaporate from around the edge of the film. Ring the coverslip with a sealing compound, such as lanolin-resin mixture (one part lanolin melted together with two parts commercial resin).

D. Electron Microscopy

Probably all observers will agree that the published electron micrographs of chromosomes, including lampbrush chromosomes, leave much to be desired. I have tried various techniques with the lampbrush chromosomes, including sections, whole mounts, replicas, and negatively stained material, all with greater or lesser degrees of disappointment. The problems—to the extent that one can analyze the failures—seem to be, first, that the chromosomes are generally too thick for useful analysis as whole mounts; second, that sections are difficult to interpret, even when the chromosomes are isolated before embedding. Electron micrographs have been published by workers from several different laboratories (Gall, 1956, 1958; Guyénot and Danon, 1953; Lafontaine and Ris, 1958; Tomlin and Callan, 1951).

Only a few general comments will be made here. Isolated chromosomes, attached to a coverslip, can be easily embedded in plastic. Attach the chromosomes as usual, run the coverslip through the appropriate dehydration technique, and invert over a gelatin capsule completely filled with the plastic monomer. After polymerization, pry the coverslip away, or freeze the coverslip and the end of the capsule on dry ice and flip the two apart with a razor blade. The chromosomes will usually though not always separate cleanly from the coverslip.

Whole mounts can be made directly on grids for the electron microscope (Fig. 11). It is probably best to use steel or platinum grids to avoid

FIG. 11. Electron micrograph of a single lampbrush loop. The chromosomes were isolated directly on a carbon-coated grid, fixed with ethanol vapor, dehydrated in an ethanol series, and dried from CO_2 by the critical point method of Anderson (1956). The bulk of the material comprising the loop appears to be fibrous in nature. From cytochemical studies it is known that this material is largely ribonucleoprotein. $\times$ 8300.

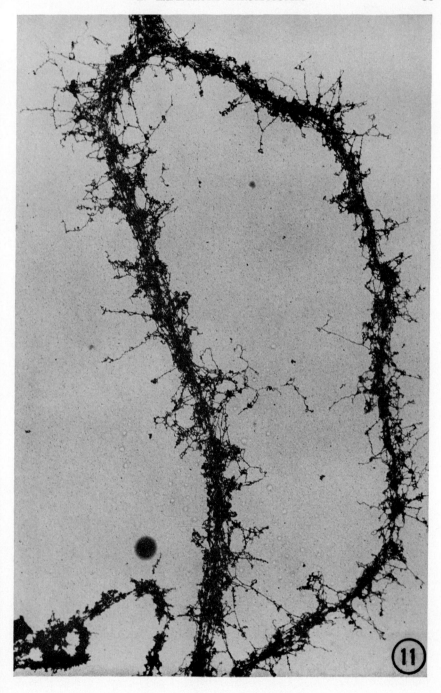

the copper in conventional grids. Place a grid, previously coated with a carbon film, in the bottom of the standard well slide. Isolate the chromosomes directly on the grid, and attach as described earlier. In order to preserve any useful structural detail in the loops, the preparations should be dried by the critical point technique of Anderson (1956; Lafontaine and Ris, 1958).

V. Materials

(1) *Triturus viridescens:* from Lewis Babbitt, Petersham, Massachusetts. Animals sexed if desired, available at all seasons.

(2) *Triturus cristatus carnifex:* from L. Haig, Ltd., Newdigate, Surrey, England. Available March-August.

(3) *Culture dishes* (finger bowls): No. A 16110, 8-inch diameter. Carolina Biological Supply Co., Burlington, North Carolina.

(4) *MS222 anesthetic:* Sandoz Pharmaceuticals, Hanover, New Jersey.

(5) *Jeweler's tweezers* (preferably not stainless): Dumont No. 4 or No. 5 obtainable from any jeweler's supply or may be ordered from B. Jadow, Inc., 860 Broadway, New York 10003. The tweezers may be used as received or ground down a little with a small Arkansas stone. Occasional sharpening is recommended. Dumont tweezers No. TW-404 and No. TW-405. Hard Arkansas polishing slips (triangle) No. ST-601 $3\frac{1}{2}$ inch $\times$ $\frac{1}{4}$ inch.

(6) *Embryological watch glasses:* No. 9844, Arthur H. Thomas Co., Vine Street at Third, Philadelphia 5, Pennsylvania.

(7) *Surgical gut:* U.S.P., Ethicon S-102, type A plain, size 000.

(8) *Pipette:* Figure 12 shows a convenient arrangement for pipette and solution.

(9) *Well slides:* The flat-bottomed depression slides used for this work are prepared by sealing a coverslip over a hole bored through a standard 3-inch $\times$ 1-inch microscope slide (Fig. 8). Holes may be bored in about 10 slides at a time. Bind slides together tightly with tape. Drill slowly, using a $\frac{1}{4}$-inch Carballoy drill with turpentine (or better still, let your glassblower do it!). Seal a coverslip over the hole as follows. Place a small piece of paraffin wax next to the hole and melt with a hot needle. Position a coverslip over the wax and hole. Place on a warming table until the paraffin melts and spreads evenly under the coverslip. Or flame with a fine bunsen flame to melt the wax. Avoid excess wax which sometimes flows into the depression.

(10) *Formaldehyde vapor chamber:* Place one half of a Petri dish in the bottom of an 8-inch finger bowl (culture bowl). Over this lay a circle

of coarse screen to serve as a support for slides. Fill the Petri dish with buffered formaldehyde (9 parts commercial 40% formaldehyde, 1 part 0.1 M phosphate buffer, pH 7.0), and add a glass plate to the top of the bowl.

(11) *Microscope:* By far the best inverted microscope which I have seen is the Zeiss plankton microscope, available from any Zeiss dealer (e.g., Brinkman Instruments, 115 Cutter Mill Road, Great Neck, L. I., New York). This microscope, with low magnification objectives, is often used in tissue culture laboratories for observation through the bottom of tubes or bottles. It may be fitted with phase-contrast optics, however, and is a very fine instrument. I would recommend the complete set of

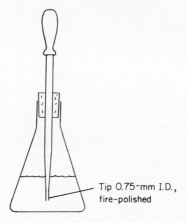

Tip 0.75-mm I.D., fire-polished

FIG. 12. Pipette and flask arrangement useful for nuclear isolation work.

Neofluar objectives: 16×, 25×, 40×, 63× (dry), and 100× (oil). The IS condenser, although of low numerical aperture, is quite good, and its long working distance is an advantage. A camera attachment and monocular tube are available on a permanent stand, and a binocular head is provided for visual observations. Of very great value is the Zeiss flash attachment (Ukatron 60), which fits into the same housing as the regular illuminator and can be kept permanently in place. Fresh or lightly fixed preparations (3–5 minutes in formaldehyde vapor) undergo Brownian motion and can be photographed only by flash. Unfortunately the flash does not provide enough light for photographs with the 63× and 100× objectives, if a slow, fine-grained film is used (Kodak high contrast copy). The complete instrument, with the attachments mentioned, costs about $2700. The cost can be substantially reduced by eliminating the camera attachment and by substituting Zeiss achromat phase objectives. The accompanying photographs (Figs. 2–6) show what can be done with the Zeiss equipment. An inverted microscope is also made by Cooke, Troughton, and Sims, Ltd., and has been used satisfactorily for many years in Professor Callan's laboratory.

Unitron Instrument Company (66 Needham Street, Newton Highlands 61, Massachusetts) sells two inverted microscopes. The first is a small model (Unitron inverted research microscope) and might suffice for many purposes. The price is about $500 with phase-contrast lenses. The large Unitron inverted microscope (Unitron phase-camera-microscope) comes with an array of accessories for only $1580, including a $3\frac{1}{4}$-inch $\times$ $4\frac{1}{4}$-inch camera, 8 phase-contrast objectives, and a binocular head.

ACKNOWLEDGMENT

Professor H. G. Callan and H. C. Macgregor supplied many helpful comments during the preparation of this manuscript. The original studies of the author have been supported by research grants from the National Science Foundation (C-10725) and from the National Cancer Institute, U. S. Public Health Service (C-3503).

REFERENCES

Anderson, T. F. (1956). Electron microscopy of microorganisms. In "Physical Techniques in Biological Research" (G. Oster and A. W. Pollister, eds.), Vol. 3, pp. 177-240. Academic Press, New York.

Battin, W. T. (1959). The osmotic properties of nuclei isolated from amphibian oocytes. Exptl. Cell Res. 17, 59-75.

Brown, G. L., Callan, H. G., and Leaf, G. (1950). Chemical nature of nuclear sap. Nature 165, 600-601.

Callan, H. G. (1952). A general account of experimental work on amphibian oocyte nuclei. Symp. Soc. Exptl. Biol. 6, 243-255.

Callan, H. G. (1956). Recent work on the structure of cell nuclei. Symposium on the Fine Structure of Cells (Leiden). Intern. Union Biol. Sci. Publ. B21, 89-109.

Callan, H. G. (1957). The lampbrush chromosomes of Sepia officinalis L., Anilocra physodes L. and Scyllium catulus Cuv. and their structural relationship to the lampbrush chromosomes of Amphibia. Pubbl. Staz. Zool. Napoli 29, 329-346.

Callan, H. G. (1963). The nature of lampbrush chromosomes. Intern. Rev. Cytol. 15, 1-34.

Callan, H. G., and Lloyd, L. (1956). Visual demonstration of allelic differences within cell nuclei. Nature 178, 355-357.

Callan, H. G., and Lloyd, L. (1960a). Lampbrush chromosomes. In "New Approaches in Cell Biology" (P. M. B. Walker, ed.), pp. 23-46. Academic Press, New York.

Callan, H. G., and Lloyd, L. (1960b). Lampbrush chromosomes of crested newts Triturus cristatus (Laurenti). Phil. Trans. B243, 135-219.

Callan, H. G., and Macgregor, H. C. (1958). Action of deoxyribonuclease on lampbrush chromosomes. Nature 181, 1479-1480.

Callan, H. G., and Spurway, H. (1951). A study of meiosis in interracial hybrids of the newt, Triturus cristatus. J. Genet. 50, 235-249.

Dodson, E. O. (1948). A morphological and biochemical study of lampbrush chromosomes of vertebrates. Univ. Calif. Publ. Zool. 53, 281-314.

Duryee, W. R. (1937). Isolation of nuclei and non-mitotic chromosome pairs from frog eggs. Arch. Exptl. Zellforsch. 19, 171-176.

Duryee, W. R. (1941). The chromosomes of the amphibian nucleus. "Cytology, Genetics, and Evolution." Univ. of Pennsylvania Press, Philadelphia, Pennsylvania.

Duryee, W. R. (1949). The nature of radiation injury to amphibian cell nuclei. *J. Natl. Cancer Inst.* **10**, 735-758.

Duryee, W. R. (1950). Chromosomal physiology in relation to nuclear structure. *Ann. N.Y. Acad. Sci.* **50**, 920-953.

Edström, J.-E., and Gall, J. G. (1963). The base composition of ribonucleic acid in lampbrush chromosomes, nucleoli, nuclear sap, and cytoplasm of *Triturus* oocytes. *J. Cell Biol.* **19**, 279-284.

England, M. C., and Mayer, D. T. (1957). The comparative nucleic acid content of liver nuclei, ova nuclei and spermatozoa of the frog. *Exptl. Cell Res.* **12**, 249-253.

Ficq, A., Pavan, C., and Brachet, J. (1958). Metabolic processes in chromosomes. *Exptl. Cell Res.* Suppl. 6, 105-114.

Finamore, F. J., and Volkin, E. (1958). Nucleotide and nucleic acid metabolism in developing amphibian embryos. *Exptl. Cell Res.* **15**, 405-411.

Flax, M., and Himes, M. (1952). Microspectrophotometric analysis of metachromatic staining of nucleic acids. *Physiol. Zool.* **25**, 297-311.

Fankhauser, G., and Humphrey, R. R. (1959). The origin of spontaneous heteroploids in the progeny of diploid, triploid, and tetraploid axolotl females. *J. Exptl. Zool.* **142**, 379-422.

Gall, J. G. (1954). Lampbrush chromosomes from oocyte nuclei of the newt. *J. Morphol.* **94**, 283-352.

Gall, J. G. (1955). Problems of structure and function in the amphibian oocyte nucleus. *Symp. Soc. Exptl. Biol.* **9**, 358-370.

Gall, J. G. (1956). On the submicroscopic structure of chromosomes. *Brookhaven Symp. Biol.* **8**, 17-32.

Gall, J. G. (1958). Chromosomal differentiation. *In* "The Chemical Basis of Development" (W. D. McElroy and B. Glass, eds.), pp. 103-135. Johns Hopkins Press, Baltimore, Maryland.

Gall, J. G. (1963a). Kinetics of deoxyribonuclease action on chromosomes. *Nature* **198**, 36-38.

Gall, J. G. (1963b). Chromosomes and cytodifferentiation. *In* "Cytodifferentiation and Macromolecular Synthesis" (M. Locke, ed.), pp. 119-143. Academic Press, New York.

Gall, J. G., and Callan, H. G. (1962). H[3]-uridine incorporation in lampbrush chromosomes. *Proc. Natl. Acad. Sci. U.S.* **48**, 562-570.

Gersch, M. (1940). Untersuchungen über die Bedeutung der Nucleolen im Zellkern. *Z. Zellforsch.* **30**, 483-528.

Guyénot, E., and Danon, M. (1953). Chromosomes et ovocytes des batraciens. *Rev. Suisse Zool.* **60**, 1-129.

Hess, O., and Meyer, G. F. (1963). Chromosomal differentiation of the lampbrush type formed by the Y chromosome in *Drosophila hydei* and *Drosophila neohydei*. *J. Cell Biol.* **16**, 527-539.

Humphrey, R. R. (1962). Mexican axolotls, dark and mutant white strains: care of experimental animals. *Bull. Philadelphia Herpetol. Soc.* **10**, 21-26.

Hunter, A. S., and Hunter, F. R. (1961). Studies of volume changes in the isolated amphibian germinal vesicle. *Exptl. Cell Res.* **22**, 609-618.

Izawa, M., Allfrey, V. G., and Mirsky, A. E. (1962). The relationship between RNA synthesis and loop structure in lampbrush chromosomes. *Proc. Natl. Acad. Sci. U.S.* **49**, 544-551.

Lafontaine, J. G., and Ris, H. (1958). An electron microscope study of lampbrush chromosomes. *J. Biophys. Biochem. Cytol.* **4**, 99-106.

Macgregor, H. C. (1962). The behavior of isolated nuclei. *Exptl. Cell Res.* **26**, 520-525.

Macgregor, H. C. (1963). Morphological variability and its physiological origin in oocyte nuclei of the crested newt. *Quart. J. Microscop. Sci.* **104**, 351-368.

Macgregor, H. C., and Callan, H. G. (1962). The actions of enzymes on lampbrush chromosomes. *Quart. J. Microscop. Sci.* **103**, 173-203.

Naora, Hiroto, Naora, Hatsuko, Izawa, M., Allfrey, V. G., and Mirsky, A. E. (1962). Some observations on differences in composition between the nucleus and cytoplasm of the frog oocyte. *Proc. Natl. Acad. Sci. U.S.* **48**, 853-859.

Pantelouris E. M. (1958). Protein synthesis in newt oocytes. *Exptl. Cell Res.* **14**, 584-595.

Ris, H. (1945). The structure of meiotic chromosomes in the grasshopper and its bearing on the nature of "chromomeres" and "lampbrush chromosomes." *Biol. Bull.* **89**, 242-257.

Rogers, M. E. (1965). Polyacrylamide gel electrophoresis of basic proteins from the nucleus and ribosomes of salamander oocytes. *J. Cell Biol.* **27**, 88A.

Rückert, J. (1892). Zur Entwickelungsgeschichte des Ovarialeies bei Selachiern. *Anat. Anz.* **7**, 107-158.

Tomlin, S. G., and Callan, H. G. (1951). Preliminary account of an electron microscope study of chromosomes from newt oocytes. *Quart. J. Microscop. Sci.* **92**, 221-224.

Wilson, E. B. (1928). "The Cell in Development and Heredity," 3rd ed. Macmillan, New York.

Wischnitzer, S. (1957). A study of the lateral loop chromosomes of amphibian oocytes by phase contrast microscopy. *Am. J. Anat.* **101**, 135-168.

The following recent references, not cited in the text, contain additional points of technical interest.

Callan, H. G. (1966). Chromosomes and nucleoli of the axolotl, *Ambystoma mexicanum. J. Cell Sci.* **1**, 85-107.

Gall, J. G. (1966). Nuclear RNA of the salamander oocyte. In "International Symposium on the Nucleolus, its Structure and Function." *Nat. Cancer Inst. Monograph* (in press).

Izawa, M., Allfrey, V. G., and Mirsky, A. E. (1963). Composition of the nucleus and chromosomes in the lampbrush stage of the newt oocyte. *Proc. Natl. Acad. Sci. U.S.* **50**, 811–817.

Macgregor, H. C. (1965). The role of lampbrush chromosomes in the formation of nucleoli in amphibian oocytes. *Quart. J. Microscop. Sci.* **106**, 215-228.

Miller, O. L. (1965). Fine structure of lampbrush chromosomes. In "International Symposium on Genes and Chromosomes, Structure and Function." *Natl. Cancer Inst. Monograph* **18**, 79–99.

Chapter 3

Micrurgy on Cells with Polytene Chromosomes

H. KROEGER

Zoologisches Institut, Eidgenössische Technische Hochschule, Zürich, Switzerland

I. Introduction

The cell physiologist is beset with a problem that also confronts the general physiologist. Both, cells and organisms, respond to experimental treatments with such a multitude of reactions that it is difficult to differentiate between primary and secondary events and to understand the relationship between them. In both cases this situation is an expression of the complex interdependence of the parts that make up the object under study; in both cases the answer should come from studies on

61

isolated systems, particularly if such studies are followed by an effort to recompose the parts into a functional whole.

This technique has been most successful at the level of the organism; it has so far been of limited value at the cell, largely because of their extremely small size. Studies of isolated cell fractions (nuclei, spindles, mitochondria) have yielded information on the composition and function of organelles, but have provided little insight into the relationship between cell parts. Thus "nucleocytoplasmic feedback" has remained a rather empty catchword: the functional significance of membranes as a means for rigid yet selective seclusion of compartments within the cell is only poorly understood; the interaction of cell organelles during cell division, the regulation of gene activities in context with the whole functioning cell, are only now coming into the range of experiment.

The ideal approach to investigating interactions in a cell would consist of dissection, separate treatment, and free recombination of cell parts. With this aim in mind a set of techniques has been developed which allows for the construction of exotic units, such as cells with a mixed cytoplasm derived from differently differentiated cells, cells containing two different sets of chromosomes in their nuclei, or cells containing more than one type of nucleus. With the development of such techniques the road is also open for interspecific combinations of this kind.

The techniques have been worked out for use with dipteran cells which contain giant chromosomes. In these cells the puffing phenomenon (see Kroeger and Lezzi, 1966) permits an evaluation of genomic activities by following puffing patterns and their changes under experimental conditions—a truly priceless asset in view of the modern concept of gene activities as the central switchboard of most or all biochemical events in the cell.

None of the techniques to be described requires either expensive and unusual equipment, other than a first-rate dissecting microscope, or more than an average degree of manual dexterity. Only microinjection into cytoplasm or cell nuclei requires a micromanipulator and more preparation and training.

II. General Procedures and Instrumentation for Micrurgy

There exists a widespread reluctance among biologists to use micrurgical techniques, caused mostly by the erroneous assumption that these techniques require extreme patience, unusual steadiness of hand, and other particular talents. While it is true that a certain amount of training is required, almost anyone should be able to learn the principles of any single technique to be described here within a few days, if taught by an expert, and within a week or two if he follows the written instructions. Any adverse disposition (lack of manual skill, shaky hands, etc.) can usually be overcome by additional training.

In contrast, the development of *new* techniques is a most laborious and time-consuming process. The student of micrurgy is taxing his skills in a world of forces far removed from everyday experience; surface tension, for instance, can exert unexpected forces; gravity, on the other hand, while predominant among the forces of everyday life, can usually be left out of consideration. The task of the micrurgist is not only to cope with these forces but to put them to use on his own behalf. However, even to the specialist with years of experience the world of small dimensions remains full of unforeseen reactions and stunning surprises. It is for this reason that schemes elaborated by a purely mental process almost invariably fail when executed physically. It is essentially by a trial-and-error process in close contact with the material that new techniques are developed.

The exactness and smoothness by which instruments perform an operation are of primary importance in this field. With the exception of microinjection, all techniques to be described are performed freehand and can only so be performed. The swift, continuous, and often complicated movements that a hand can perform, and the simplicity of procedure, an important mark of a truly useful technique, can hardly be achieved by an experimental device such as a micromanipulator. However, it should be stressed again that it is not difficult to train one's hand to perform the movements to be described.

Steadiness and control over fine movements of the hand depend to a high degree on a relaxed and comfortable position of the rest of the body. Even a slightly strained position of the feet, for instance, can substantially interfere with the free movement of the fingers. The first prerequisite is, therefore, that maximum comfort be provided for the whole body. The general position of the seated micrurgist should be such that the eyepieces of the dissecting microscope are directly before his eyes when a most comfortable position is achieved. The arms, from the elbow to the tip of the small finger, rest firmly on the table, the armrests of the dissecting microscope, and the microscope stage, respectively. A Styrofoam rest for the arms and hands gives maximum support to all parts of the lower arm. It is advisable to make these adjustments before actually starting work, since first successes in a less than perfect position are likely to keep the experimenter from improvements and swiftly carry him into a habit difficult to break. Strong muscular efforts should be avoided during the last hours before surgery since they increase the normal muscular tremor to an extent that can interfere with operations.

Most important is the positioning of the hands. All five fingers should be in contact over as large an area as possible with one another and with the microscope stage. The surgical instrument is held like a pencil; the full length of the small finger, the first member of the ring finger (and possibly also the tip of the middle finger), are all in firm contact with the microscope stage. It is important that this particular position be instilled as a habit from the beginning. The optical equipment is a dissecting microscope giving magnifications up to $\times$ 200. It can sometimes be useful to increase magnification beyond the point where resolution is improved, as the impression of "nearness" of the field of observation can aid in directing movements. Provisions for cooling the stage can be useful.

Most experiments are performed with transmitted light and therefore require understage. A white plane serves as a reflector. The white plane is brought into a very steep position (not far from 90° to the direction of the light), giving the same effect as a diaphragm in a compound microscope, i.e., a partial darkfield. Delicate adjustments of the reflector in this range allow one to discern many details otherwise invisible. A foot-driven focusing device can be useful but is not obligatory. It should be power-driven or else adjustable with a minimum of strain, as movements of the leg are apt to be transmitted to the arms and hands.

The following instruments are used: (a) one pair of iridectomy scissors, the blades of which must close tightly; (b) a mouth pipette containing a trap for saliva and ending with a rubber fitting which allows rapid and simple exchange of capillaries with tapered tips; (c) several pairs of fine watchmaker forceps; they are sharpened on a fine grinding stone under the dissecting microscope; and (d) tungsten and glass needles mounted on holders. Tungsten needles are sharpened by repeated immersion of the tip in a mixture of 8 parts KNO_3 and 7 parts $NaNO_2$, which is kept molten in a crucible over a gas flame. Glass needles are drawn from soft glass capillaries of diameter 1–2 mm (melting point capillaries are a good material) over an alcohol or a gas microburner. The pulling is done quickly after removal from the flame by a swift sideways movement. The taper should make the tip very fine (far beyond the resolution of the dissecting microscope), yet provide the needle with enough rigidity to perform its function. On the other hand, if the needle is too rigid it breaks easily when it touches the glass surface of the slide.

The microscope slides on which all surgery is carried out are siliconized in such a way that the oil which covers the preparation does not spread beyond a certain zone, yet the watery phase under the oil can contact the glass surface without being repelled. The mode of preparation of such slides is described in Appendix, 1. Henceforth in this chapter "slide" refers to microscope slides treated as described there.

All operations are performed through an oil layer which covers the preparation and prevents evaporation. Kel-F halofluorocarbon oils have proved to be outstandingly useful for this technique. These oils were introduced to micrurgy by Kopac (1955a,b), who found them to be particularly nontoxic in contact with cytoplasm: while most other oils induce a zone of denaturation in the surrounding protein if injected into amoebae, Kel-F oils[1] did not show such an effect. For the experiments to be described, No. 10 "heavy" oil has proved to be most useful. By mixing different grades with the wax, any viscosity can be easily produced. As a less expensive substitute, a mixture of 2 parts heavy mineral oil with 1 part Oronite Polybutene 128[2] was found to be satisfactory for most purposes. Henceforth in this chapter "oil" refers to either one of these mixtures.

It is often necessary to prepare solutions with a specific titer of materials available only in minute quantities (e.g., hemolymph, hormone solutions, etc.). Two methods can be employed.

a. Mixing two solutions of known composition in a specific proportion. Two drops of oil are placed side by side on a slide. A small quantity of solution I is transferred into the oil and pushed under the surface where it forms a perfect sphere. The diameter of this sphere is measured by an ocular micrometer. Subsequently it is pushed down through the oil and brought into contact with the glass surface. A small amount of solution II is transferred to the second oil drop and pushed under the surface. By quickly opening and closing the prongs of a forceps several times, the sphere is divided into several dozens or hundreds of spheres of different sizes. By using an ocular micrometer a sphere is selected with a volume such that, if combined with the measured amount of solution I, the desired titer will be produced in the mixture. Once the sphere with the desired size is found, it is separated from the cluster of spheres and transferred between the prongs of forceps surrounded by a small volume of oil into the other oil drop, where it is combined with the measured amount of solution I. This technique has the advantage of simplicity and swiftness, particularly if tables are prepared from which the sphere diameter of solution II can be read directly. The accuracy of this method

[1] Kel-F oils are supplied by the Minnesota Mining and Manufacturing Co., Jersey City, New Jersey, in various viscosities including a wax.
[2] Oronite is supplied by the California Chemical International Inc., 200 Bush Street. San Francisco. California.

is limited in that the volume bears a cubic relationship to the diameter measured; the inaccuracy of the method can be reduced by measuring with a compound microscope under high magnification.

b. *Drying out solution I and redissolving the dry material in solution II.* A certain amount of solution I is sucked up into the fine taper of a capillary fixed to the mouth pipette, and the meniscus is marked on the glass. Next, the content of the capillary is forced out and deposited on a slide where it is allowed to dry down; solution II (containing material to be added to solution I in the final titer) is sucked up into the same capillary to the meniscus mark. The content is forced out, deposited on the dry matter on the slide, and immediately covered by oil. The dry material of solution I dissolves in the droplet of solution II. By a repetition of this process more than two solutions can be combined: several solutions are deposited on the dry matter of solution I and each is left to dry, only the last one providing the solvent for the sum of materials deposited.

III. Cell Types with Polytene Chromosomes; Their Handling and Culture

Polytene giant chromosomes have been reported from the macronucleus of ciliates (Ammermann, 1964; Alonzo and Perez-Silva, 1965) and from the following tissues of various dipteran insects: (a) in larvae: salivary glands, Malpighian tubules, various sections of the gut; some degree of polyteny has also been described in muscle, tracheal, and adipose tissue and in isolated ganglia by Makino (1938); (b) in pupae: pulvilli of the pretarsus (foot pads, so far described only for some cyclorrhaphous flies), heart wall, trichogen and tormogen cells of the hypodermis (Whitten, 1963, 1964); (c) in the adult: ovarial nurse cells (Bier, 1960).

Cells containing polytene chromosomes generally seem to have some type of secretory function. As a cell becomes more and more polytene, it usually enlarges in over-all size in a more or less constant proportion to its chromosomes, the eventual size attained varying from tissue to tissue and from species to species (see also Beermann, 1962). While the measurement of cell volumes is often difficult because of an irregular outline, the diameter of the spherical nuclei and the proportions of the polytene chromosomes can easily be determined. Table I gives a list of such measurements in tissues and species most often used in experimental analyses.

TABLE I

APPROXIMATE SIZES REACHED BY NUCLEI AND POLYTENE CHROMOSOMES[a]

Species	Tissue	Diameter of nucleus (μ)	Width of polytene chromosomes (μ)	Remarks
Chironomus tentans	Salivary gland	90-100	18-20	
	Rectal portion of gut	—	4	Calculated from photograph in Beermann (1962)
	Malpighian tubules	40-50	6-9	
Chironomus thummi	Salivary gland	90-100	14-18	
	Malpighian tubules	30-40	5-8	
Drosophila hydei	Salivary gland	30-40	6-10[b]	
Drosophila melanogaster	Salivary gland	30-35	4-6[b]	
Rhynchosciara angelae	Salivary gland	—	8-10	Calculated from photograph in Sirlin and Schor (1962a,b)
Sarcophaga bullata	Foot pads	60-70	6-8	Measured in permanent preparations (kindly provided by Dr. J. M. Whitten, Northwestern University, Evanston, Illinois)

[a] The values given should approximately represent the maximal sizes attained in the particular species and tissues. Stage of development: *Sarcophaga* foot pads: mid-pupal stage; all other measurements: late last larval instar. Values obtained by measuring 3-4 objects or by calculating from photographs.
[b] Measured in fixed material.

Due to the small size of most Diptera, tissues with polytene chromosomes can be collected only in small quantities and are therefore not amenable to gross chemical techniques. In addition, the tissues with the largest chromosomes, those of the salivary glands, surround a lumen filled with secretory product (see Section A, below). It is difficult, if not impossible, to separate the cells from this secretion after dissection. On the other hand, the secretion can be collected in somewhat larger quantities and in reasonable purity by permitting aquatic larvae (for instance, chironomids) to secrete into sterile sand in depression slides moistened with distilled water (Defretin, 1951; Laufer and Nakase, 1965), or by keeping such larvae in a pilocarpine solution (0.2% by volume) which causes them to convulsively express large amounts of their salivary secretion.

While the large size and the ease by which patterns of genomic activity can be determined make tissues with polytene chromosomes a most favorable experimental material, they have a number of disadvantages. Several of the tissues have been shown to exhibit "ionic coupling" (see Section III,B). Further, the culturing techniques for these tissues are rather imperfect (see Section B, below). If wounded, they exhibit very little regenerative or healing power; they do not undergo cell division, and so can be used only for studies on deoxyribonucleic acid (DNA) synthesis and not for studies on mitosis.

A. Morphology

Salivary glands of Diptera are sac-like structures consisting of rows or layers of cells which surround a lumen. In the more advanced stages of chironomid larvae (late last larval instar and prepupa), the nuclei sit in columns of cytoplasm which connect the upper and lower margins of the slightly flattened gland (Fig. 1). In chironomids the salivary gland contains a lobe which, in most species, produces a special type of granular secretion (Beermann, 1961). In sciarids the salivary gland is sharply divided into two zones—a proximal one, which is flattened and has a straight lumen, and a distal one, which is cylindrical, is thinner, and has a lumen which passes through this portion of the gland in a zigzag line.

Since the evaluation of some of the experiments to be described can best be done with the electron microscope, a short description of some cell types with polytene chromosomes, as revealed by this instrument, seems warranted. In chironomids the cell wall toward the hemolymph (Fig. 2A) consists of two layers. The outer (basement) membrane (BM) is thick and resists puncturing by a needle with some strength, the inner (plasma) membrane (PM) is thin and contains large infoldings stretching into the layer of mitochondria that lines the cell wall toward the hemolymph all around the outer margin of the gland. Below the mitochondrial layer the cytoplasm is filled with an endoplasmic reticulum (ER) densely seeded with ribosomes. In this region the nucleus and many Golgi bodies are located. The border to the glandular lumen (i.e., the secretory surface, Fig. 2B) is densely covered with microvilli (MV) projecting into the lumen (L). There is also visible in Fig. 2B the septate membrane (SM) which separates adjacent cells (discussed in Section B, below). The morphology of nucleus and polytene chromosomes has been described by Beermann (1962) *in extenso*.

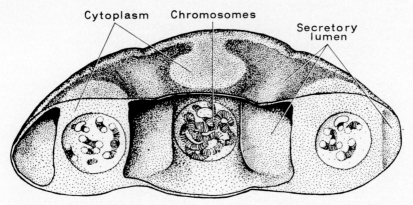

Fig. 1. Cross section through chironomid salivary gland. (From Kroeger, 1964, by permission of Springer-Verlag, Berlin.)

Malpighian tubules are long, blind-ended sacs, which project from the gut into the body cavity. They are subdivided into several portions with distinctly different morphological and physiological features. Electron micrographs of Malpighian tubules of *Drosophila melanogaster* show three distinct regions in each cell, a basal one with many infoldings of the cell membrane, an intermediate region with an endoplasmic reticulum and many Golgi bodies, and an apical region characterized by microvilli projecting into the lumen (Wessing, 1962).

The foot pad cells of higher flies secrete extensive areas of adult cuticle during the pupal stage. They are derived from unspecialized small hypodermal cells and are superficially located (Whitten, 1963).

Temporary or permanent squash preparations of polytene chromosomes are made as described in Appendix, 3. Such preparations are particularly well suited for evaluations of puffing patterns (see Section I, and Kroeger and Lezzi, 1966).

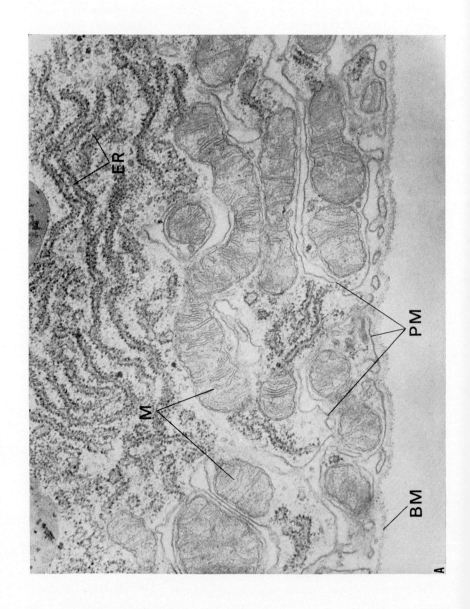

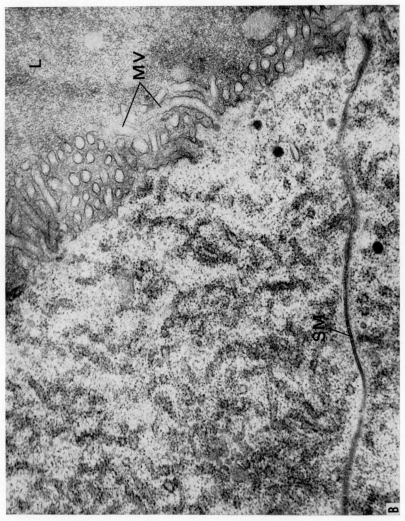

FIG. 2. See legend on p. 72.

B. Culture

Most types of tissue with polytene chromosomes can be kept without difficulty in short-term (up to 24 hours) tissue culture. Prolonged *in vitro* culture of these tissues has so far—with the possible exception of *Drosophila*—been unsatisfactory. Salivary glands or Malpighian tubules can be kept alive for rather extended periods, but their cells soon fail to show signs of active metabolism or of growth. However, no extensive efforts to culture these tissues have come to the attention of the author; possibly only moderate effort is required to adapt general culturing experience to these tissues.

Short-term *in vitro* culture is done either with the hanging drop technique on a cover slip inverted over a depression slide and sealed with oil, or directly on the glass surface of a slide under a cover of oil. With the latter technique, culturing beyond approximately 2 hours (depending on the type and quantity of the medium) requires provision for a sufficient oxygen supply. This can be achieved by blowing air or pure oxygen through a pipette into the medium after it is covered with oil; if the tip diameter of the pipette is small enough, the medium is transformed into a foam which contains approximately equal amounts of liquid and gas. The formation of a foam is dependent on trace amounts of protein or other large molecules in the medium. However, even in an artificial medium devoid of protein, a sufficient quantity of protein has usually leaked out from the gland within a few minutes after its explantation.

As a general rule, hemolymph is by far the best medium for any type of culture; it is greatly superior to any artificial medium so far tested. This is particularly true for wounded cells. In some Diptera, the unbalancing of the phenoloxidase dehydrogenase system after withdrawal of hemolymph constitutes a certain problem, but the hemolymph tyrosinase can be either blocked by phenylthiourea (Schmidt and Williams, 1953) or inactivated by heat treatment at 60°C for 5 minutes (Wyatt, 1956). In chironomids the reaction is so weak that such precautions are not necessary.

FIG. 2. Electron micrographs of cell boundary (A) toward hemolymph, and (B) toward secretory lumen, of a *Chironomus thummi* salivary gland cell. BM, basement membrane; PM, plasma membrane with large infoldings into the mitochondrial layer; ER, endoplasmic reticulum, seeded with ribosomes; M, mitochondria; L, secretory lumen; SM, septate membrane, the border between adjacent cells; MV, microvilli, projecting from the secretory surface into the lumen. Osmium fixation. A: × 36,000; B: × 28,800. (Courtesy of Dr. Barbara J. Stevens, University of Chicago.)

For artificial media the requirements have proved to be remarkably different from one dipteran group to the next. For tissue culture of *Drosophila melanogaster* an extensive literature exists and several media are in use (see Jones, 1962). It seems likely that these media can be used with tissues containing polytene chromosomes, although long-range *in vitro* culture of such tissues has not been reported. A particular kind of *in vivo* culture has been devised by Hadorn *et al.* (1963): by repeatedly transferring salivary glands back into more juvenile hosts, salivary glands from *Drosophila melanogaster* were stimulated to grow beyond their normal size, some nuclei reaching a DNA content 3.5 times higher than the controls. For short-term storage of *Drosophila* tissues, "drosophila-Ringer" as devised by Ephrussi and Beadle (1936) is satisfactory: NaCl 7.5 gm; KCl 0.3 gm; CaCl$_2$ 0.21 gm; distilled water 1000 gm.

For chironomids, the artificial medium devised by Jones and Cunningham (1961) is the most satisfactory one. The medium of choice for sciarids is Morton, Morgan and Parker's "Medium 199" (Glaxo Laboratories Ltd., Greenford, England). It is used undiluted.

One of the principal problems in culturing cells with polytene chromosomes (and possibly a major handicap in general insect tissue culture) is the recently detected phenomenon of "ionic coupling": Loewenstein and Kanno (1964) have determined that dipteran salivary gland cells have high ionic barriers toward the hemolymph and toward the glandular lumen, but practically no such barriers between themselves, a condition brought about by a particular type of membrane (septate membrane, SM in Fig. 2B) which separates adjacent cells (see also Kanno and Loewenstein, 1964; Kroeger, 1966). The whole organ is practically devoid of internal ion barriers. This means: if any one cell in this organ (and the same is probably true for many other organs) is leaky against the hemolymph or against the lumen, the whole gland is opened to the milieu and rapidly loses various small molecules while gaining other molecules from the medium. In view of the recently established regulatory function of the internal ionic milieu on gene activities (Kroeger, 1963, 1966; Kroeger and Lezzi, 1966), it is therefore imperative to use tissues without cellular wounding if prolonged culturing effects are to be achieved. This is most difficult with some tissues; salivary glands of chironomids are held in place by a salivary duct and fine ligaments, these ligaments having—at least in some species—a tendency to be torn out of their anchorage by the process of explantation, leaving large cellular wounds. Tearing can be prevented by using forceps to control the release of glands from the decapitated larva, and carefully cutting the ligaments well away from the gland (see Kroeger, 1966).

IV. Isolation of Nuclei and Chromosomes

A. Nuclei

Obtaining isolated cell nuclei with their biochemistry as undisturbed by the isolation procedure as possible is one of the principal problems of the cell physiologist. Because of the strictly controlled, purely mechanical forces employed, micrurgy yields isolated nuclei which are very clean and biochemically intact; on the other hand, due to size limitations, only salivary gland cells and Malpighian tubule cells seem amenable to these techniques and only small quantities can be prepared. The predominant problem in this technique is posed by the fact that micrurgical means can be employed for nuclear isolation only where there is no firm attachment of the nucleus to the cytoplasm, a condition which varies from one dipteran group to the next and between developmental stages. The Diptera of choice for this technique are the sciarids, as in these the nucleus is very loosely connected with the cytoplasm at all stages. The second choice is *Drosophila* salivary gland cells just prior to puparium formation. Foot pads of higher cyclorrhaphous flies are also a good source for nuclei with giant chromosomes. In chironomids, the anchorage of the nucleus in the cytoplasm is firm at all times, having so far allowed only semiisolation of nuclei. The experience with the various dipteran groups and tissues is summarized in Table II.

Two types of culture medium have been employed with success.

a. Diluted contents of Drosophila eggs. In insect eggs the cleavage nuclei are not encased in cells until blastoderm formation; therefore egg contents of this stage would seem to be a most natural environment for isolated nuclei. Pursuing a lead given by von Borstel (1959), who found that cell division in eggs of *Habrobracon juglandis* can continue for at least 8 generations after the eggs have been opened under oil, semiisolated nuclei from salivary glands of *Drosophila busckii* (Kroeger, 1959) and of *Chironomus thummi* (Lezzi, 1961) have been cultured in egg contents up to 4 hours, a time amply sufficient for observations on changes of puffing patterns and for tracer studies of various kinds. It should be noted that the contents of eggs from different stages of development were found to induce different puffing patterns in the semiisolated nuclei (Kroeger, 1959). The egg content had too high a tonicity for somatic nuclei from later stages of development; it was adjusted by the addition of distilled water, using technique (a) described in Section II.

b. An artificial sugar medium. The artificial sugar medium was devel-

TABLE II

ISOLATION AND CULTURE OF CELL NUCLEI

Species	Tissue	Stage, when isolation is possible	Culture medium	Author
Bradysia mycorum	Salivary gland	Last larval instar	Sugar medium (see Appendix)	Rey (1963)
Rhynchosciara angelae	Salivary gland	Last larval instar		Sirlin and Schor (1962a,b)
Drosophila melanogaster	Salivary gland	Immediately before and after puparium formation	*Drosophila* egg content	Kroeger, unpublished data (1958)
Drosophila busckii	Salivary gland			Kroeger (1959)
Sarcophaga bullata	Foot pads	Mid-pupa	No culturing described	Whitten (1964)
Chironomus thummi	Salivary gland	No isolation possible, semiisolation possible during last larval instar	*Drosophila* egg content	Lezzi (1961)

oped by Frenster *et al.* (1960) and is replenished with large synthetic molecules (see Appendix, 4). Isolated nuclei of salivary glands from *Rhynchosciara angelae* have been cultured successfully in this medium for 3 hours (Sirlin and Schor, 1962a,b), nuclei from *Bradysia mycorum* for 2 hours (Rey, 1963).

Nuclei are isolated and cultured in the following way: (1) Donor tissue is placed on a slide and covered with oil. Salivary glands should contain as few cellular wounds as possible, since damage to the organ causes salivary secretion to leak from the lumen; the medium around the tissue quickly turns into a sticky mass and all further procedures become impossible. (2) From a second oil drop containing culture medium, two spheres of medium are transferred to the oil drop which covers the donor tissue, one approximately 1/5–1/10 the volume of donor tissue, the other with a volume 10–20 times larger. (3) The smaller sphere is brought into contact with the donor tissue. The ratio between tissue and fluid should be such that the tissue is surrounded by a rim of medium but does not float.

From this point on, the techniques for isolation and semi-isolation (where a rim of cytoplasm is left to surround the nucleus) are different:

a. Procedure for semi-isolation. By means of a tungsten needle bent into a microhook, the cytoplasm around a nucleus is excised. The hook is held sideways and pressed down on the cytoplasm so that it cuts out a piece; the piece is removed by withdrawal of the tungsten hook sliding along the surface of the glass. Cuting stepwise nearer and nearer to the nucleus (without ever touching it), this procedure is repeated until the nucleus is surrounded by only a rim of cytoplasm. A number of nuclei are isolated in this way from each gland. Since the above mentioned leakage of glandular secretion cannot be avoided entirely, it is imperative to work fast at this stage.

Except for the semi-isolated nuclei, all tissue remnants are removed through the oil. Then a narrow connection to the larger drop of medium (see step 2, above) is produced along the glass surface by rubbing one prong of a forceps against the glass surface, thereby gradually enlarging the contact between medium and glass in the desired direction. As soon as the two droplets are interconnected by a liquid bridge, the semi-isolated nuclei are pushed over into the larger droplet one by one (Fig. 3), preferably by creating small waves in the medium with a needle; the nuclei should not be directly touched. Next the liquid bridge is broken by rubbing the tip of a needle across its path, the needle having previously been dipped into slightly wetted lecithin. It is important that the connection between the droplets be kept narrow, as otherwise a large amount of medium tends to move over into the smaller droplet.

Furthermore it is essential that no remnants of the tissue be passed to the larger droplet along with the nuclei, as such remnants, cytoplasm or secretion, prove deleterious to the nuclei in culture.

b. *Procedure for isolation.* A glass needle is inserted into a cell; by a swift movement a slit is ripped through the cellular membrane. Subsequent gentle squeezing of the cell causes the nucleus to pop out

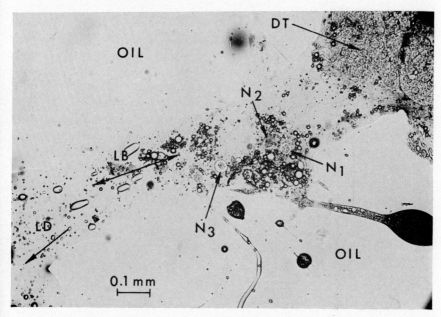

FIG. 3. Isolation procedure for salivary gland nuclei. A small drop, containing donor tissue (DT, *Drosophila melanogaster* salivary gland, immediately after puparium formation; nuclei *in situ* are visible), is connected by a liquid bridge (LB) to a large drop (LD), everything covered with oil (OIL). The watery phase consists of the sugar medium described in Appendix, 3. N_1–N_3, 3 nuclei in various stages of separation from cytoplasm. While the photograph was taken the nuclei were left near cell debris; they therefore show the typical clumping of chromosomes concomitant with a swelling of the nucleus, which reflects the deleterious influence of cytoplasmic fragments. N_1 and N_2, early stage of chromosome clumping; N_3, advanced stage of clumping.

through the slit without cytoplasm adhering to it. From this point onward, the procedure is the same as for semi-isolated nuclei. Isolated nuclei are particularly sensitive and should never be touched directly by a needle. The nuclei can be observed continuously during the culture period by phase contrast. The general appearance of the chromosomes should not perceptibly change during this period; shrinkage and gain of refractoriness of the chromosomes are typical signs either of a wounded nuclear envelope or of an adverse condition in the medium. Abortive preparations are most often caused by a deviation of the medium from isotonicity.

If only the chromosomes and their puffing pattern are of interest, squash preparations can be prepared directly on the slides. At the end of the incubation period a sphere of acetic-orcein (see Appendix, 2) is transferred to the oil and brought into contact with the incubation medium. If the droplet of stain is small, the medium along with the nuclei is transformed into a compact mass that adheres firmly to the glass surface. A larger drop of stain causes strong currents at the moment of admixture with the medium, which prevents the production of a continuous mass.

After a staining period of 20 minutes to 4 hours (depending on the stain, the species, and the type of tissue) the slide is placed perpendicularly in a staining dish with acetone. Within minutes the oil is dissolved and the slide is transferred to distilled water. After 30 seconds it is withdrawn and the adhering water is blotted up except for a small amount covering the preparation. A drop of acetic-orcein is placed on top of the preparation for further staining and for a softening of the tissue; after a few minutes a cover slip is placed over the preparation. If many preparations are lost from the slide in either the acetone or the water, the degree of siliconizing of the slide should be decreased.

B. Chromosomes

Techniques for the isolation of functional polytene chromosomes are not as well developed as those for the isolation of nuclei. Fixed chromosomes can easily be obtained: 45% acetic acid (with or without orcein) or 5% formalin renders the chromosomes rigid. They can be easily freed from the cell with the help of glass or tungsten needles, recognized, and collected separately. Fixed chromosomes can be of interest for investigations on their chemical composition (see, for instance, Edström and Beermann, 1962), but cannot be used for physiological studies.

Unfixed salivary gland chromosomes can be collected by an unusually simple and rapid technique: if a glass needle is inserted into the nucleus of a salivary gland cell of *Chironomus thummi* explanted under oil, the nuclear content can be almost completely scooped out of the gland and brought into the surrounding oil with one swift movement that moves the needle sideways and out of the cell. The nuclear content adhering to the needle is left to float in the oil, or is wiped off on the surface of the slide. Once the gland is explanted, one nuclear content can be gathered every 6–8 seconds; the technique lends itself particularly to experiments where larger numbers of chromosomes or nuclear sap are to be amassed.

The nuclear contents collected by this technique are, due to the extremely swift passage through fairly stiff cytoplasm, not much contaminated with extranuclear material and probably little altered in their chemical composition. However, they have one most decisive disadvantage: they have been subjected to sufficient stress to disfigure their giant chromosomes; no studies on chromosome configuration or puffing pattern can be made on them.

Three techniques have been described which yield unfixed, structurally intact chromosomes: (1) Karlson and Löffler (1962) have treated chironomid salivary glands with pronase, homogenized and subjected them to differential centrifugation; they produced fairly clean chromosome preparations. (2) Buck and Malland (1942) found that immersion of salivary glands in 0.25% solution of dried egg-white for 2–3 hours renders polytene chromosomes sufficiently stiff to allow their isolation by micrurgy. (3) Glancy (1946) explanted chironomid salivary glands, tore them, and ruptured their nuclei with fine steel needles; the chromosomes were pressed out of the nucleus and immediately floated away into the surounding medium. Lezzi (1965) made use of this technique but employed a sugar medium (see Appendix, 4) developed by Frenster et al. (1960).

Although the chromosomes obtained by these methods are structurally intact, it is doubtful that, with the probable exception of the sugar medium, they are in a native state. Extensive efforts to liberate chromosomes from their nuclei in nonaqueous media were to no avail,—the chromosomes proved to be extremely flabby and gel-like and their disfiguration could not be avoided (for the elasticity, consistency, and other properties of polytene chromosomes under various conditions, see Glancy, 1946). However, it seems probable that most or all the alterations induced during isolation with either one of these techniques are reversible, and that the chromosomes can be rendered functional once they are reintroduced into an appropriate milieu.

As far as ribonucleic acid (RNA) synthesis is concerned, the minimum requirements for physiological activity of a chromosome should be: (a) four nucleotides (as triphosphates), (b) RNA polymerase, (c) an energy-regenerating system, and (d) certain ionic titers serving as regulators for gene activity (see Kroeger and Lezzi, 1966). However, certain proteins (for instance, an "acidic puff protein," see Kroeger and Lezzi, 1966) are probably necessary in addition.

No such medium has so far been described. However, it is felt by this author that with some effort the present limitations could be overcome to allow true *in vitro* culture of polytene chromosomes.

V. Removal of Chromosome Segments and Whole Chromosomes

In salivary gland cells of chironomids, it is possible to make cells highly deficient by micrurgical means. This technique produces cells devoid of part or all of their genetic complement, a situation most intriguing for the cell physiologist intent on studying the contribution of various parts of the genome to the metabolism of the cell or to intranuclear regulatory processes.

The salivary gland is placed on the slide with as little adhering hemolymph as possible and quickly covered with oil. A glass needle, which has been dipped into concentrated siliclad (Clay-Adams Inc., New York, N.Y.) is introduced into the nucleus, preferably into one that is superficially situated (Fig. 4A). The needle is then withdrawn vertically with one or more chromosomes looped around it. Basically the movement is similar to that used to scoop out nuclear contents (see Section IV,B), but is done slowly and in a controlled manner. It is difficult to remove one chromosome singly, as the other chromosomes have a tendency to move with it. Therefore one chromosome piece is pulled slowly, while the amount of chromosome material remaining in the nucleus is continuously controlled. If the desired fraction is removed from the cell, the needle is wiped over the surface of the gland to free it from the adhering chromosomes. The chromosome parts outside the cell can be cut off by two glass needles sliding against one another like the blades of a pair of scissors, or can be left dangling into the thin layer of hemolymph around the gland, where they quickly degenerate (Fig. 4B).

The withdrawal of chromosome parts is done at random. Since any chromosome part, which was outside the nucleus, stains with orceine as a blackish mass devoid of a banding pattern, the part of the chromosome set actually removed from the nucleus can be determined upon termination of the experiment in a squash preparation. The determination is accurate down to a few bands in a chromosome. Random removal plus retrospective evaluation is the most economical technique: 4–8 nuclei can be made deficient per minute. However, to a limited extent the removal of specific pieces of the chromosome set is also possible. In chironomids the nucleolus is so large that it resists removal from the nucleus through the small opening in the nuclear envelope more than do other chromosome pieces. It is therefore possible to remove all chromosomes except for the nucleolus and closely adjacent chromosome segments; in *Chironomus thummi* this retains all of the small 4 chromosome, which also carries the Balbiani rings. It is also possible to remove the nucleolus alone; a variety of other combinations can be achieved.

Prolonged culture of deficient cells has not yet been accomplished; the

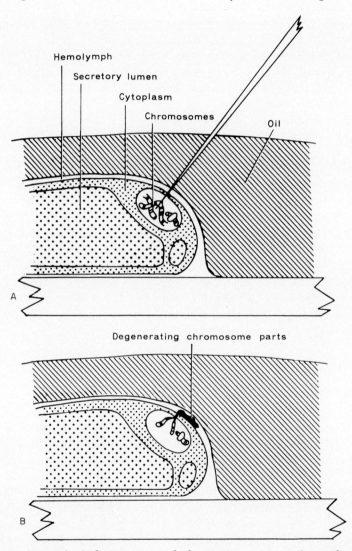

FIG. 4. Removal of chromosomes and chromosome segments from *Chironomus* salivary gland nuclei. A: A glass needle is inserted into a nucleus and is brought into contact with a chromosome. B: Part of the chromosome set has been withdrawn and left to hang into the hemolymph, where it degenerates quickly. (From Kroeger, 1964, by permission of Springer-Verlag, Berlin.)

cell wound inflicted by the withdrawal procedure does not heal, and in time material is exchanged through it until the cell eventually dies. Increasing the viscosity of the medium (preferably hemolymph) by the addition of methyl cellulose considerably retards the onset of alterations in the cell; efforts to close the external wounds by using epoxy resin systems have led to a further increase in survival time of wounded cells. However, the technique is still in an imperfect state.

If left under oil without the addition of culture media, the cells carry enough nutritional reserves to keep them metabolically active for 2–4 hours. Their puffing activity is not significantly lower then when they are surrounded by hemolymph.

VI. Transplantation of Cytoplasm and Chromosomes

One of the truly challenging goals of the micrurgist is the construction of viable cells composed of parts from different origin. Thus, cells containing cytoplasm from different types of tissue or from different species, cells with more than one nucleus, each donated by a different tissue, or cells whose nucleus or chromosome set has been replaced by the respective part from another cell type are of particular interest for the student of somatic cell differentiation and for the student for interactions between cell organelles. Most of these composite units can be composed by micrurgy.

A. Cytoplasm

Large pieces of cytoplasm from any source can be implanted into the cytoplasm of salivary gland cells of *Chironomus thummi;* in fact, cells with ⅔ of their cytoplasm of foreign origin can be prepared. A gland is placed on a slide with as little hemolymph adhering as possible, and covered with oil. The donor tissue is transferred into the same oil droplet, and brought into contact with the glass surface near the gland but not in liquid connection with it. With two needles the tissue is torn into pieces and, if possible and desired (the feasibility depending on the size of donor cells), the nuclei are removed. If large cells (e.g., Malpighian tubule cells) are used as donor material, the plasma membrane of the donor cells should be torn so that the plasma fragments to be transferred do not contain large, uninterrupted areas of plasma membrane; otherwise such areas may wrap themselves around the transplant, forming a barrier between host and donor cytoplasm.

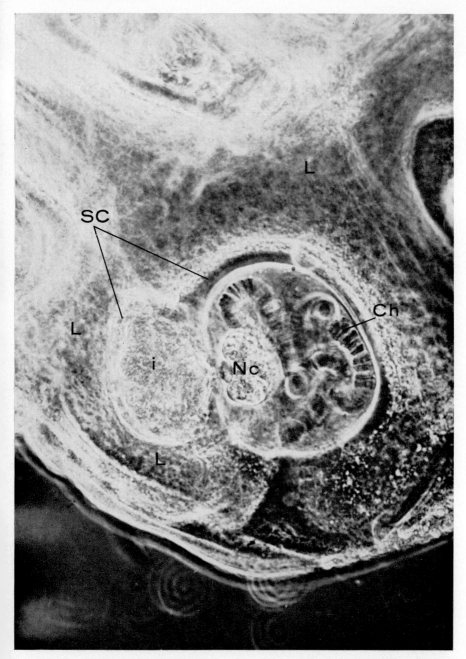

Fig. 5.   Cytoplasm from Malpighian tubule cell implanted into a salivary gland cell of *Chironomus thummi*. Note the rim of salivary cytoplasm which surrounds the implant at its luminar side. I, implant; SC, salivary cytoplasm; Ch, chromosomes; Nc, nucleolus; L, secretory lumen. Phase-contrast photograph. ca. × 500.

A glass needle is introduced into the host cell near a nucleus. By a fast movement the needle is removed vertically, cutting a plane through the cytoplasm and a slit through the cell membrane. The nucleus must not be touched during this process. Since after withdrawal of the needle the cut through cytoplasm and plasma membrane is invisible, it is necessary to remember its length and position in relation to the cell nucleus and the general outline of the cell.

A piece of donor cytoplasm is taken up with the tip of a needle, pushed through the oil, and laid down on the surface of the gland directly over the slit. It is pressed through the slit into the host cytoplasm until the slit closes over it. The final appearance of a medium-sized transplant in a salivary gland cell is shown in Fig. 5. It is often difficult to determine whether all the transplant is actually in the host cell and whether the slit is well closed. This can be ascertained by moving the very tip of the needle over the slit, touching the surface very slightly: projecting parts of the transplant and irregularities of the surface show up by being specifically moved by the needle.

The major difficulty with this technique is a tendency of the transplant to reappear at the surface of the cell when the needle is retrieved. If the needle is very strongly siliconized (using silicone vapor plus prolonged curing), and if it is withdrawn in a direction strictly parallel to its long axis, this can be partly eliminated. Nevertheless, withdrawal of the needle is the point where transplantations most often fail.

One, two, or more chunks of cytoplasm can be introduced either through the same cut or through two cuts. Cytoplasm from any source can be used as a donor material. Malpighian tubule cells are particularly easy to handle due to their large size; in addition, if transplanted into a salivary gland cell, Malpighian cytoplasm can readily be identified in electron micrographs since its endoplasmic reticulum is devoid of ribosomes (B. J. Stevens, personal communication, 1964). It is also possible to introduce whole, unbroken cells of small diameter (for instance, nerve cells from the same organism) into the cytoplasm, if the reaction of cells surrounded by an intracellular milieu is of interest.

The introduction of cytoplasm into the salivary gland nucleus is also possible. It leads, however, within minutes to drastic pathological reactions on the part of the chromosomes. The procedure follows that for the introduction of foreign chromosomes into the nucleus (see below).

B. Chromosomes

The transfer of chromosomes into the cytoplasm follows essentially the same lines as the transplantation of cytoplasm. The chromosomes are obtained by the scooping-out technique (described in Section IV,B). They are rolled into bundles, placed over the slit in the cell membrane, and pressed into the cell. There they degenerate quickly, but the subsequent distribution of their components (DNA, RNA, proteins) over the cell can be followed by autoradiography and is of interest as it follows specific patterns (see Kroeger *et al.*, 1963). A most challenging technique, taxing the skill of the micrurgist to a considerable degree, is the introduction of chromosomes (or cytoplasm, see above) into the nucleus of *Chironomus* salivary gland cells. Again, essentially the same technique is used as for transplantation into the cytoplasm, but here the slit is made directly over the nucleus without actually piercing the nuclear envelope. Once the transplant is positioned in the cytoplasm above the nucleus, it can be pushed further down, the needle tip moving in advance of the transplant and cutting the necessary hole into the nuclear membrane. Sometimes it is easier to select superficially located nuclei where the transplant can be pressed into the nucleus in one step. It is to be stressed, however, that this technique is not easy to learn and is one of the very few described in this chapter which must be designated as "tricky."

The transplanted chromosomes are disfigured and not amenable to cytological investigations (see Section IV,B). Since they are probably intact chemically and physiologically, their effect on the set of host chromosomes and on the host cytoplasm can be studied. Furthermore, it is possible to study the effect of transplanted chromosomes in a cell devoid of its own genome. To this end, salivary gland cells are used as hosts from which all chromosomes have previously been withdrawn by the technique for the production of deficient cells (see Section V).

As for short- or long-term culture of cells carrying transplants in their nucleus or cytoplasm, the situation is the same as with cells made mechanically deficient (see Section V): for short-term experiments the cells carry enough nutritional reserves to stay active without a medium, for long-term culture the techniques are as yet not satisfactory.

VII. Microinjection

A. Injection into Larvae

The injection of small, measured amounts of liquids into insect larvae requires a system where the injection pressure can be delicately controlled. A glass needle with a sharp tip (bore at the tip: around 100μ) is internally siliconized by directing dry air through silicone oil (one of the brands listed in Section II) and subsequently through the needle.

It is then connected by polyethylene or hard rubber tubing to a mouth pipette or a mounted syringe driven by a micrometer.

Dipteran larvae to be injected should be anesthetized to avoid contractions which lead to a loss of hemolymph and injected material upon withdrawal of the needle. To avoid their giving way to the injecting needle, some types of larva must be fastened. Nonaquatic larvae are placed on the sticky side of a piece of Scotch tape, and firmly held there. A drop of water liberates them immediately. Aquatic larvae with a wet surface do not stick well to Scotch tape. If anesthetized, they do not need fastening if placed on a slide. Injection without anesthesia is possible by rolling them into wetted cellulose wadding or cotton wool, which is fastened by insect pins to paraffin or cork, leaving the site of injection uncovered (Pelling, 1964).

The hemolymph of most nonaquatic larvae coagulates within minutes and it is usually not necessary to cover the wound. If special circumstances call for it, the wound can be covered by a droplet of molten paraffin. In contrast, wounds of aquatic larvae, such as chironomids, do not close and paraffin does not stick well to their wet body surface. Injections are therefore made through body appendages (see Pelling, 1964) and the appendages are ligated immediately afterward. Pelling (1964) has also described a foot-controlled machine, which is used to pull the ligature tight while the injecting needle is withdrawn.

If particulate matter such as tissue fragments or imaginal discs are to be injected, a needle with a constriction must be used to prevent the particle from being sucked into the injection apparatus proper during aspiration (Ephrussi and Beadle, 1936).

B. Injection into Cytoplasm and Cell Nuclei

Injection into the cytoplasm or into cell nuclei requires much more refined techniques. Although originally developed for use with chironomid salivary glands (Robert, 1964), the technique to be described can doubtless be employed with various other cell types. As compared with other microinjection techniques (injection by controlled heating, by micrometer-driven microsyringes, etc.), its principal advantage is the simplicity of procedure; its disadvantages lie in imperfect quantitation (see below) and a tendency of the needle tip to get clogged if material to be injected is aspirated.

The intruding injection needle inevitably wounds the cell. To keep the wound small, while providing an opening at the tip sufficient for injection without application of extreme pressure, the tip should be as thin-walled as possible. Tubes with very thin walls, however, are difficult to handle, so the wall thickness is made to decrease toward the tip. For this purpose, thick-walled soft glass tubing (2.3–3 mm diameter, 0.3–0.4 mm bore; such glass tubing is provided by thermometer manufacturers) is cut into pieces 10–20 cm in length, which are closed at both ends over a flame (Fig. 6A). Next, the tube is heated over its entire length, the air within creating an internal pressure. While it is still hot, the tube is heated intensely over a short segment of its length; the internal pressure causes the softened glass to bulge out into a bubble, the walls of which are proportionately much thinner than the rest of the tube (Fig. 6B). A much higher internal pressure is achieved if the tube is partially filled with water by means of steel tubing of 0.3-mm external width before it is closed. However, the resulting vapor pressure is more

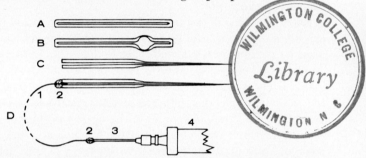

FIG. 6. Instrumentation for microinjection into *Chironomus* salivary gland cytoplasm or cell nuclei. A: Piece of thick-walled glass tubing closed at both ends. B: Tubing with thin-walled bulge. C: Thin-walled portion pulled out into a fine taper. D: Blunt end connected to a syringe (1, steel tubing; 2, epoxy resin mount of steel tubing to capillary and hypodermic needle; 3, hypodermic needle; 4, syringe).

difficult to control. The tube is then opened at both ends, and the bubble is heated again and pulled out into a fine point, the tip being strong enough to support the advancing point, although it tapers so slowly that the hole produced at the cell membrane remains small (Fig. 6C). If the first effort to achieve such a taper fails, the pulling procedure can be repeated with the same tube, using forceps to pull small pieces of glass from the tip immediately after removal from the flame.

The tip is strongly siliconized by dipping it into concentrated silicone oil (one of the brands listed in Section II), and 15 cm of the steel tubing mentioned above is glued into the blunt end of the glass tube by an epoxy resin cement. Previously or at the same time, the other end is glued into a hypodermic needle which just fits over it (Fig. 6D). After the glue has hardened, the hypodermic needle is attached to an ordinary 1-ml syringe which is greased so as to give a stiff movement and which contains the liquid to be injected. Once filled, the system should be entirely or almost free of air bubbles. The movable stage of a compound microscope has been successfully used as a substitute of a micromanipulator.

A salivary gland is explanted under oil, and the tip of the injection needle is brought into the same optical plane. It is then advanced into the cell, its tip either remaining in the cytoplasm or piercing the nuclear membrane and resting in the nuclear sap. If the needle does not slide into the cell smoothly but pulls on the plasma membrane, it is insufficiently siliconized. By applying moderate pressure to the plunger, part of the material held in the injection needle is transferred into the cell.

This injection method is based on the fact that a system of the type and dimensions described above offers a resistance that happens to be in the range conveniently overcome by an amount of force the hand is used to exert in everyday life.

Entry of material into the nucleus can be recognized by the movement of chromosomes or sudden increase in the nuclear diameter, if a somewhat larger amount is injected. This increase should be barely perceptible and can be kept fairly uniform from one cell to the next; differences in material injected are achieved by diluting the material, not by altering the volume. Injections into the cytoplasm usually result in small reversible or irreversible changes in the optical properties of the cytoplasm around the tip of the injection needle. Keeping size and intensity of such changes constant from injection to injection can be learned. However, these observations afford nothing more than a rough estimate of quantities; for this technique the problem of quantitation is not yet adequately solved.

VIII. Conclusions

It seems unnecessary to indicate any of the many problems that can be approached by the techniques described in this chapter; obviously their number is large, and many experimental pathways point directly toward some of the most pressing questions of modern biology.

As for the evaluation of cell reactions to various operational procedures, three techniques seem particularly well suited: (a) autoradiographic tracer techniques, allowing the identification of molecules in small amounts and often requiring only short-term experiments, (b) electron microscopy, which can reveal submicroscopic reactions of nucleus and cytoplasm to experimental influences, and (c) immunological techniques, specifically the use of fluorescent antibodies.

The three unsolved problems which now most severely hamper further extension of micrurgical techniques with cells containing polytene chromosomes are (a) the difficulty of closing wounds at the cellular and nuclear membrane, (b) the insufficiency of methods for prolonged culture of tissues containing polytene chromosomes, and (c) the lack of media suitable for chromosomes culture. In the case of tissue culture, the problem probably does not consist so much of elaborating suitable culture media and culture conditions, but rather in the development of methods yielding explants without wounds or with their cellular wounds artificially closed (see Section V). Neither obstacle seems to be insurmountable and should yield to a determined effort.

When describing the single steps of a technique, care has been taken to present sufficient background knowledge of the rationale and experience behind any one step to make the description useful to tissues other than those tested. It should therefore be possible to apply some of the techniques wherever cells of sufficient size and properties are encountered.

Appendix

1. *Preparation of weakly siliconized slides.* The slides are submerged for 5 minutes in a solution of 8 drops silicone oil (SC-87, a product of General Electric, Silicone Products Department, Waterford, New York, or Rhodorsil-240, a product of Société des Usines Chimiques, Rhône-Poulenc, France) in 250 ml acetone, and allowed to dry for 24 hours at room temperature. If particular needs call for it, the water repellent effect can be increased by a higher silicone titer in the acetone, by a longer exposure to the silicone solution, or by a curing at higher temperatures.

2. *Squash preparations of cells with polytene chromosomes.* Dissected tissues are placed on a slightly siliconized slide (see this Appendix, 1) and covered by a drop of one of the following solutions:

   a. Acetic-orcein:
      45% acetic acid
      55% distilled water
      ca. 2% orcein
      (keep barely boiling under reflux, filter when entirely cooled)
   b. Lactic-orcein:
      50% acetic acid
      50% lactic acid (85%)
      ca. 2% orcein
      (shake well, filter)

Acetic-orcein stains faster, but must not be left unattended during the staining period since it dries out; it must be replenished periodically. Lactic-orcein does not require replenishment. However, with both stains it is imperative that fresh stain be added a few minutes before squashing since, if the droplet is not covered, acetic acid evaporates from both stains, yet has to be present during the moment of squashing to keep the tissue soft. The tissues are preferably squashed on strongly siliconized slides with strongly siliconized cover slips (see Nicoletti, 1959).

Lowering the pH by increasing the acetic acid content of the stain leads to solutions which stain faster, but have a tendency to cause disintegration of cells upon prolonged staining.

Clever (1961) has described a counterstain to be used in addition to orcein; it is freshly prepared before use by mixing the following two solutions in a proportion of 55:45 (for chironomid salivary glands; other tissues require different proportions):

Solution I: light green FS, 0.1% in 96% alcohol
Solution II: orange G, 0.2% in 70% alcohol

By an addition of 1–2 drops acetic acid to the mixture the pH is brought to about 5. Preparations are kept in the stain for 2 days, and subsequently embedded. In good preparations the cytoplasm appears pale green; the puffs, Balbiani rings, and nucleoli appear bright green.

3. *Permanent squash preparations.* Two techniques are in use for obtaining squash preparations which are not under cover:

a. Tissue is fixed, stained in acetic-orcein (see above), and squashed on a nonsiliconized, albuminized slide with a strongly siliconized cover slip. The preparation is placed on dry ice (cover slip down). After a while the slip can be lifted off by a razor blade. The use of lactic-orcein is inadvisable, as it is difficult to get this solution into the frozen state (see Pelling, 1964).

b. Tissue is fixed, stained in acetic-orcein or lactic-orcein, and transferred to an albuminized, nonsiliconized slide in a small droplet of any solution which contains about 45% acetic acid. The tissue is covered by a strip of cellophane, on top of which a piece of filter paper is placed. Squashing is carried out by rolling a vial over the filter paper, applying a little pressure with the hand. The squashed preparation is immersed in water or buffer solution, in which in a minute the cellophane strip wrinkles and detaches itself from the preparation (Slizynski, 1952).

From this point onward the slides obtained by either technique can be made into permanent preparations by dehydrating and embedding them, or can be used for autoradiography.

4. *Medium for culturing isolated salivary gland nuclei of sciarids* (Frenster *et al.*, 1960).

64.1805 gm saccharose
3.3741 gm glucose
1.7866 gm $MgCl_2$
1.6659 gm NaCl
0.5 liter Tris buffer (3.025 gm Tris + 20.7 ml 1 N HCl)

The medium is supplemented with 2.8% of either polyvinylpyrollidone of molecular weight 40,000 (Sirlin and Schor, 1962a,b), or Luviskol K-90 (Badische Anilin- und Sodafabrik, Ludwigshafen, Germany) (Rey, 1963).

ACKNOWLEDGMENTS

Aided in part by grants from the "Schweizerischer Nationalfonds zur Förderung der wissenschaftlichen Forschung" and the Damon Runyon Memorial Fund for Cancer Research (DRG-668). I gratefully acknowledge the help of Dr. John S. Edwards, Western Reserve University, Cleveland, in the preparation of this manuscript.

REFERENCES

Alonso, P., and Perez-Silva, P. (1965). *Nature* **205**, 313.
Ammermann, D. (1964). *Naturwiss.* **51**, 249.
Beermann, W. (1961). *Chromosoma* **12**, 1.
Beermann, W. (1962). *Protoplasmatologia* **6D**, 1-161.
Bier, K. (1960). *Chromosoma* **11**, 335.
Buck, J. B., and Malland, A. M. (1942). *J. Heredity* **33**, 173.
Clever, U. (1961). *Chromosoma* **12**, 607.
Defretin, R. (1951). *Compt. Rend.* **233**, 103.
Edström, J. E., and Beermann, W. (1962). *J. Cell Biol.* **14**, 371.
Ephrussi, B., and Beadle, G. W. (1936). *Am. Naturalist* **70**, 218.
Frenster, J. H., Allfrey, V. G., and Mirsky, A. E. (1930). *Proc. Natl. Acad. Sci. U.S.* **46**, 432.
Glancy, E. (1946). *Biol. Bull.* **90**, 71.
Hadorn, E., Gehring, W., and Staub, M. (1963). *Experientia* **19**, 530.
Jones, B. M. (1962). *Biol. Rev. Cambridge Phil. Soc.* **37**, 512.
Jones, B. M., and Cunningham, I. (1961). *Exptl. Cell. Res.* **23**, 368.
Kanno, Y., and Loewenstein, W. R. (1964). *Science* **143**, 959.
Karlson, P., and Löffler, U. (1962) *Z. Physiol. Chem.* **327**, 286.
Kopac, M. J. (1955a). *Trans. N.Y. Acad. Sci.* [2] **17**, 257.
Kopac, M. J. (1955b). *Trans. N.Y. Acad. Sci.* [2] **18**, 22.
Kroeger, H. (1959). *Chromosoma* **11**, 129.
Kroeger, H. (1963). *Nature* **200**, 1234.
Kroeger, H. (1964). *Chromosoma* **15**, 36.
Kroeger, H. (1966). *Exptl. Cell Res.* **41**, 64.
Kroeger, H., and Lezzi, M. (1966). *Ann. Rev. Entomol.* **11**, 1-22.
Kroeger, H., Jacob, J., and Sirlin, J. L. (1963). *Exptl. Cell Res.* **31**, 416.
Laufer, H., and Nakase, Y. (1965). *Proc. Natl. Acad. Sci. U.S.* **53**, 511.
Lezzi, M. (1961). Diploma Thesis, E.T.H., Zürich.
Lezzi, M. (1965). *Exptl. Cell Res.* **39**, 289.
Loewenstein, W. R., and Kanno, Y. (1964). *J. Cell Biol.* **22**, 565.
Makino, S. (1938). *Cytologia* **9**, 272.

Nicoletti, B. (1959). *Drosophila Inform. Serv.* 33, 181.

Pelling, C. (1964). *Chromosoma* 15, 71.

Rey, V. (1963). Diploma Thesis, E.T.H., Zürich.

Robert, M. (1964). Diploma Thesis, E.T.H., Zürich.

Schmidt, E. L., and Williams, C. M. (1953). *Biol. Bull.* 105, 174.

Sirlin, J. L., and Schor, N. A. (1962a). *Exptl. Cell Res.* 27, 165.

Sirlin, J. L., and Schor, N. A. (1962b). *Exptl. Cell Res.* 27, 363.

Slizynski, B. M. (1952). *Drosophila Inform. Serv.* 26, 134.

von Borstel, R. C. (1959). *Federation Proc.* 18, 164.

Wessing, A. (1962). *Protoplasma* 55, 264.

Whitten, J. M. (1963). *Proc. 16th Intern. Congr. Zool., Washington, D.C., 1963* Vol. 2, p. 276.

Whitten, J. M. (1964). *Science* 143, 1437.

Wyatt, S. S. (1956). *J. Gen. Physiol.* 39, 841.

Chapter 4

A Novel Method for Cutting Giant Cells to Study Viral Synthesis in Anucleate Cytoplasm

PHILIP I. MARCUS AND MORTON E. FREIMAN

Department of Microbiology and Immunology, Albert Einstein College of Medicine, Bronx, New York

I. Introduction

The continuous development of new and exacting methods to probe the nature of regulatory mechanisms in the mammalian cell provides the investigator with an ever increasing range of sophisticated tools; still, the extent to which programming of biochemical events in the nucleus and cytoplasm is interrelated persists as a problem of fundamental concern to the cell biologist. Within the time-space confines of the individual cell, biochemical events have been resolved or "dissected" by numerous ingenious means, for example: chemically, with agents like actinomycin D and puromycin; physicochemically, with high energy radiations directed at the whole cell or its various parts and with microbeams of ultraviolet light or of coherent light from lasers; with autoradiography, immunofluorescence, viral infection, and micurgy through micromanipulation. Questions involving the biochemical autonomy of the major portions of the cell, nucleus and cytoplasm, are perhaps answered most directly by the last named technique, where the dissection of individual cells into nucleate and anucleate cytoplasm provides a straight forward means of separating biochemical reactions requiring nuclear participation from those that do not. In spite of the directness of the micrurgical approach it is used with great hesitancy by most investigators, probably because of the elaborate specialized equipment and technical dexterity required to tool for and perform the simplest of operations (*cf.* Goldstein and Eastwood, 1964). For these reasons we sought to simplify the micrurgical procedures, first, by working with significantly larger cells than normally available and, second, by modifying the cutting technique to the extent that greater amounts of nucleate and anucleate cytoplasm could be produced, under conditions where they may be manipulated readily for biochemical or virological studies.

This chapter describes our experience along these lines, and illustrates how we have applied these modifications to a simple virological problem —defining the role of the nucleus in viral synthesis. This description includes the preparation and use of X-ray–induced giant HeLa cells as hosts, and the novel means of micrurgy we have used on these enormous cells, dissection with an ordinary glass-cutting wheel.

II. X-Ray–Induced Giant HeLa Cells

The development of giant cells in populations of HeLa cells that have been exposed to X-rays has been reported previously (Tolmach and Marcus, 1960a; Puck and Marcus, 1956) and only a brief résumé of the essential features for obtaining large numbers of giant cells will be elaborated here. The work of Tolmach and Marcus (1960a) provides a detailed description. Monolayer cultures of HeLa S3 cells containing approximately 10^7 cells per bottle were exposed to 1200 r of 230-kV X-rays while still attached to the surface of the bottle. Following irradiation, cells were detached from the bottle by treatment with 0.05% trypsin (Ham and Puck, 1962), washed in growth medium (Tolmach and Marcus, 1960a), and dispersed into 35-mm plastic Petri dishes (Falcon Plastics #3001) at 5×10^5 cells in 1.5 ml of growth medium. This cell density allows for adequate spacing between cells following maturation to essentially full giant status by 10 days, and permits unambiguous resolution of individual cell fragments following cutting. The medium was replaced routinely every 3 or 4 days and particularly on the day prior to testing. Within 8–10 days following irradiation and incubation at

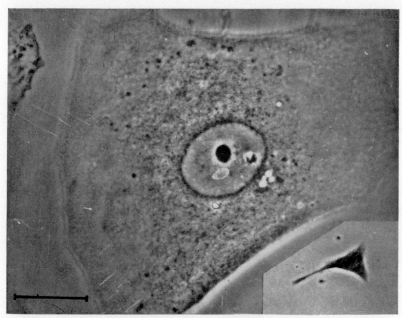

Fig. 1. Mononucleate giant HeLa cell 9 days after X-irradiation. Insert shows normal size HeLa cell. Phase-contrast photomicrograph. Marker = 100μ.

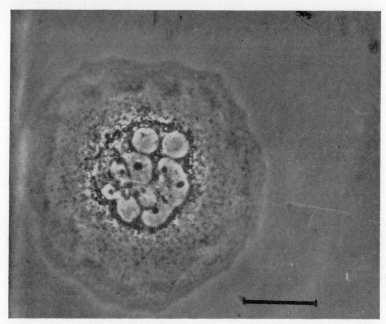

Fig. 2. Multinucleate X-ray–induced giant HeLa cell 9 days after irradiation. Phase-contrast optics. Marker = 100μ.

37°C, the average cell volume may increase 20–40 times the normal value of $2.4 \times 10^3\mu^3$ per cell. At this time individual giant cells may achieve spread diameters 10–15 times that of normal cells, the former measuring 500–1000μ, as illustrated in Fig. 1.

Although X-ray–induced giant cells are not stable indefinitely, with medium changed regularly they may persist in culture for at least 3 or 4 weeks. Most of our experiments have been conducted with giant cells used 9–12 days after irradiation.

The use of X-ray–induced giant cells as suitable nucleate and anucleate hosts in virus studies requires that these cells respond normally to infection. Our previous experience with giant HeLa cells revealed that they were more than adequate as hosts, since the giant cell behaves like a normal unirradiated cell whose dimensions have simply increased to the extent that they produce an amount of myxovirus proportional to their volume, in some cases as high as 50 times greater than normal (Tolmach and Marcus, 1960b). This seems true also for poliovirus production in X-ray–induced monkey kidney giant cells (Levine, 1963).

One basic objection to the use of giant cells for obtaining anucleate cytoplasm is the presence of fragmented nuclei. However, by adhering strictly to optimal growth conditions, including frequent medium changes, it is possible to obtain populations of giant cells in which this aberrant characteristic is seen in very few cells. Furthermore, it is possible to choose for study anucleate cytoplasm derived from cells with single intact nuclei, as exemplified by the giant cell illustrated in Fig. 1. Giant cell populations are obtained in which 50% of the cells are mono-, bi-, or trinucleate, with the remainder of the cells consisting primarily of multinucleate cells of the type illustrated in Fig. 2. Occasionally nuclei fragment more extensively, migrate from each other, and are present in various parts of the cytoplasm. This cell type is scarce under optimal growth conditions and is easy to identify with phase-contrast optics.

III. The Glass-Cutting Wheel Technique for Micrurgy

Because of its enormous size, the X-ray–induced giant cell lends itself to a novel means of micrurgy. We have found that an ordinary glass-cutting wheel possesses a unique characteristic as a micrurgical instrument when rolled gently over giant HeLa cells contained in plastic Petri dishes. Within 20 seconds a single Petri dish can be traversed over 100 times with the cutting wheel, which when pressed very gently

leaves behind well-defined lines, as illustrated in Fig. 3. A silhouette of
the cutting wheel is also shown. The lines are permanently impressed
into the soft plastic surface of the dish. This procedure, carried out
aseptically on giant cells bathed with growth medium, produces large
numbers of cells cleaved randomly into nucleate and anucleate cytoplasm, as illustrated in Figs. 4 and 5. These phase-contrast photographs
were taken 5 minutes after cutting, and reveal the nucleate and anucleate
portions separated by the furrow produced with the cutting wheel. This
furrow is usually about 20μ in width. Characteristically within 1 hour

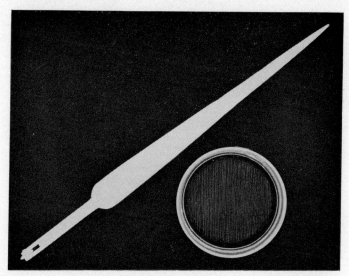

Fig. 3. Silhouette of an ordinary "Red Devil" glass cutter, which has been
trimmed of excess metal for use as a micrurgical instrument, and a 35-mm plastic
Petri dish scored with the cutter and photographed to reveal the permanent furrows
created as the cutting wheel traverses the plate.

of cutting most nucleate portions of the cell retract from the line of
cleavage, as shown in Fig. 6, whereas anucleate fragments usually remain undisturbed. Not all cells "survive" the trauma of cutting, and
within hours the cytoplasm of some cell fragments takes on a granular
appearance which we have come to associate with "nonfunctional"
cytoplasm. We have noted that many anucleate fragments react normally
to trypsinization, and have been observed to reattach to the surface of
a plastic Petri dish.

The ease with which large numbers of anucleate fragments can be
obtained from cells cleaved at random by the cutting wheel should
make available the harvest of anucleate cytoplasm relatively free of

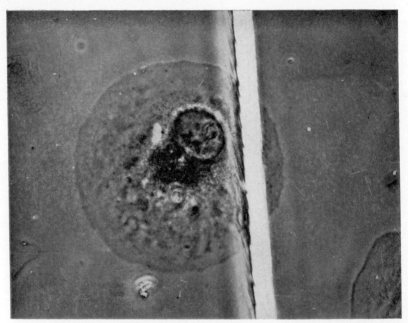

Fig. 4. Giant HeLa cell dissected with the cutting wheel into nucleate and anucleate cytoplasm, separated by the furrow left in the plastic by the cutting wheel. Phase-contrast photomicrograph taken 5 minutes after cutting. Width of furrow = ca. 20μ.

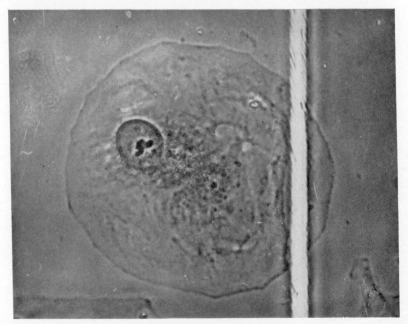

Fig. 5. Giant HeLa cell dissected into nucleate and anucleate cytoplasm with the furrow of the cutting wheel far removed from the nucleus. Phase-contrast photomicrograph taken 5 minutes after dissection. Furrow width = ca. 20μ.

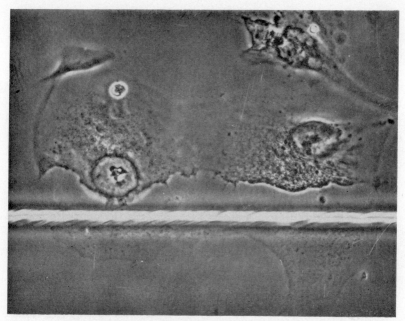

FIG. 6.   Two giant HeLa cells dissected with the cutting wheel and photographed 1 hour later by phase-contrast optics. Nucleate portions of both cells retract from the furrow, whereas anucleate cytoplasm (shown in lower part of figure) remains fully attached. Width of furrow = ca. 20μ.

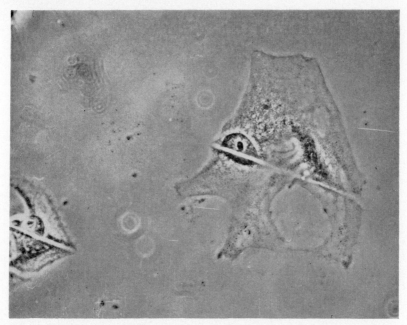

FIG. 7.   Giant HeLa cell grown on a *glass* Petri dish surface and dissected through the nucleus by the cutting wheel. Phase-contrast optics. Width of the furrow on glass = ca. 10μ.

nuclei-containing material by centrifugation in appropriate density gradients.

The simultaneous action of rolling and cutting accomplished with the glass-cutting wheel appears to possess desirable characteristics for cutting cells. We have not observed any of the tearing action seen when the usual scraping stroke is made with a fine scalpel or microneedle. Furthermore, the division between cell parts is always clean even when cutting takes place on cells bound to glass, as illustrated in Fig. 7. Here, the nucleus of one cell has been cleanly cut although no record of the gently pressed cutting wheel is apparent on the hard glass surface.

As they age, cultures of giant HeLa cells accumulate considerable numbers of the dumbbell-shaped cells considered by Goldstein and Eastwood (1964) as especially favorable for cutting. These cells attain such great lengths that the cytoplasmic strand connecting the nucleate and anucleate portions often exceeds the field diameter of our 20× phase-contrast objective (600μ). Cells of this type are readily severed with the hand-held cutting wheel, viewed under a 5× objective, and respond with a rapid contraction of both ends of the strand—as if a stretched rubber band has been cut. The cut ends swing freely in the medium and tend to act as a lever to dislodge the cell fragment. For this reason few of our successful experiments involve this type of cell.

Our experience with the hand-held cutting wheel and large giant cells seen with a 5× objective shows that it is relatively simple to cut individual preselected cells and, if desired, to scrape away the nucleate fragments. Perhaps the principle of the glass-cutting wheel would prove advantageous in normal cell micrurgy and thus justify fabrication of miniaturized cutting wheels.

IV. Viral Synthesis in Anucleate Cytoplasm

We have chosen in the remainder of this chapter to demonstrate the usefulness of combining the attributes of large cell size with the favorable micrurgical characteristics of the cutting wheel by describing our experience with viral synthesis in anucleate cytoplasm. Briefly, the purpose of the experiments was to define the role of the nucleus in the synthesis of virus specific material. For many viruses there is a definite nuclear involvement, and readily demonstrable synthesis of viral nucleic acid or protein in the nucleus. However, for some of the myxoviruses and for the enteroviruses, most of the evidence points to an autonomous viral synthesizing system in the cytoplasm, independent of nuclear involvement. In this connection, the short-pulse uridine-H[3] labeling experiments of Franklin and Baltimore (1962), designed to detect Mengo virus RNA synthesis autoradiographically, support the model of viral ribonucleic acid (RNA) synthesis exclusively in the cytoplasm, as do the cell fractionation studies of Holland and Bassett (1964), but do not totally rule out synthesis in the nucleus and extremely rapid transport to the cytoplasm, as pointed out by Dulbecco (1962) in his search for a unifying concept of patterns of viral RNA synthesis. Our data showing viral RNA synthesis in metaphase-arrested cells infected at high multiplicities with poliovirus (Marcus and Robbins, 1963) indicate that an intact nucleus per se is not required for the replication of certain RNA viruses; however, the mitotic state must be considered a rather unique type of "physiological" anucleation. The demonstration of viral RNA or protein synthesis in anucleate cytoplasm would resolve this question. These experiments, involving two very different RNA viruses, are described below.[1]

A. Newcastle Disease Virus Synthesis: Detection by Hemadsorption

The single cell hemadsorption technique (Marcus, 1962), a refinement of the Vogel and Shelokov (1957) hemadsorption procedure, can detect and localize viral specific hemagglutinin upon its earliest appearance in the plasma membrane of myxovirus-infected cells. Hence, viral hemagglutinin, synthesized and incorporated into the surface of anucleate cytoplasm, would be revealed by the specific adsorption of erythrocytes. The following experimental procedure was used to determine the viral synthetic capacity of anucleate cytoplasm: Giant HeLa cells were randomly cleaved into nucleate and anucleate cytoplasm with the glass-cutting wheel, allowed to "heal" for 15 minutes at 37°C in the original growth medium, and then infected with a high multiplicity of Newcastle disease virus to insure virus contact with all cell fragments. Following a 30-minute period for virus adsorption and entry, and a brief wash, the preparation of cut cells was exposed to viral antiserum to

[1] This work was presented at the 64th meeting of the American Society for Microbiology (abstract in *Bacteriol. Proceedings*, 1964). An independent study of this type, using conventional micromanipulative micrurgy on normal size cells, was reported simultaneously at the Federation Meetings by Crocker, Pfendt, and Spendlove (abstract in *Federation Proceedings*, 1964) and later in detail (Crocker *et al.*, 1964). The results with both normal and giant cell anucleate cytoplasm as hosts for poliovirus RNA synthesis are identical, but in addition Crocker and his co-workers used immunofluorescent techniques to demonstrate the synthesis of specific poliovirus antigen.

neutralize any nonengulfed virus (Marcus and Carver, 1965) and lower the "background" hemadsorption to zero, washed again, and incubated for 10–12 hours to allow maximum synthesis of viral hemagglutinin and incorporation into the cell membrane (Marcus, 1962). Following incubation, the cut and infected cells were subjected to the single cell hemadsorption test by flooding the plates with bovine erythrocytes, adsorbing for 15 minutes at 4°C, and finally washing off nonadsorbed red

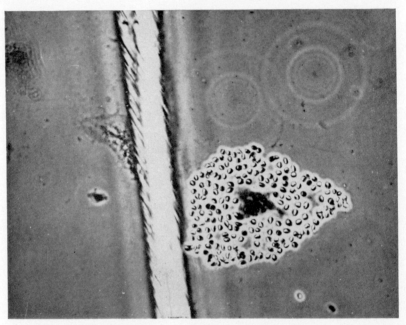

Fig. 8. Nucleate and anucleate cytoplasm of a giant HeLa cell 10 hours after exposure to Newcastle disease virus and just after completion of the single cell hemadsorption test to reveal viral hemagglutinin on the plasma membrane. The nucleate portion of the cell (right of the cut) is HAD$^+$, whereas the anucleate cytoplasm is HAD$^-$. Notice that the upper left-hand edge of the nucleate cytoplasm has retracted significantly from the furrow made by the cutting wheel. Phase-contrast optics. Bovine red blood cells attached to the nucleate portion are 6µ in diameter.

blood cells with cold phosphate-buffered saline (pH 6). The cells were examined immediately under phase-contrast optics, or fixed with OsO_4, and searched for hemadsorption-positive (HAD$^+$) anucleate cytoplasm. The photomicrograph shown in Fig. 8 represents a typical finding, namely, that no HAD$^+$ anucleate cytoplasm was ever found among scores of fragments observed, although nucleate cytoplasm invariably showed a strong HAD$^+$ reaction characteristic of Newcastle disease virus infection in uncut normal or giant size cells. Figure 8 also reveals a strong HAD+ state of the plasma membrane even along the cut and retracted edge of the cell, demonstrating that the severed portion of the nucleate cytoplasm can "heal" and function normally with respect to incorporation of viral hemagglutinin. The negative results do not disclose whether viral hemagglutinin was not synthesized, or whether this molecule was made but never incorporated into the surface of the anucleate cytoplasm, the latter situation resulting perhaps from a loss of

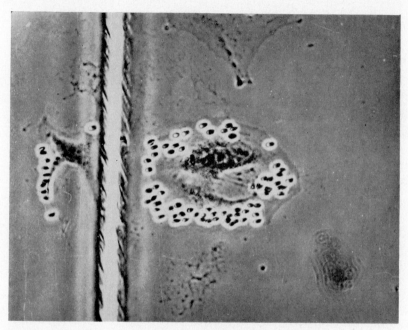

FIG. 9. Giant HeLa cell dissected with the cutting wheel 3 hours *following* exposure to Newcastle disease virus. Both nucleate and anucleate portions of the cell are HAD+ 10 hours after infection, i.e., 7 hours after cutting. Notice retraction of nucleate cytoplasm from the furrow, and its HAD+ state. Phase-contrast optics. Bovine erythrocytes are 6μ in diameter.

functional capacity following micrurgy. To resolve this question, a series of experiments were carried out in which cell cutting *followed* virus adsorption by 3 or 4 hours, an interval just encompassing the latent period for this virus in giant cells (Tolmach and Marcus, 1960b). Under these conditions, several HAD+ anucleate fragments were observed of the type illustrated in Fig. 9. Hence, we conclude that anucleate cytoplasm does possess the degree of functional integrity needed to support synthesis of viral hemagglutinin and incorporation of this molecule into the cell surface, once some early stage(s) of viral development is initiated prior to isolation of the infected cytoplasm from the nucleus. It is not clear whether these findings reflect a definite role of the nucleus per se in initiating synthesis of Newcastle disease virus, or a minimal requirement of functional integrity not met in the micrurgically prepared anucleate fragments some 3 or 4 hours after separation from the nucleus. The published accounts of essentially normal synthesis of this virus in actinomycin D-treated cells (Kingsbury, 1962; Marcus and Robbins, 1963) direct preference to the latter proposal but do not rule out the former.

B. Poliovirus RNA Synthesis: Detection by Autoradiography

As a member of the myxoviruses, Newcastle disease virus matures at the cell surface and might require a degree of plasma membrane function not present in anucleate cytoplasm 3 or 4 hours post-cutting, i.e., when new viral hemagglutinin would begin to appear on the cell surface. The same may not be true for poliovirus, which appears to mature intracytoplasmically. Other features point to poliovirus as a more likely candidate to manifest an autonomous existence in anucleate cytoplasm: a relatively short growth cycle of 5–6 hours, and much higher maximum levels of viral RNA synthesis than are reached with Newcastle disease virus (Marcus and Robbins, 1963). The following experiments were carried out: Giant HeLa cells are cleaved with the glass-cutting wheel, incubated for 15 minutes at $37°C$ to "heal," treated with actinomycin D at 10 μg per milliliter for 20 minutes to suppress cellular RNA synthesis in nucleate cytoplasm (used as controls on viral RNA synthesis) infected with poliovirus at high multiplicities, and incubated for 5 hours in growth medium with uridine-H^3 at 0.25 microcuries per milliliter. Following this procedure, cells are fixed and processed to detect incorporation of acid-insoluble label autoradiographically. The 35-mm plastic Petri dishes can be used for autoradiography after fixation, once two small holes are drilled in the bottom at the edges of each dish to permit drainage of water and an even settling of the circularly cut stripping film over the dish surface.

Cut giant cells infected with poliovirus and several controls were processed as above and searched microscopically for evidence of label incorporation. Representative examples of each test and control experiment are illustrated in pairs as viewed by phase-contrast microscopy and dark-field optics to reveal silver grains of the autoradiograph. Figure 10 illustrates the first control, normal cellular RNA synthesis in the uninfected X-ray–induced giant HeLa cell, and shows the characteristic intense labeling over the nucleoli and in the nucleus after 5 hours' la-

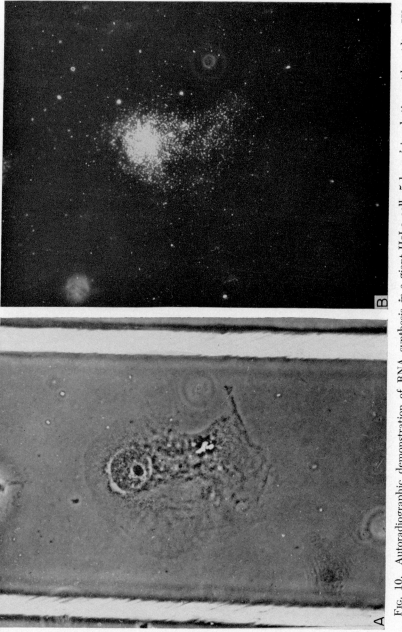

FIG. 10. Autoradiographic demonstration of RNA synthesis in a giant HeLa cell: 5 hours' incubation with uridine-H³. Phase-contrast (A) and dark-field (B) optics.

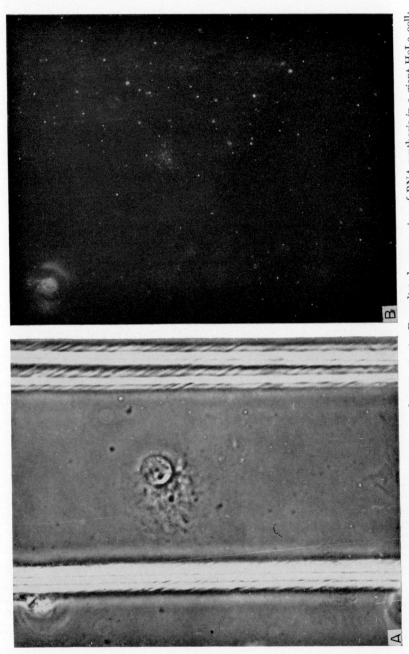

FIG. 11.   Autoradiographic demonstration of actinomycin D-mediated suppression of RNA synthesis in a giant HeLa cell: 5 hours' incubation with uridine-H³. Phase-contrast (A) and dark-field (B) optics.

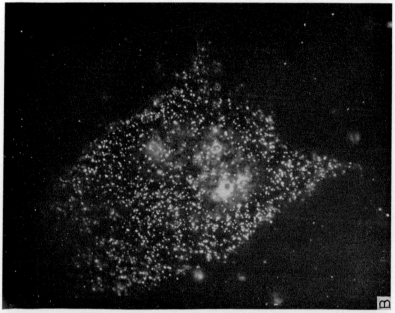

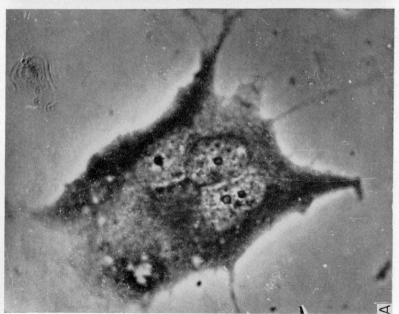

FIG. 12. Autoradiographic demonstration of poliovirus RNA synthesis in an actinomycin D-treated trinucleate giant HeLa cell: 5 hours' incubation with uridine-H³ and poliovirus. Phase-contrast (A) and dark field (B) optics.

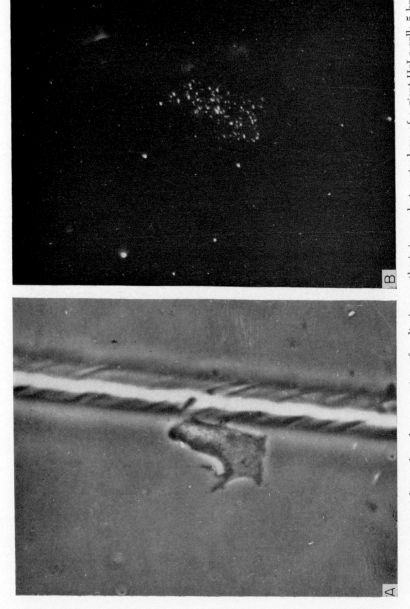

FIG. 13. Autoradiographic demonstration of poliovirus synthesis in anucleate cytoplasm of a giant HeLa cell: 5 hours' incubation with uridine-H³ and poliovirus. Phase-contrast (A) and dark-field (B) optics.

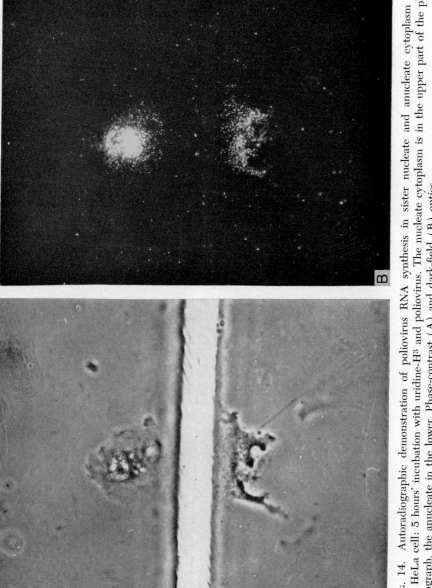

FIG. 14. Autoradiographic demonstration of poliovirus RNA synthesis in sister nucleate and anucleate cytoplasm of a giant HeLa cell: 5 hours' incubation with uridine-H[3] and poliovirus. The nucleate cytoplasm is in the upper part of the photomicrograph, the anucleate in the lower. Phase-contrast (A) and dark-field (B) optics.

beling. Cellular RNA synthesis was suppressed with actinomycin D (10 μg per milliliter for 20 minutes) in order to reveal poliovirus-induced RNA synthesis as a control on the synthetic capabilities of the giant cell. Figure 11 illustrates the lack of cellular RNA synthesis in such uninfected actinomycin D-treated cells, as shown by the absence of all but background silver grains, whereas Fig. 12 shows an infected cell and demonstrates poliovirus-induced RNA synthesis under similar conditions. The relatively uniform distribution of label throughout the cytoplasm is evident, as is the customary absence of label concentrated over the nucleus—in this particular cell, over the three nuclei. In the test experiments, several anucleate cytoplasmic fragments were observed to contain the uniform distribution of silver grains characteristic of the actinomycin-treated, poliovirus-infected cell (cf. Fig. 12). An anucleate fragment is seen without its sister nucleate part in Fig. 13. Figure 14 illustrates a giant cell cut into sister nucleate and anucleate cytoplasm. The nucleus is visible in the upper portion of the figure, and the anucleate portion in the lower; the latter shows a relatively uniform distribution of label.

V. Conclusions

The giant HeLa cells that develop following X-irradiation are so large as to be suitable for certain aspects of conventional micrurgy carried out with hand-held instruments; moreover, their enormous size permits a novel method for micrurgy—cell cleaving with an ordinary glass-cutting wheel. If operations are carried out on plastic Petri dishes, the wheel leaves a permanent impression of the swath cut through the cell. Extensive furrows through the cytoplasm of giant cells—produced as the glass-cutting wheel simultaneously rolls, presses, and cuts the cell—yield cytoplasmic edges which "mend" rapidly. The plasma membrane at the site of micrurgy becomes functional with respect to maintaining the morphological integrity of the cell fragment and sustaining incorporation of viral specific molecules.

The cutting wheel technique possesses several favorable characteristics suggesting that its principle may be profitably applied, in a scaled-down version, to problems involving conventional micromanipulative procedures.

The virus studies reported here provide an example of functional anucleate cytoplasm in a mammalian cell, and demonstrate that the cell genome or the nucleus per se is not required to initiate or maintain the program of events that lead to the production of poliovirus RNA, and presumably infectious virus,[2] whereas the more complex surface maturing myxovirus might require nuclear involvement or, what is considered more likely, simply an additional degree of functional integrity to incorporate specific molecules into the plasma membrane. Additional studies should distinguish between these alternatives.

Our experience with X-ray–induced giant HeLa cells as ideal subjects for large-scale micrurgy by a hand-held cutting wheel, the availability of most cell types as X-ray–inducible giant cells, and the suitability of these cells as hosts in virus-cell studies portend a favorable prognosis of experiments concerned with mammalian cells cultured *in vitro*. Combined with the powerful tools of autoradiography and immunofluorescence, these semimacro-micrurgical techniques should provide further impetus in defining the role of cell genome and nuclear function in the total economy of the cells' physiology.

ACKNOWLEDGMENTS

This investigation was aided by National Institutes of Health grants AI-03619-04 and 2-K3-GM-15,461-04, the latter as a Research Career Development Award from the National Institute of Allergy and Infectious Diseases, U.S. Public Health Service, to the senior author.

REFERENCES

Crocker, T. T., Pfendt, E., and Spendlove, R. (1964). *Science* **145**, 401.
Dulbecco, R. (1962). *Cold Spring Harbor Symp. Quant. Biol.* **27**, 519.
Franklin, R. M., and Baltimore, D. (1962). *Cold Spring Harbor Symp. Quant. Biol.* **27**, 175.
Goldstein, L., and Eastwood, J. M. (1964). *In* "Methods in Cell Physiology" (D. M. Prescott, ed.), Vol. 1, Chapter 21. Academic Press, New York.
Ham, R. G., and Puck, T. T. (1962). *In* "Methods in Enzymology" (S. P. Colowick and N. O. Kaplan, eds.), Vol. 5, Chapter 9. Academic Press, New York.
Holland, J. J., and Bassett, D. W. (1964). *Virology* **23**, 164.
Kingsbury, D. W. (1962). *Biochem. Biophys. Res. Commun.* **9**, 156.
Levine, S. (1963). *Progr. Med. Virol.* **5**, 127.
Marcus, P. I. (1962). *Cold Spring Harbor Symp. Quant. Biol.* **27**, 351.
Marcus, P. I., and Carver, D. H. (1965). *Science* **149**, 983.
Marcus, P. I., and Robbins, E. (1963). *Proc. Natl. Acad. Sci. U.S.* **50**, 1156.
Puck, T. T., and Marcus, P. I. (1956). *J. Exptl. Med.* **103**, 653.
Tolmach, L. J., and Marcus, P. I. (1960a). *Exptl. Cell Res.* **20**, 350.
Tolmach, L. J., and Marcus, P. I. (1960b). *Bacteriol. Proc.* p. 91. (Abstract).
Vogel, J., and Shelokov, A. (1957). *Science* **126**, 358.

[2] See footnote 1 on p. 101.

Chapter 5

A Method for the Isolation of Mammalian Metaphase Chromosomes

JOSEPH J. MAIO[1] AND CARL L. SCHILDKRAUT[2]

*Department of Cell Biology and the Unit for Research in Aging,
Albert Einstein College of Medicine, Bronx, New York*

I. Introduction

As opposed to the nucleochromatin material extracted from interphase cells, isolated metaphase chromosomes offer unique possibilities for studying the genetic material in a state perhaps more closely approximating that of the genetic material in the living organism. Thus, we might expect that the isolated metaphase chromosomes would preserve in separate and distinct structures not only the specific deoxyribonucleic acid (DNA) sequences corresponding to genetic linkage groups, but also the association of specific types of histone and ribonucleic acid (RNA)

[1] Postdoctoral trainee supported by U.S. Public Health Service Grant GM 876-03 from the National Institute of General Medical Sciences.
[2] Kennedy Scholar.

with specific chromosomes, if such an association does indeed exist in nature.

In recent years, several attempts to isolate metaphase chromosomes *en masse* from mammalian cells have been described (Prescott and Bender, 1961; Chorazy *et al.*, 1963; Somers *et al.*, 1963; Lin and Chargaff, 1964). In some instances, the methods employed removed the chromosomal histones (Busch, 1965), or precluded returning the extracted chromosomes to aqueous media without destroying their characteristic morphology. In other instances, the preparations apparently contained large amounts of cytoplasmic debris, broken chromosomes, or chromosomes that tended to aggregate.

A method is described here for obtaining large numbers of metaphase chromosomes from several types of mammalian cells grown *in vitro*. This method, we believe, offers several advantages in that (1) the isolated chromosomes are morphologically intact and stable over extended periods of time, (2) the isolation is performed at neutral pH and exploits the high specific gravity of the chromosomes, circumventing many of the difficulties encountered in attempting to remove cytoplasmic debris, (3) the use of enzymes, acids, or organic solvents in the purification is avoided, (4) contamination by nuclei and visible cytoplasmic debris is very slight, and (5) the extraction may be easily completed in 1 day.

II. Preparation of Cells

A. Cell Cultures

Suspension cultures of HeLa strain S3 cells, mouse strain L-cells, Chinese hamster cells (strain V-79-379A, female lung) and a strain of Syrian hamster cells transformed by SV-40 virus are grown in Eagle's medium supplemented with nonessential amino acids and 5% fetal calf serum. The generation time of the Chinese hamster cells is about 12 hours; the generation time of the other cell lines is about 22–24 hours.

B. Metaphase Arrest

Vinblastine sulfate (Eli Lilly & Co.) is added to logarithmically growing cultures of HeLa cells and Chinese hamster cells to a final concentration of 0.01 µg per milliliter. Incubation at 37°C is continued for 15 and 8 hours, respectively. Syrian hamster cells and L-cells are incubated in the presence of 0.5 µg per milliliter vinblastine sulfate for 11 hours.

The highest numbers of metaphase figures can be accumulated in the

HeLa cell populations; on occasion, as many as 90% of the HeLa cells are in metaphase arrest at the time of harvesting. The number of cells in metaphase arrest in cultures of the other cell lines seldom exceeds 50% of the population and usually varies from 20 to 40%. In general, HeLa cells give the most satisfactory results not only because of the high proportion of metaphase cells, but also because they show no tendency to form micronuclei during the period of incubation with vinblastine sulfate. However, these advantages are offset by the considerable morphological uniformity that characterizes most of the HeLa chromosomes. This is an important consideration if the scope of the research includes attempts to fractionate and identify chromosomes according to individual types.

It is advisable to determine the optimum length of the incubation period with the metaphase arrest agent for each cell line. This is especially important when working with cell lines that have a tendency to form micronuclei during prolonged incubation with colchicine or vinblastine sulfate (e.g., Chinese hamster cells). Micronuclei are very difficult to separate from the chromosomes by differential centrifugation and represent a loss in chromosome yield.

III. Chromosome Extraction

A. Hypotonic Treatment and Homogenization of Cells

From 4 to 12 liters of cell culture, containing 2×10^9 to 6×10^9 cells, are harvested for each extraction. The cells are sedimented by centrifugation at $500 \times g$ for 15 minutes in the horizontal head of a refrigerated centrifuge, and washed twice in Earle's balanced salts solution. Throughout all subsequent operations the cells and extracts are maintained near 0°C by using ice buckets.

The pellet of washed cells is resuspended in $CaCl_2$, $MgCl_2$, and $ZnCl_2$ (Frenster, 1963), each at a concentration of 0.001 M in 0.02 M Tris, pH 7.0 (hereafter, this solution will be referred to as TM). The ratio of volumes of homogenization medium to packed cell volume should be at least 10 to 1. Hypotonic swelling and intracellular dispersion of metaphase figures occurs over a period of about 20 minutes. Five per cent saponin (Fisher Scientific Co.) previously filtered through two layers of Whatman No. 1 paper is then added to a final concentration of 0.05%. Five minutes after the addition of saponin, aliquots of the cell suspension are transferred to a 40-ml capacity Dounce homogenizer equipped with a small clearance pestle. The cells are broken and the chromosomes liberated by several strokes with the homogenizer. Phase-contrast microscopy is nearly indispensable for following the extent of cell breakage and the effectiveness of the subsequent steps in the chromosome purification.

B. Preparation of Crude Chromosome Suspension

The homogenate is diluted in TM containing 0.05% saponin to triple its original volume, and distributed into centrifuge tubes. Best results are obtained by adding a volume of homogenate to each tube so that the distance from the surface of the liquid to the bottom of the tube does not exceed 3 cm. The tubes are centrifuged in the horizontal head of a refrigerated centrifuge for 5 minutes at $120 \times g$. Most of the nuclei and unbroken cells will sediment, while the chromosomes and fine cellular debris remain in suspension. The supernate, containing the chromosomes, is poured off and saved (care must be taken not to pour off any of the loosely packed pellet and thereby contaminate the supernate with nuclei). Chromosomes entrapped among the nuclei are extracted by resuspending the pellet in each tube to the original volume with TM containing 0.05% saponin and repeating the centrifugation at $120 \times g$. The second supernate is combined with the first, and the pellets extracted yet another time if significant quantities of chromosomes are still present. The final pellet may then be discarded. Rather vigorous pipetting should be used for resuspending the pellets in order to disperse clumped and sedimented chromosomes, but the production of foam during the pipetting must be avoided.

If nuclei or unbroken cells are still present in the combined supernatant fractions, they must be removed by repeating the low-speed centrifugation as described above, until only chromosomes and fine, particulate debris remain in suspension.

C. Purification of Chromosomes

The chromosomes are sedimented by centrifugation in an angle-head centrifuge for 10 minutes at $2500 \times g$. The pellets are washed once in $0.02 M$ Tris, pH 7.0, containing 0.1% saponin. Divalent metals are omitted during this step since their presence prevents the loosening of chromosomes from attached debris and the solubilization of much amorphous material in the extracts. Furthermore, massive amounts of debris will sediment with the chromosomes during the high-speed centrifugation in dense sucrose (see below) if these metals are present. The chromosomes themselves expand somewhat at this time, but resume their former configuration as soon as they are returned to a medium containing divalent metal ions at sufficiently high concentration.

The pellets obtained from the last washing step are thoroughly resuspended in 2.2 M sucrose (sp. gr. ca. 1.28) in 0.02 M Tris, pH 7.0, and 0.1% saponin. The suspension is layered over 10 ml of the dense sucrose solution in each of three 1 × 3-inch cellulose nitrate tubes suitable for the SW 25.1 rotor of the Model L ultracentrifuge. The interfaces are blended by stirring with a glass rod, but care is taken to leave about 1 cm of sucrose solution undisturbed at the bottom of the tubes. The tubes are centrifuged at 50,000 × g for 1 hour. Chromosomes are very dense structures (we estimate their specific gravity to be near 1.35) and will sediment. Nuclei will also sediment (Chauveau et al., 1956) if they have remained as contaminants of the crude chromosome suspension. Cytoplasmic debris remains in suspension or rises to the top of the tubes to form a pellicle.

The pellets of purified chromosomes are resuspended in TM and examined under the phase-contrast microscope. If cytoplasmic debris is still present, a second cycle of washing in Tris-saponin buffer and centrifugation in dense sucrose may be required. Sucrose is removed from the purified preparation by washing the chromosomes at least 3 times (2500 × g for 10 minutes) in TM.

As determined by counts in a phase-contrast hemacytometer, 30–40% of the chromosomes in HeLa metaphase cells are recovered in the final preparations. The average DNA content of the isolated HeLa chromosomes is 0.5 μμg per chromosome, indicating that less than 10% of the DNA is lost from the chromosomes during the isolation procedure. Contaminating nuclei contribute less than 1% of the dry weight of these chromosome preparations.

In TM, the chromosomes may be stored in suspension for weeks at 0°C in ice buckets. For long-term storage at −20°C, we recommend the addition of glycerol to a final concentration of 20%.

A buffer composed of 0.005 M $CaCl_2$ and 0.05% saponin in 0.02 M Tris, pH 7.0, is an effective medium for maintaining the chromosomes both in a contracted state and in monodisperse suspension for experimental purposes. In this medium, clumps of aggregated chromosomes may be dispersed by drawing the suspension up and down several times through a No. 22 spinal tap needle attached to a syringe.

Permanent slide mounts of the isolated chromosomes may be prepared by adding an equal volume of methanol-acetic acid (3:1) fixative to a portion of the chromosome suspension, followed by thorough mixing with a pipette and centrifugation at 2500 × g for 3 minutes. The chromosomes are resuspended in fixative only, dispersed by pipetting, and again centrifuged. This step is repeated and the final pellet resuspended in a small volume of fixative. The concentrated chromosome suspension is spread over a clean coverslip and left to dry. After hydrolysis in HCl (this step may be omitted), the coverslip is washed in tap water, and stained with any of the usual chromosome stains, such as Unna methylene blue, Giemsa, or aceto-orcein.

IV. Some Properties of the Isolated Chromosomes

Figures 1 and 2 show fixed and stained metaphase chromosomes isolated *en masse* from four different cell lines. The morphological characteristics of chromosomes as seen in karyotype preparations of individual cells are well preserved in the isolated chromosomes and there is little evidence of extensive breakage. It should be noted that the isolated chromosomes stain strongly with the Feulgen stain.

At present it is uncertain whether all chromosomes of the individual karyotypes are proportionally represented among the isolated chromosomes. Among the isolated chromosomes of the Chinese hamster cells, a chromosome of an individual type might be expected to appear in the general population with a frequency of about 9%. This expectation is fulfilled when counts are made of easily recognizable chromosomes such as the longest metacentric and acrocentric chromosomes of the set. The two pairs of smallest chromosomes of the set are found less frequently than expected (about 6%). Although these chromosomes often stain poorly and are difficult to distinguish from background, it is conceivable that they are selectively lost during the several centrifugation steps of the extraction.

Phase-contrast photomicrographs of suspensions of purified chromosomes are shown in Figs. 3 and 4. Individual chromatids and centromeric constrictions are easily visible as in fixed and stained preparations, but in addition the chromosomes appear here as plump, refractive, and rather rigid bodies. Occasionally there is some suggestion of fine differentiation in the region of the centromere (Fig. 4). The plump appearance of chromosomes in suspension is probably related to the fact that they have not been stripped of their associated histones by the fixation procedures used in ordinary cytological preparations. In this condition, the chromosomes are excellent material for studying the effects of changes in ionic environment and enzyme treatment on chromosome structure. Responses to these changes are often dramatic and swift.

By using the following simple technique it is possible to observe certain interesting structural properties of the isolated chromosomes. To 0.1 ml of a suspension of chromosomes in TM is added 10 µl of 1% aqueous solution of sodium lauryl sulfate (recrystallized from ethanol). A drop of the suspension is placed under a coverslip and observed with phase-contrast optics. The chromosomes expand to many times their original volume and the chromatids assume a nearly spherical shape. Two round dots may regularly be seen at each side of the longitudinal midline of the chromosomes in the region of the centromere (Fig. 5). If a solution of pronase (Calbiochem, grade B) at a concentration of 1 mg per milliliter in 0.02 M NaCl is allowed to flow under the coverslip, the expansion and lengthening of the chromosomes continue until they are hardly visible in the unstained preparations. This process of chromosome dissolution may be stopped at any moment by allowing methanol to flow under the coverslip, followed by methanol-acetic acid fixative. With a little practice, this can be accomplished so that many chromosomes can be made to adhere to the coverslip. The coverslip is floated off the slide in fixative, dried, and stained. The chromosomes appear as filamentous structures whose shape frequently suggests a figure eight (Fig. 6A, B, and C). They may also appear as very long double strands that sometimes exceed 100µ in length (Fig. 6D and E).

Incubation for 1 hour at 37°C in the presence of 25 µg per milliliter of pancreatic deoxyribonuclease or trypsin does not greatly alter the appearance of the chromosomes as observed in phase contrast; however, only disorganized fibrous material is observed after such treated chromosomes are fixed and stained. Incubation under the above conditions with pancreatic ribonuclease or pepsin has no obvious effect on chromosome morphology in either phase-contrast or stained preparations.

The average chemical composition of the HeLa chromosomes in preparations isolated by the procedure described here is 16.4% DNA, 11.6% RNA, and 72.0% protein. We find similar chemical compositions for chromosomes isolated from Chinese hamster, Syrian hamster, and mouse L-cells, and for HeLa cells using the recently developed procedures of Cantor and Hearst (1966) and J. Huberman and G. Attardi (manuscript in preparation). Cantor and Hearst have obtained similar values for the composition of chromosomes isolated from mouse ascites tumor cells.

The DNA and RNA associated with the HeLa chromosomes have guanine + cytosine contents of 41% and 63% respectively. At least 80% of the RNA liberated from the chromosomes by the presence of 1% sodium lauryl sulfate sediments in the 16 S and 28 S regions during sucrose density-gradient centrifugation. There is approximately twice as much 28 S as 16 S RNA, the same proportion that is found in HeLa cell ribo-

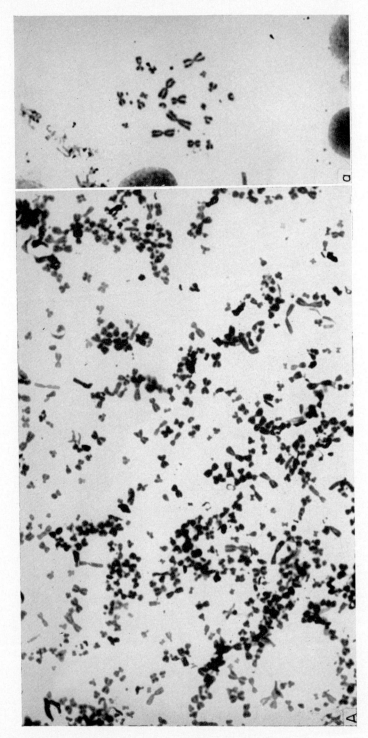

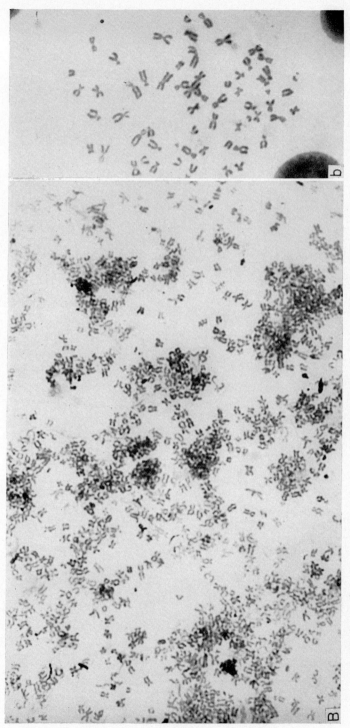

Fig. 1. Isolated chromosomes from Chinese hamster cells (A) and HeLa cells (B). To the right of each photograph of the isolated chromosomes is a representative karyotype for comparison (a and b, respectively). Unna methylene blue. × 860.

121

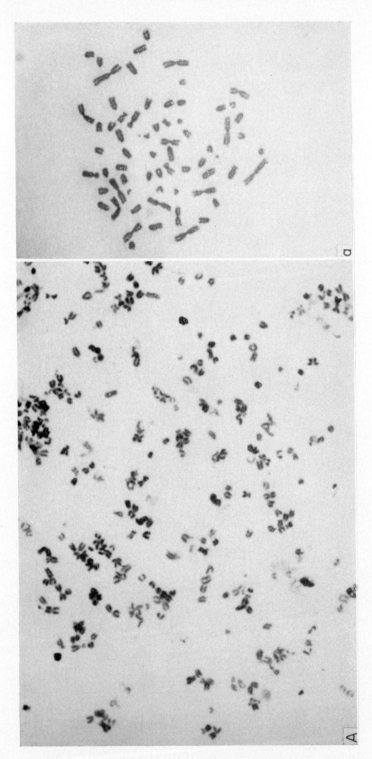

122

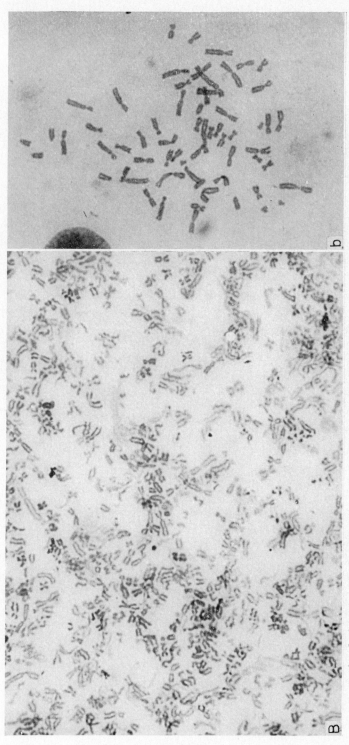

Fig. 2. Isolated chromosomes from L-cells (A) and Syrian hamster cells (B). To the right of each photograph of the isolated chromo-somes is a representative karyotype for comparison (a and b, respectively). Unna methylene blue. × 860.

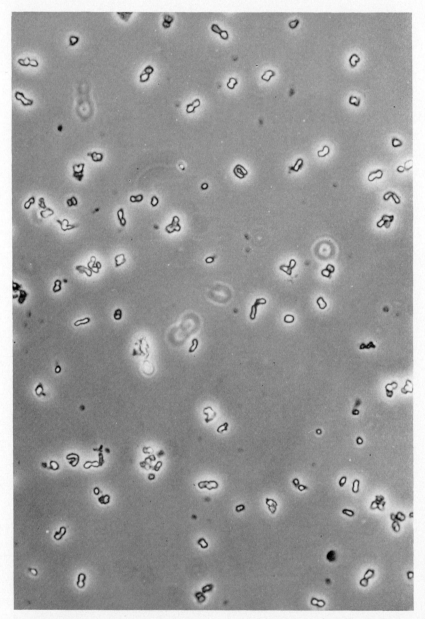

FIG. 3. Phase-contrast photomicrograph of isolated Chinese hamster chromosomes in suspension. × 860.

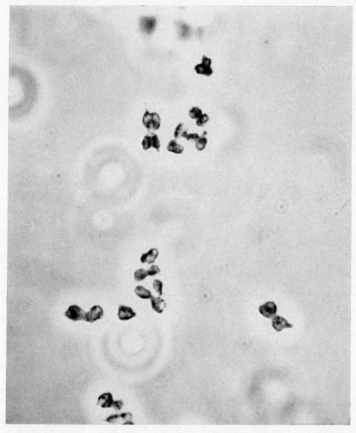

FIG. 4. Phase-contrast photomicrograph of isolated Chinese hamster chromosomes in suspension. × 2100.

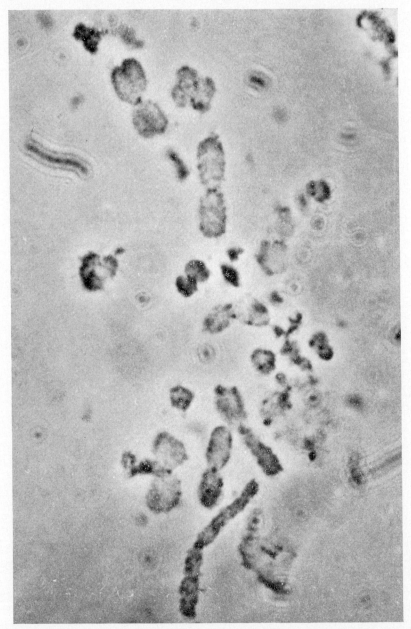

FIG. 5. Phase-contrast photomicrograph of isolated Chinese hamster chromosomes in TM and 0.1% sodium lauryl sulfate. × 1700.

somes. An appreciable quantity of RNA is present in these chromosome preparations (about 12% of the dry weight), and since this RNA resembles ribosomal RNA, it is conceivable that ribosomes adsorb to the chromosomes during extraction, and are not removed during the washing steps and sucrose sedimentation. This possibility requires further exploration. Most of the chromosomal proteins (ca. 50–78% in preparations from different cell types) are soluble in 0.2 N HCl and show acrylamide gel electropherograms similar to that of histones extracted from the nuclei of the respective cells.

Since the isolated chromosomes of Chinese hamster cells may be separated according to size into at least three classes by low-speed centrifugation in steep sucrose gradients (Maio and Schildkraut, 1966), it may be possible to determine whether there are differences in the species of nucleic acids and proteins associated with morphologically distinct chromosomes.

V. Discussion

The method of extracting and purifying mammalian metaphase chromosomes described here represents only one of many possible approaches to the problem of obtaining unequivocally recognizable chromosomes for studies of their physicochemical properties. Experiments in this laboratory have shown that not only many different divalent metal ions may be employed to maintain chromosome structure during isolation at neutral pH, but also polyamines and even added histones. Furthermore, our criteria of purity of the final preparations rest primarily upon observations in the phase-contrast microscope—e.g., absence of nuclei, membrane fragments, and cytoplasmic debris—and cannot yet take into account the possibilities that extraneous materials have adsorbed to the chromosomes during the extraction, or that materials that are intrinsic components of the chromosomes have been lost. It is obvious, then, that the value of this or any method for extracting and purifying chromosomes can be assessed only when more information is available concerning the influence of the isolation procedure on their final composition.

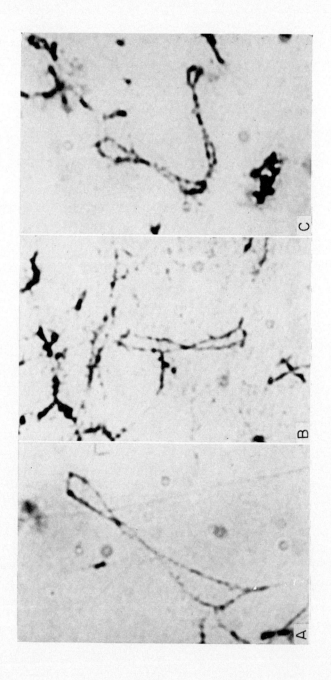

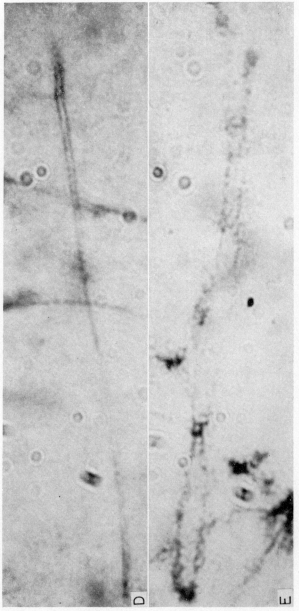

Fig. 6. Chinese hamster chromosomes treated with sodium lauryl sulfate and pronase. Unna methylene blue stain. A, B, and C: × 1700. D and E: × 2100.

ACKNOWLEDGMENTS

We gratefully acknowledge the capable assistance of Mrs. Irene Bossert and Mrs. Vera Loeffler.

These studies were supported in part by grants from the National Science Foundation (GB-2030) and from the National Institutes of Health (HD 00674).

REFERENCES

Busch, H. (1965). "Histones and Other Nuclear Proteins." Academic Press, New York.
Cantor, K., and Hearst, J. (1966). *Proc. Natl. Acad. Sci.* **55**, 642.
Chauveau, J., Moulé, Y., and Rouiller, C. (1956). *Exptl. Cell Res.* **11**, 317.
Chorazy, M., Bendich, A., Borenfreund, E., and Hutchison, D. J. (1963). *J. Cell Biol.* **19**, 59.
Frenster, J. H. (1963). *Exptl. Cell Res.* Suppl. **9**, 235.
Lin, H. J., and Chargaff, E. (1964). *Biochim. Biophys. Acta* **91**, 691.
Maio, J. J., and Schildkraut, C. L. (1966). *Federation Proc.* **25**, 707.
Prescott, D. M., and Bender, M. A. (1961). *Exptl. Cell Res.* **25**, 222.
Somers, C. E., Cole, A., and Hsu, T. C. (1963). *Exptl. Cell Res.* Suppl. **9**, 220.

Chapter 6

Isolation of Single Nuclei and Mass Preparations of Nuclei from Several Cell Types[1]

D. M. PRESCOTT,[1a] M. V. N. RAO,[2] D. P. EVENSON,
G. E. STONE, AND J. D. THRASHER[3]

*Department of Anatomy, University of Colorado Medical Center,
Denver, Colorado*

I. Introduction

In order to assay the syntheses of ribonucleic acid (RNA), deoxyribonucleic acid (DNA), acid-soluble protein, and acid-insoluble protein in single nuclei and in mass preparations of nuclei, a method using the detergent, Triton X-100, and spermidine or spermine has been developed for nuclear isolation that eliminates all cytoplasm and avoids loss of

[1] Supported by NSF Grant GB-1635 to D. M. Prescott.

[1a] Present address: Institute of Developmental Biology, University of Colorado, Boulder, Colorado.

[2] Permanent address: Department of Zoology, Andhra University, Waltair, India.

[3] Present address: Department of Anatomy, The Center for the Health Sciences, University of California, Los Angeles, California.

macromolecules. The use of Triton was originally suggested to us by the abstract of Hymer (1963). Various staining and isotope tracer experiments have indicated that no RNA, DNA, or protein is lost during the treatment. The spermidine stabilizes the nucleus with remarkable effectiveness, and even such extremely fragile structures as the *Euplotes* macronucleus or the macronuclear anlagen are easily and cleanly isolated free of cytoplasm.

The procedure has been used successfully with single and mass isolation of macronuclei from the ciliates *Tetrahymena* and *Euplotes*, interphase and division stages of the micro- and macronuclei of *Paramecium*, the developing macronuclear anlagen of *Euplotes*, macronuclei from various other ciliates, nuclei from *Amoeba proteus*, and mass isolation of nuclei from HeLa cells. Very likely the technique could be applied to practically any cell type that lacks a cell wall.

II. Isolation of Individual Macronuclei of
Euplotes eurystomus

Using a dissecting microscope, the cell or cells are transferred from the culture to a dish containing an isolation medium of Triton X-100 (Rohm and Haas, Philadelphia) diluted 2000-fold with distilled water and containing 100 mg per liter of spermidine phosphate trihydrate. The pH of the medium is 6.8 without buffering. Cells are easily handled using a braking pipette (Stone and Cameron, 1964) or some other type of micropipette. Within a minute in the isolation medium *Euplotes* becomes fragile and the macronucleus can be freed by gentle agitation of the cell with a braking pipette. The process is easily observed with a dissection microscope. When left in the medium for several minutes the macronuclei will often float free without agitation.

Isolated nuclei can be pipetted directly onto a subbed slide (Caro, 1964) and air-dried. Normal cytology is better preserved by fixation in alcohol before drying. This is achieved by micropipetting the free nuclei into 70% alcohol in a Petri dish with a subbed slide on the bottom (McDonald, 1958). The nuclei will rapidly settle onto the slide and adhere firmly within a few minutes. The best adhesion occurs with a slide that has been freshly subbed. The slide can be rinsed in 100% alcohol and air-dried. If the nuclei are to be assayed for radioactivity with a conventional tube counter, the coverslip of a size that will fit into a planchet can be substituted for the slide. Figure 1 shows an isolated macronucleus of *Euplotes* stained with methyl green-pyronin.

Mass isolation of *Euplotes* macronuclei can be achieved by concentrating the cells by centrifugation and resuspending in the Triton-

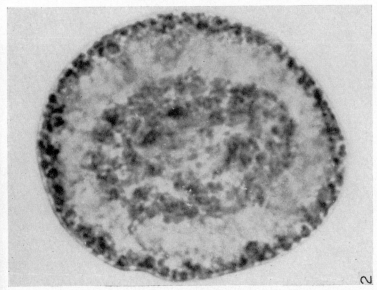

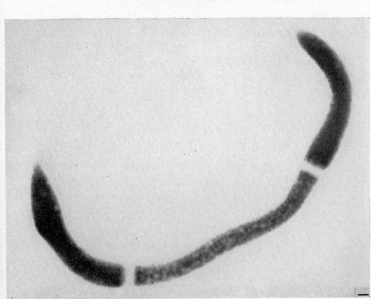

FIG. 1. Isolated macronucleus of *Euplotes eurystomus* stained with methyl green-pyronin. The positions of the replication bands indicate about 50% completion of DNA synthesis.

FIG. 2. Isolated nucleus of *Amoeba proetus* stained with methyl green-pyronin. Multiple nucleoli line the inner side of the nuclear membrane. The clumping of material in the central region of the nucleus is caused by alcohol fixation.

spermidine solution. Gentle pipetting breaks the cells and repeated centrifugation and resuspension of the pellet in Triton-spermidine result in a clean macronuclear preparation. Such preparations have been used for electrophoretic analysis of basic nuclear proteins of *Euplotes*.

III. Isolation of Individual Nuclei of *Amoeba proteus*

Amoeba proteus is more resistant to breakage in Triton and several minutes' exposure to the dilute Triton is necessary before the cells become fragile enough for disruption with a pipette. A number of concentrations of Triton and spermidine have been tried and the most successful has been a solution containing Triton diluted 1 : 200 with distilled water and 4 mg/100 ml of spermidine. Figure 2 shows an amoeba nucleus stained with methyl green-pyronin. Nuclei are deposited on slides or coverslips by dropping through 70% alcohol as described for *Euplotes*.

IV. Isolation of Macronucleus of *Stentor coeruleus*

When the macronucleus of *Stentor coeruleus* is fully extended, the internodal connections become extremely tenuous; consequently, the isolation of the intact nucleus becomes relatively more difficult than with some other organisms. The most successful method employed has been essentially the same as described above for amoeba with a few minor exceptions.

One to several *Stentor* are placed in a Syracuse watch glass containing a 1 : 200 dilution of Triton X-100 plus 4.0 mg/100 ml of spermidine phosphate trihydrate. Using a dissecting scope with a dark field (created by placing a black glass filter or a piece of black tape on the scope stage) and reflected light, the macronucleus is readily recognized as a bright bluish, monofiliform structure.

The ease with which the macronucleus can be removed depends greatly upon the nutritional state of the animal. The extraction is easily accomplished from a semistarved *Stentor*; if well fed, the cytoplasm adhering to the macronucleus is more difficult to detach. To remove the intact macronucleus from a well-fed *Stentor*, the organism must be left in the Triton-spermidine medium for about an hour. The cytoplasm can be freed by gently agitating the macronucleus with a braking pipette whose bore tip diameter is slightly smaller than the diameter of the *Stentor*. The macronucleus is then drawn into the pipette and placed in the 70% alcohol over the subbed slide. For those animals deprived of abundant food organisms, a 5–10-minute treatment in the isolation medium is usually sufficient. Figure 3 shows an intact macronucleus stained with the Feulgen reaction.

If it is not required that the beaded structure of the nucleus be left intact, the isolation of the entire nucleus, although broken up to individual nodes or groups of nodes, may be easily and quickly accomplished. After several minutes in the isolation medium, the *Stentor* are broken up with a braking pipette whose bore tip diameter is about one half the

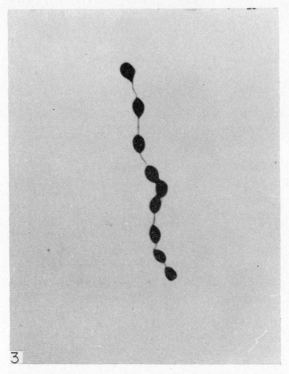

FIG. 3. Isolated macronucleus of *Stentor coeruleus* stained with Feulgen nuclear reagent.

diameter of the organism. The nodes are then selected out, agitated free of cytoplasm, and placed in the alcohol over the subbed slide.

V. Isolation of Macronuclear Anlagen of *Euplotes woodruffi*

The macronuclear anlagen of *Euplotes woodruffi* 4 hours after the separation of the exconjugants appear as a translucent disc against a dark background of cytoplasm. It measures about 20–25μ in diameter (Fig. 4). To isolate the anlagen the exconjugants are transferred to a

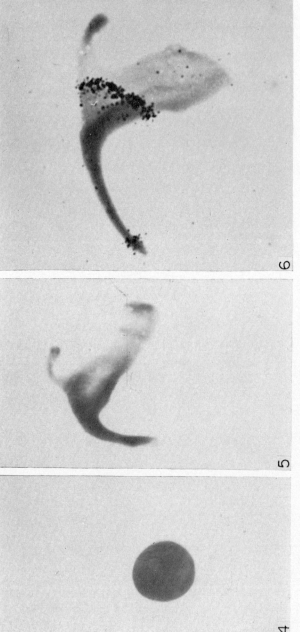

Fig. 4. Isolated macronuclear anlagen of *Euplotes woodruffi*, 4 hours after separation of the conjugants. Stained with acetocarmine.

Fig. 5. Isolated macronuclear anlagen of *Euplotes woodruffi*, 36 hours after conjugation. Stained with acetocarmine.

Fig. 6. Macronuclear anlagen of *Euplotes woodruffi* labeled with tritiated thymidine for 30 minutes before isolation. Radioactivity is restricted to the band regions. Stained with toluidine blue.

Syracuse watch glass placed under a dissecting microscope and the isolation medium (a 1 : 2000 dilution of Triton with 10 mg/100 ml of spemidine) added. The cells become very fragile in 1–2 minutes. Gentle agitation by pipetting disrupts the cells and liberates the macronuclear anlagen free of cytoplasm. The anlagen is swirled around in a fresh medium to remove any attaching material.

The macronuclear anlagen assumes the shape of the letter T by 24 hours after separation of the conjugants at 22°C. The cells at this time are indistinguishable morphologically from vegetative individuals. The anlagen is very fragile and breaks at several points, but can be freed by gentle agitation of the cell in the Triton-spermidine solution. Figure 5 shows the macronuclear anlagen 36 hours following conjugation. Figure 6 shows the macronuclear anlagen from an exconjugant after 30 minutes of labeling with tritiated thymidine.

VI. Isolation of Micro- and Macronuclei of *Paramecium caudatum*

The micronucleus and macronucleus are isolated with a solution of Triton diluted 1 : 2000 with distilled water containing 100 mg/1000

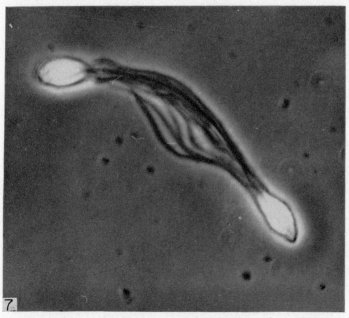

Fig. 7. Phase-contrast photograph of division figure of the micronucleus of *Paramecium caudatum* mounted in 70% alcohol.

ml of spermidine. Figure 7 shows an isolated micronucleus in division. The food vacuoles of well-fed animals often contain brownish discoidal bodies of the shape and size of the micronucleus. Gentle agitation, however, results in the dissolution of these bodies, leaving the micronucleus in the debris. The micro- and the macronuclei may be washed in fresh medium to remove any attaching material.

Transferring to the subbed slide by the usual procedure of dropping into 70% alcohol, as described for *Euplotes, Stentor,* and *Amoeba,* often results in wide dispersal of the nuclei on the slide. Such scattering can be lessened by gently releasing the nuclei very close to the surface of the subbed slide in 70% alcohol, rather than letting the nuclei drop through the alcohol. This procedure is applicable generally to nuclei of smaller size, like the micronucleus of *Paramecium* and the early anlagen stages of *Euplotes woodruffi.*

VII. Isolation of Macronuclei of Other Ciliates

In the case of larger ciliates with a ribbon or beaded nucleus, like *Spirostomum ambiguum* (Fig. 8), *Blepharisma japonicum* (Fig. 9), and *B. undulans* (Fig. 10), the cells are left undisturbed in the isolation medium for an hour. The cytoplasm disintegrates during this time, leaving the macronucleus intact. The nuclei are drawn into a pipette and ejected with a slight force into the Petri dish containing 70% alcohol. This force usually extends the macronucleus by the time it reaches the surface of the subbed slide at the bottom of the Petri dish.

Prolonged treatment with the isolation medium in the case of these larger ciliates releases the macronucleus. However, such long treatment of *Euplotes woodruffi* and *Paramecium caudatum* renders it impossible to isolate either the macronuclear anlagen or the micronucleus.

VIII. Mass Isolation of *Tetrahymena* Nuclei

The solution used to obtain mass quantities of nuclei from *Tetrahymena* differs from that described in that the Triton X-100 (0.1%, v/v) and spermidine (0.001%, w/v) are dissolved in 0.25 M sucrose. The procedure that gives the best results in obtaining intact nuclei is as follows. The cells are centrifuged and resuspended twice in 50 ml 0.15 M KCl. To 1 ml or less of packed cells, 9 volumes of the lysing solution are added and the cells suspended and allowed to stand for 5 minutes. The cells are then agitated by drawing them into a Pasteur pipette with forceful expulsion. The agitation is continued (5–10 additional minutes) until no whole cells are observed when examined under the light microscope.

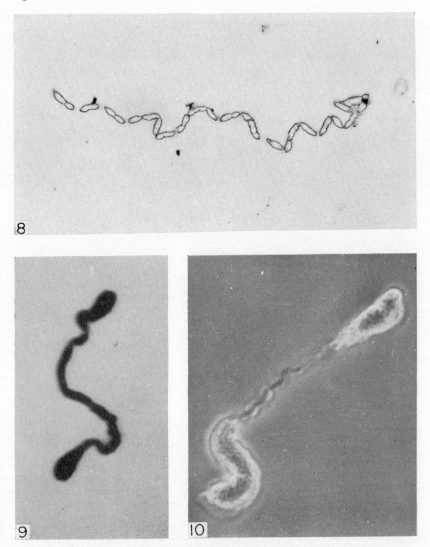

Fig. 8. Phase-contrast photograph of isolated macronucleus of *Spirostomum ambiguum* mounted in 70% alcohol.

Fig. 9. Isolated macronucleus of *Blepharisma japonicum* stained with acetocarmine.

Fig. 10. Phase-contrast photograph of isolated macronucleus of *Blepharisma undulans* mounted in 70% alcohol.

When lysis is complete the preparation is brought to 50 ml with 0.25 *M* sucrose. This solution is centrifuged at 700 × *g* for 30 minutes. The resulting pellet contains the nuclei. The nuclear pellet is washed by resuspending and centrifuging two more times in either 0.25 *M* sucrose or isotonic KCl. With the light microscope no significant contamination can be detected. The micronuclei appear to accompany the macronuclei in the isolation procedure either in close association with the macronucleus or as free micronuclei.

The best results are obtained when the total volume of packed cells is 1 ml or less. If more than 1 ml cells is lysed in the tube the resulting pellet upon centrifugation becomes a gelatinous mass, and the nuclear pellet is contaminated with cytoplasmic debris.

IX. Nuclear Isolation in S3 HeLa Cells

Nearly confluent monolayers of S3 HeLa cells, in 500-ml Blake bottles grown in F-10 medium supplemented with 10% bovine calf serum containing penicillin (50 units per milliliter) and streptomycin (50 µg per milliliter), are trypsinized with 5 ml 0.25% trypsin in saline D-2 (pH 7.2–7.4) (Ham, 1963). After incubation for 5–10 minutes at 36.7°C the trypsinized cells are suspended in an additional 5 ml saline D_2. The cells are centrifuged at 1000 *g* at 0–2°C for 5 minutes and the trypsin removed. The pellet is washed 3 times in 50 ml 0.154 *M* KCl and centrifuged at 1000 *g* for 5 minutes. It is necessary to remove protein and traces of medium before lysing.

The cells are lysed and nuclei isolated by a slight modification of the method described for *Tetrahymena*. Packed cells volumes of 0.1 ml are resuspended in 4 ml 0.1% (v/v) Titron X-100 containing 0.001% (w/v) spermine dissolved in double distilled water. The cells are continuously agitated with a Pasteur pipette to facilitate rupturing of the cell membrane. The nuclei are free of cytoplasmic contamination as determined by Carnoy fixation, toluidine blue staining, and light microscopy observation at 1250 magnification within 2–4 minutes after addition of the Triton solution (Fig. 11). Packed cell volumes larger than 0.1 ml have proven difficult to lyse. Prophase and telophase nuclei appear normal, while metaphase figures have not been observed.

After the cells have been lysed, 0.25 *M* sucrose containing 0.001% spermine is added to a final volume of 50 ml and centrifuged at 700 *g* (0–2°C) for 20 minutes. The resulting nuclear pellet is resuspended in 0.25 *M* sucrose with spermine and centrifuged. The sucrose washing pro-

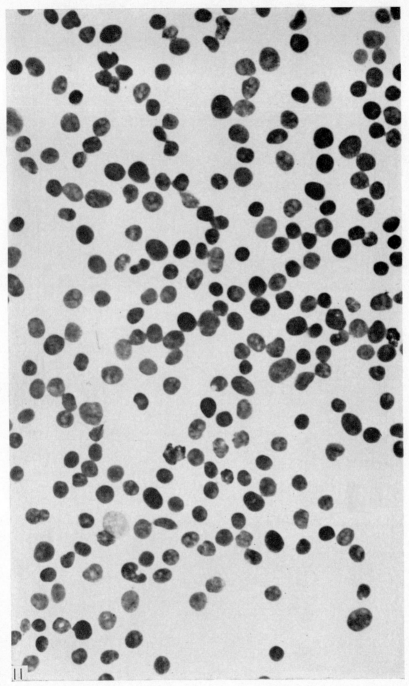

Fig. 11. HeLa nuclei after treatment of whole cells with Triton X-100 and 0.25 M sucrose. Note the absence of cytoplasmic contamination in the nuclear preparation. × 1000.

cedure is repeated twice. The spermine prevents excess nuclear-membrane rupturing and tends to keep the nucleolar apparatus intact. However, it appears at present that saline D_2 (Ham, 1963) with spermine added may give better nuclear preparations than sucrose.

REFERENCES

Caro, L. G. (1964). *In* "Methods in Cell Physiology" (D. M. Prescott, ed.), Vol. I, pp. 327–363. Academic Press, New York.

Ham, R. G. (1963). *Exptl. Cell Res.* 29, 515.

Hymer, W. C. (1963). *Federation Proc.* 22, 473.

McDonald, B. B. (1958). *Biol. Bull.* 114, 71.

Stone, G. E., and Cameron, I. L. (1964). *In* "Methods in Cell Physiology" (D. M. Prescott, ed.), Vol. I, pp. 127–140. Academic Press, New York.

Chapter 7

Evaluation of Turgidity, Plasmolysis, and Deplasmolysis of Plant Cells

E. J. STADELMANN

Department of Plant Pathology and Physiology, University of Minnesota, St. Paul, Minnesota[1]

[1] This work is supported by the Minnesota Agricultural Experiment Station. Miscellaneous Journal Series, Paper No. 1243. The author wishes to thank Professor T. T. Kozlowski, University of Wisconsin, Madison, Wisconsin, for his help in revising this article.

I. Introduction

Very few (mostly physical) methods are suitable for research on living cells, when little or no damage is allowable (cf. Stadelmann, 1955, p. 77f). Of available procedures, only the experimental exploitation of the osmotic properties of the cell has brought about significant progress toward understanding the functioning of the living plant cell. This approach led to (1) an evaluation of the osmotic quantities of the cell (cf. Ursprung, 1935; Blum, 1958), (2) discovery of the differential permeability of the protoplasmic envelope (Pfeffer, 1877, p. 12lf), and (3) the first data on the composition of cell membranes (cf. Overton, 1895), which were confirmed by studies with the electron microscope (Sitte, 1963; Whaley et al., 1964). These results are also in good agreement with some first chemical analyses of the outer cell membrane (cf. Mitchell and Moyle, 1951, p. 388, Table V; Macfarlane, 1964).

Improvement of basically very simple techniques of microscopy resulted in increasingly elaborate procedures used in the discovery of the permeability series. Further, significant differences in these permeability series were found for various cell types and for the same material when conditions were different (cf. Höfler, 1942; Hofmeister, 1938). In addition to these findings, information on other qualities of the protoplasm (e.g., viscosity and wall attachment) was obtained from plasmolysis experiments. Also the investigation of alterations in the protoplasm and its components (cf. Küster, 1956, p. 17f; Stadelmann, 1956a, p. 85f), induced by plasmolysis, provided interesting insight into the reactivity of the cell and demonstrated the dynamic state of the living protoplasm, which is not appreciated on the basis of electron microscope or chemical investigations.

The equipment needed usually for these experiments is relatively simple compared with the elaborate apparatus ordinarily used in other fields of cell research. However, increased experience and alertness of the investigator are required here to recognize all the fallacies which may derive from inadequacies of the procedures used and the evaluation of the cell reaction under special conditions.

The importance of data on the osmotic quantities and the protoplasmic permeability of individual cells is evident. Only detailed analysis of water relations of plants under different environmental conditions will explain the interaction of the factors involved in their water economy. The osmotic quantities of the cells which control the water flow are key factors in recognizing the quantitative water relationships.

The most significant accomplishment of comparative permeability research on plant cells was the discovery of differences in the protoplasm that depend upon the tissue or plant species. Also, the development of methods which allow one to measure quantitatively such a quality of the living protoplasm is an important step in obtaining further information on the living protoplasm and its structure.

Voluminous data are available on changes in turgidity and plasmolysis (cf. Stadelmann, 1966a), but information on osmotic behavior of the cell and of the qualities of the living protoplasm is still insufficient. Therefore much additional work is needed for a more comprehensive insight concerning these important functions of the living cell.

A. History

Experimentally induced changes of cell size were first observed by Nägeli (1855) after immersion in external solutions with sufficiently high osmotic activity. The retraction of the protoplasmic envelope from the cell wall was described first by Braun (1852) and subsequently by Pringsheim (1854) and Nägeli (1855, p. 1, referring to his experiments from 1849 and 1850). Important work was done later by De Vries (1877), who introduced the distinction of two phases after application of a hypertonic solution: (1) decrease of cell volume, and (2) separation of the protoplast from the wall. De Vries coined the term "plasmolysis" and defined it as the separation of the living protoplasmic envelope from the cell wall, caused by the action of an external water-withdrawing solution. Experimental evidence that the turgor pressure found in the plant cell is caused by osmosis and differential permeability of the protoplasmic layer was given by Pfeffer (1877), who in this way established the concept of the osmotic mechanism for cell turgor.

After a subsequent period when some misunderstanding occurred, especially in interpreting results of plasmolysis experiments (cf. Blum, 1958, p. 5f), the classical work of Ursprung and Blum (1916) and Höfler

(1920) gave a basis for correct interpretation of the results derived with the different methods. These concepts, further developed and propagated especially by Meyer (1938) and Kramer (cf. 1956), are now widely recognized and adopted for interpretation of data concerning the water relations of cells (cf. Walter 1963, 1965; Kozlowski, 1964).

Passive permeability was first investigated by Janse (1887) and De Vries (1889), who explained in this way the slow deplasmolysis observed in experiments with certain nonelectrolytes as external hypertonic solutions. Later observations, especially on deplasmolysis time, increased our knowledge of the permeability of the protoplasmic layer to a certain group of substances. After initial attempts to calculate the absolute permeability constant (Lepeschkin, 1909), the techniques introduced by Fitting (1915) and especially the *plasmometric method* (Höfler, 1918) were the start for a broader investigation on *permeability series* and calculation of the absolute permeability constant for harmless nonelectrolytes.

Plasmolysis time and plasmolysis form were first used for determination of protoplasmic viscosity and wall attachment by Weber (1921, 1924). Detailed investigations of cytoplasmic alterations caused by plasmolysis are of more recent date and concern a variety of particular effects on the living protoplasm, i.e., systrophe, partial protoplasts, cap plasmolysis, regeneration membranes (cf. Stadelmann, 1956a).

B. Basic Principles and Definitions

There is wide disagreement on appropriate terminology for describing water relations of the plant cell (cf. Eyster, 1943; Levitt, 1951; Walter, 1952; Kozlowski, 1964, p. 33). In contrast there is almost complete agreement about the dynamics and interplay of the factors involved in influencing cell water relations. The essential relationship was first described by Ursprung and Blum (1916, p. 530) with the formula:

$$Sz = Si - W \qquad (1)$$

Meyer (1938, p. 535) expressed the same formula with different symbols, which were widely adopted by American workers:

$$DPD = OP - WP \qquad (2)$$

Formula (1), which shows better agreement with the established rules for mathematical symbols, will be used here in its extended form:

$$Sz = Si - W \pm A \qquad (3)$$

These symbols and other terms should be defined as follows:

$Si =$ suction potential (sometimes called suction force) of the cell contents (this includes protoplasm and central vacuole; when a large central vacuole is present, Si is almost entirely composed of the suction potential of the cell sap). Si is equal to the osmotic pressure, which could be measured in an osmometer filled with the cell contents. Si is acting in the direction of water uptake into the cell. Si is measured in atmospheres.

Sz = suction potential of the cell (sometimes called suction force or diffusion pressure deficit = DPD). Sz is the excess hydrostatic pressure to which the cell must be subjected to bring it into equilibrium with pure water (Spanner, 1952, p. 380). Sz is acting in the direction of water uptake into the cell. Sz is measured in atmospheres.

W = wall pressure, resulting from the elastic stretch of the cell wall. W is acting in the direction of release of water from the cell. The wall pressure is equal in magnitude and opposite in direction to the turgor pressure (P): $W = -P$. W and P are measured in atmospheres.

A = any eventual external pressure working either in the direction of water release (e.g., suction potential of a surrounding medium) or of water uptake (e.g., elastic forces from cell walls of neighboring cells, hindering the volume decrease of the cell). A is measured in atmospheres.

O (osmotic value) = the concentration of an ideal nonelectrolyte (e.g., glucose, saccharose) which produces the same suction potential as the cell content. O is measured in moles per liter or moles/cm^3.

Osmotic quantities (German: *Osmotische Zustandsgrössen*): Si, Sz, W, P, and O as described above.

Osmotic pressure: the maximum hydrostatic pressure developed in an osmometer filled with the solution under consideration, measured in atmospheres.

Recently a thermodynamic terminology for the relationship described in formula (1) was introduced by Taylor and Slatyer (1961, p. 344; cf. Boyer, 1965). It was intended to emphasize the description of the energy state of the water in the plant cell and in the soil. The new concept should also lead to a unification of the terms used for plant and soil water relationship.

The basic relation is given with the formula (Taylor and Slatyer, 1961, p. 344):

$$\psi = P + \pi + \tau$$

where $\psi = - Sz^* = - DPD$,† water potential of the cell, numerically equal but opposite in sign to suction potential of the cell* and to diffusion pressure deficit†; $P = W^* = WP$,† wall pressure (erroneously called

---

† As used in formula (2).
* As used in formula (3).

turgor pressure by Taylor and Slatyer, 1961, p. 340f); $\pi = -Si^*$ $= -$ OP,† solute (osmotic) potential, numerically equal but opposite in sign to suction potential of the cell contents* or osmotic pressure†; $\tau =$ matric potential, considered to be important for cells with small vacuole or with cell contents occupied largely by hydrophilic colloidal substances (Taylor and Slatyer, 1961, p. 345). τ may correspond to A in formula (3), which accounts for any additional pressure developed under special conditions.

All these quantities are indicated as pressure:

$$\text{Pressure} = \frac{\text{force}}{\text{area}} = \frac{\text{energy}}{\text{volume}}$$

and measured in atmospheres or bars (1 atm $= 1.013$ bar; 1 bar $= 10^6$ dyne cm^{-2}).

However, no significant gain of insight into the water relations of the plant results from these new terms (cf. Walter, 1963, p. 41f), since they are based upon the same experimental data as the former osmotic quantities.

For the average plant cell, three characteristic stages must be distinguished sharply in describing the osmotic situation:

a. Maximal water saturation: When the cell is transferred to pure water an equilibrium will be established between the wall pressure and the suction potential of the cell content, while the suction potential of the cell becomes zero. This state of the cell is characterized by the subscript s:

$$Sz_s = Si_s - W_s; \quad \text{when} \quad Sz_s = 0, \quad Si_s = W \qquad (4)$$

b. Incipient plasmolysis: An external solution is applied, which is concentrated enough to cause only slight, just visible separation of the protoplast from the cell wall. When there is no interference with wall tensions of neighboring cells, the cell wall is usually under no elastic stretch by turgor pressure and, therefore, W is zero. This state is indicated by the subscript g:

$$Sz_g = Si_g - W_g; \quad \text{when} \quad W_g = 0, \quad Sz_g = Si_g \qquad (5)$$

c. Normal state: A condition between incipient plasmolysis and maximal water saturation and corresponding to the cell under normal living conditions. This state is indicated by the subscript n. Osmotic characteristics of the normal state are the most valuable, since they indicate the water relation of the cell in its natural situation:

$$Sz_n = Si_n - W_n \qquad (6)$$

Since the cell sap concentration changes according to the cell size, its osmotic value also varies and the same subscripts are applied as above to specify a certain state. The relationship

$$O_g > O_n > O_s \qquad (7)$$

can be established and for O_g the term "osmotic ground value" (*osmotischer Grundwert*, Höfler, 1920, p. 290) is often used.

Peculiar situations are found in aquatic algae in their natural habitat. There they exhibit the state of maximal water saturation (cf. Hoffmann, 1935). For fresh water with no appreciable concentration of osmotically active substances the formula becomes

$$Si_n = W_n \qquad (8)$$

For brackish and salt water the osmotic potential A of the water must be taken into consideration:

$$Si_n = W_n + A \qquad (9)$$

For some of these organisms the cell wall is at the normal state practically under no stretch [$W = O$; "isotonobiontes" (Höfler, 1963, p. 26; Höfler and Höfler, 1963)]. In this case the osmotic suction potential of the cell contents is equal to the osmotic potential of the milieu:

$$Si_n = A$$

Practically no change of cell volume takes place in diatoms due to their special wall structure. Therefore:

$$Si_s \approx Si_n \approx Si_g \qquad (10)$$

Inward folding of a more or less supple cell wall, swelling of the cell wall, positive or negative elastic pressure from the neighboring cells, and strong wall attachment of the protoplasm may interfere with the experiments and influence the results, as discussed later.

Assuming that none of these complications interferes, the changes in osmotic quantities of the cell in relation to water saturation and cell volume are shown diagrammatically in Fig. 1A and B. Figure 1A shows the relationship for a cell with thin, considerably elastic and stretchable cell walls; in Fig. 1B a cell with relatively thick cell walls and little possibility of stretching is assumed. The continuation of the curves left of the line for V_g indicates the situation in a wilting cell.

A cell with stretchable elastic cell walls may show a noticeable increase in cell sap concentration by a decrease of volume until incipient plasmolysis is reached. Therefore, cell sap concentration in incipient plasmolysis will be higher than under natural conditions.

A

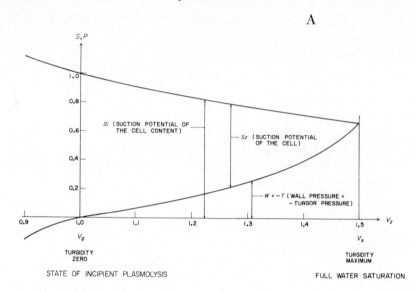

STATE OF INCIPIENT PLASMOLYSIS   FULL WATER SATURATION

B

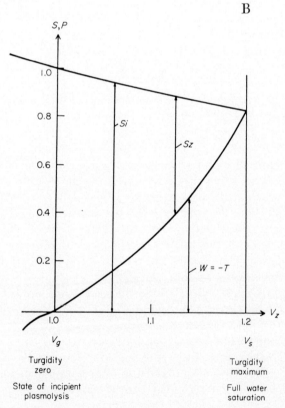

Some terms used in permeability studies:

Degree of plasmolysis: the numerical ratio of the volume of the plasmolyzed protoplast and the inner volume of the turgorless (released) cell. $G = V_p/V_z$; G is a pure number without dimension. It becomes 1 for the cell in incipient plasmolysis and decreases with increasing degree of contraction.

Deplasmolysis: the process of dilatation of the plasmolyzed protoplast (cf. Unger, 1855, p. 262) and is also called recovery (cf. Meyer and Anderson, 1947, p. 135).

Diffusion: a process of equalization resulting from the thermal agitation of molecules. In a solution system with inequalities in concentration, diffusion causes a net migration of the solute from a region of higher concentration to one of lower concentration.

Hypertonic solution (or concentration): a solution of an osmotically active substance which has a higher suction potential than the cell sap of the particular cell material in the state of incipient plasmolysis.

Hypotonic solution (or concentration): a solution of an osmotically active substance which has a lower suction potential than the cell sap of the particular cell material in the state of incipient plasmolysis.

Partial concentration: the concentration of one constituent of a solution which has also other dissolved substances.

Partial suction potential: suction potential developed by a partial concentration.

Permeability: a quality of a membrane, its passability for certain substances. In a broader sense it describes the capacity of the cell for any kind of uptake and exchange of material with the milieu (cf. Höfler, 1959, p. 236).

Permeability constant: a measure of the permeability of a membrane for a particular substance, derived from an equation relating some factors determining the permeation process. *Absolute permeability constant:* a permeability constant derived from a formula which takes into account all (and only those) factors which determine the permeation process (cf. Stadelmann, 1961, 1963, p. 663).

Permeability series: list of permeating substances tested, arranged in

---

FIG. 1. Diagrammatic presentation of the relationship of cell volume and osmotic quantities in a cell between full turgidity and turgor pressure 0 showing (A) cell with highly stretchable walls, and (B) cell with thick and less stretchable walls. Abscissa: relative cell volume, V_z, (volume of the cell at the state of incipient plasmolysis, V_g, is taken as unity). Ordinate: suction potential, S, and wall pressure, P, (measured in atmospheres) in relative values (suction potential of the cell content at the state of incipient plasmolysis, Si_g, is taken as unity). (After Höfler, 1920, p. 290; cf. also Tamiya, 1937, p. 558; Bennet-Clark, 1959, p. 173; Kozlowski, 1964, p. 31.)

the order of increasing or decreasing permeability constant with its values.

Permeation: the process of migration of molecules through a membrane.

Plasmometric method: a procedure for quantitative evaluation of the degree of plasmolysis of a cylindrical cell, where geometrical conditions allow quite accurate calculation of the volume of the cell and the protoplast.

Protoplast: the protoplasmic envelope with the central vacuole in a differentiated cell.

Protoplast length: length of the protoplast in a cylindrical cell when the protoplast has its final form, i.e., the form of a cylinder with two hemispheres at the ends. It is the distance from the top of one hemisphere to the top of the other hemisphere (see Fig. 3).

II. Methods for Measurement of the Osmotic Ground Value (O_g)

This osmotic quantity may be determined (a) by direct evaluation of incipient plasmolysis or, (b) by immersion in a sufficiently high hypertonic solution and calculating from geometrical data of the cell. Injuries resulting from the plasmolysis procedure (cf. Stadelmann, 1956a, p. 101) are unlikely, since the time needed here for evaluation is too short for cells to develop noticeable damaging effects.

A. Method of Incipient Plasmolysis

The most frequently used procedure to determine O_g is the direct determination of the concentration provoking incipient plasmolysis. This method can be used for cells of any shape and consists essentially in testing the same cell, or preferably different pieces of the same tissue, in a series of stepwise increased concentrations. These steps must be sufficiently narrow to allow accurate determination of O_g. For a given kind of tissue cell the average value of O_g is defined as corresponding to the external concentration which caused incipient plasmolysis in 50% of the cells. The best way to determine the exact value of O_g is by interpolation from the ratios of the number of plasmolyzed and unplasmolyzed cells of the two concentrations closest to O_g.

1. MATERIALS

To avoid alteration in osmotic quantities only fresh material should be used for the experiments. When parts of a plant are tested, the preparation must be made immediately before the experiment. The cuttings should be thick enough to allow a sufficiently high number of intact cell layers on both sides of the cells to be tested. For epidermal cells that do not readily separate from the cell layer underneath, at least one sub-epidermal layer should be kept intact in the cutting. Cells near the edge of the cutting may behave differently (due to the shock of the cutting process) and should not be evaluated.

Small entire organs (e.g., moss leaves) or samples of filamentous algae should bring only a small amount of water with them when transferred to the testing solution. The diluting effect of this water has to be neglected.

2. EXPERIMENTAL PROCEDURES

In most cases a series of small glass bottles with ground-glass stoppers should be used and the concentration steps prepared with great accuracy. When each glass bottle is filled with one of the concentrations a piece of the sample to be tested is placed in each bottle. These pieces, of course, should be as uniform as possible. Extreme care must be taken that these sample pieces are not injured mechanically, which could result in stimulation plasmolysis or other effects. Therefore a brush or an inoculating needle loop is most appropriate for the transfer procedure of the sample pieces.

Since each of the pieces must be tested after the same immersion time, it is convenient to proceed in intervals of 1 or 2 minutes. Depending upon the speed of plasmolysis of the cells, reading may start after approximately 10 minutes of immersion. Thereafter the specimen from the first concentration is taken out of the glass bottle and examined on a microscopic slide in a large droplet of the same solution. For the evaluation the smallest degree of separation of the protoplast from the cell wall, which can be definitely recognized, must be considered (cf. Fig. 2). Small plasmolytic contractions sometimes are difficult to recognize. Noncolored cell sap and small differences in refractive index between protoplasm, cell sap, and medium may cause special difficulties, so that some experience in recognizing incipient plasmolysis in these cells will be needed before starting the experiment proper to determine O_g.

3. ACCURACY OF THE DETERMINATION

Accuracy of the determination of O_g with this method depends greatly on the experience of the investigator and can be as high as $\pm$ 0.015–0.03 M saccharose (Höfler, 1918, p. 64). *Deplasmolysis* of the cells after evaluation is recommended for all material where harmful effects of the experimental procedures are expected.

4. Sources of Errors

Although experimental procedures are not difficult, some sources for error and misinterpretation exist (cf. Oppenheimer, 1932). Concerning the experimental setup, it is important to avoid conditions which may lead to a *change of the concentration* of the test solutions. Also, harmful

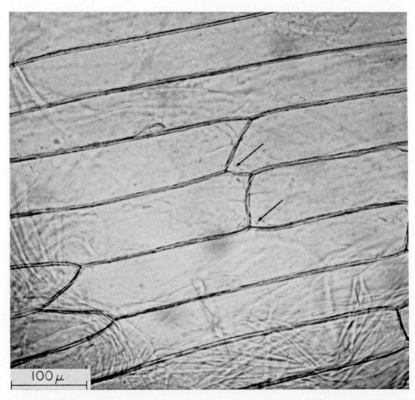

100 μ

Fig. 2. Cells near or in the state of incipient plasmolysis. Material: inner epidermis of the bulb scale of *Allium cepa*. Concentrations applied: 0.30–0.40 M mannitol. Arrows indicate the places where the protoplast had just separated from the cell wall. When these areas are very small and just recognizable the cell is considered to be in the state of incipient plasmolysis.

substances must be avoided in the preparation of the test solution. Whenever possible, glucose or saccharose should be used. Polyethylenes or their derivatives have recently shown to be harmless in plasmolysis experiments of plant cells (cf. Pirson and Schaefer, 1957; Walter, 1963, p. 56). Salt solutions may permeate into the vacuole and therefore change the value of O_g.

Factors which may lead to error:

a. Wall attachment of the protoplasm layer. This is very high in red algae (Höfler, 1932b) but in the majority of other plant cells very low. Its intensity may be inferred from a second plasmolysis and determination of O_g with the same sample. The value of O_g for the second plasmolysis should be only slightly lower (cf. Ursprung, 1939, p. 1179).

b. Permeability to sugars. Many diatoms have an appreciable permeability to sugars and must be handled with rapidity so that permeation has only negligible effects on O_g.

c. Impermeability of the cell wall. In some Musci the external cell wall is impermeable to molecules as big as saccharose. In other species even smaller molecules such as formamide penetrate the cell wall less readily (Biebl, 1954). Access to the testing solution may be given in these cases by incisions, since the cell walls in the interior of the leaf show higher permeability to the plasmolytic agents.

d. High protoplasmic viscosity. This may delay attainment of osmotic equilibrium and separation of the protoplasm from the cell wall. When this occurs, the method of incipient plasmolysis may even become impracticable.

e. Stimulation plasmolysis and vacuolar contraction. These processes (cf. Stadelmann, 1956a, pp. 83, 90) sometimes lead to confusion with true plasmolysis. Normally they are rarely encountered under the conditions of these experiments.

f. Osmoregulation. This would lead to an alteration of the concentration of the cell sap. However, its possible effect should be small enough during the time needed for measuring O_g so that it may be neglected.

g. Swelling of the cell wall. This interferes seriously with evaluation of plasmolysis in some marine algae (cf. Hoffmann, 1932).

h. Plasmolysis-like contractions of dead protoplasm. These are easily distinguished from plasmolysis of intact cells and should not cause serious difficulties.

The intensity of these effects may have to be taken into account in evaluating the plasmolysis experiments, and sometimes may even preclude the application of the plasmolytic method.

However, the majority of cell types found in higher plants will show none of these difficulties to such extent and it is therefore unrealistic to overemphasize these errors (Ernest, 1935).

5. EXAMPLE OF A TYPICAL EXPERIMENT

After some preliminary experiments to familiarize the investigator with the material, the experiment itself may be performed.

Material: leaf of *Elodea* sp., middle third of the lower leaf side (cells of the "vein" were not evaluated). Plasmolyticum: saccharose. Entire and undamaged leaves (except at the cut surface) were used. For every concentration a different leaf was placed in the testing solution. Leaves

TABLE I

NUMBER OF TOTAL AND PLASMOLYZED CELLS AT DIFFERENT TIMES
IN THE CONCENTRATION SERIES

Time elapsed after transfer to the test solution (min)	Concentration (moles of saccharose per liter); number of cells plasmolyzed (Pl.) and total number of cells evaluated (T)												
	0.28		0.30		0.32		0.34		0.36		0.38		0.40	
	Pl.	T	Pl.	T	Pl.	T	Pl.	T	Pl.	T	Pl.	T	Pl.	T
5	0	48	0	52	2	47	2	56	12	58	32	61	45	56
15	0	75	0	74	3	61	4	60	23	64	61	66	68	68
30	0	68	1	71	4	58	8	56	62	62	68	68	70	70
50	0	65	2	78	5	74	8	80	All cells plasmolyzed					
70	0	63	2	81	5	72	11	73	All cells plasmolyzed					
100	0	66	2	80	5	79	12	65	All cells plasmolyzed					

of three consecutive whorls were used. Table I (Ursprung, 1939, p. 1232) shows the frequency of plasmolysis.

From the relation

$$\frac{O_g - C_1}{C_2 - C_1} = \frac{50 - P_1}{P_2 - P_1} \qquad (11)$$

where P_1 is the frequency of plasmolyzed cells in the concentration C_1 in per cent, P_2 the frequency of plasmolyzed cells in the concentration C_2 in per cent, C_1 the highest concentration of the series where the frequency of the plasmolyzed cells is less than 50%, and C_2 the lowest concentration of the series where the frequency of the plasmolyzed cells is higher than 50%, the value of O_g is calculated.

When P_2 in formula (11) = 100%, the formula can be rewritten:

$$\frac{O_g - C_1}{C_2 - C_1} = \frac{\dfrac{N_1}{2} - N_2}{N_1 - N_2} \qquad (12)$$

where N_1 is the total number of cells examined in the concentration C_1, and N_2 the number of plasmolyzed cells from the total number N_1 in concentration C_1.

Formula (12), which involves fewer calculations, can be applied to the above experiment for $t = 30$ minutes, $C_1 = 0.34$ M, and $C_2 = 0.36$ M:

$$\frac{O_g - 0.34}{0.36 - 0.34} = \frac{28 - 8}{56 - 8}$$

$$O_g = 0.348 \text{ mole per liter}$$

In the same way, for $t = 50$ minutes, the value $O_g = 0.349$ mole per liter; for $t = 70$ minutes, $O_g = 0.348$ mole per liter; and for $t = 100$ minutes, $O_g = 0.348$ mole per liter. Their high degree of conformity shows clearly the reliability of the plasmolytic method for suitable material and conditions.

B. Plasmometric Method

This method (Höfler, 1917) is based upon the geometry of a cylindrical cell. In the state of final plasmolysis the protoplast will have the form of a cylinder with two (almost) hemispheric ends (= "plasmometric shape"; see Fig. 3). The volume V for this protoplast is

$$V_p = \frac{\pi \cdot b^2}{4} \cdot \left(L - \frac{b}{3}\right) \tag{13}$$

where L = length of the protoplast from one end to the other (see Fig. 3), b = inner width of the cell, and h = inner length of the cell. L, b, and h are measured in relative micrometer units.

For cells with protoplast ends which have less curvature than a hemisphere (see Fig. 3B), instead of $b/3$ the height of the meniscus m and a correction factor α must be introduced in the above formula; the factor α depends upon the ratio m/b and can be found in Table II. In this case formula (13) becomes

$$V_p = \frac{\pi \cdot b^2}{4} \cdot (L - \alpha \cdot m) \tag{14}$$

The volume change of the protoplasm during contraction will be very small when the protoplasmic layer is thin (as is usually the case) and no swelling of the protoplasm occurs. In many cells the small volume of the protoplasmic envelope can be neglected, and the volume of the protoplast may be taken as the volume of the vacuole. For the osmotically active material in the vacuole it can be assumed that the laws of Dalton, Van't Hoff, and Boyle-Mariotte can be applied, i.e., that the concentration is inversely proportional to the volume. Although this proportionality law is valid for many cell types, deviations also have been reported (cf. Höfler, 1918, p. 113).

Applying the reciprocity law, the concentration of the cell sap at the time of incipient plasmolysis can be calculated from its known final volume in equilibrium with any concentration of the plasmolyticum. This concentration, of course, must be strong enough to produce sufficient contraction of the protoplast to bring about the plasmometric shape and

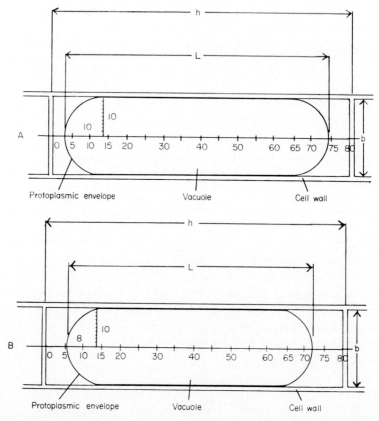

Fig. 3. Diagram of a plasmolyzed cylindrical cell showing (A) with hemispheric menisci, and (B) with menisci having the shape of a spherical segment. $L =$ length of the protoplast. In (A), $L = 74 - 3.5 = 71.5$ micrometer units (MU); in (B), $L = 72.0 - 5.5 = 66.5$ MU. $h =$ inner cell length $= 80$ MU; $b =$ inner cell width $= 2 \times 10$ MU $= 20$ MU; $m =$ height of the segment, in (B) $= 8$ MU. (After Höfler, 1918, p. 112.)

to free the two ends of the protoplast so that they do not touch the transverse cell wall.

The concentration of the osmotic ground value O_g is then derived from the external concentration C used and the volume V_p of the protoplast in osmotic equilibrium with C:

TABLE II

CORRECTION FACTOR α FOR DIFFERENT RELATIVE HEIGHT (m/b) OF THE
MENISCUS OF THE PROTOPLAST IN FINAL PLASMOLYSIS[a]

m/b	α
0.00	1.000
0.05	0.997
0.10	0.987
0.15	0.970
0.20	0.961
0.25	0.917
0.30	0.880
0.35	0.837
0.40	0.787
0.45	0.730
0.50	0.667

[a] After Höfler (1918, p. 105).

$$O_g = C \cdot \frac{V_p}{V_z} \tag{15}$$

where V_z is the volume of the cell lumen at turgor pressure $= 0$. V_z is calculated with the formula for the cylinder:

$$V_z = \frac{\pi \cdot b^2}{4} \cdot h \tag{16}$$

where h is the inner length of the cell measured in the cell lumen. The ratio V_p/V_z is called "degree of plasmolysis" (G), so that O_g becomes

$$O_g = C \cdot \frac{V_p}{V_z} = C \cdot G = C \cdot \frac{L - (b/3)}{h} \quad \text{or} \quad C \cdot \frac{L - \alpha \cdot m}{h} \tag{17}$$

The degree of plasmolysis G becomes unity for the state of incipient plasmolysis and decreases with the protoplast volume V_p. It should not be confused with the degree of contraction which, of course, increases when V_p becomes smaller.

The great advantage of the plasmometric methods is that they make possible the determination of O_g for an individual cell using only one single external concentration of a plasmolyticum. Using the method of incipient plasmolysis for this determination would result in the testing of several concentrations, and it is most likely that the cell would be damaged by the procedures involved in making such a series of concentration changes. Another advantage is the elimination of possible effects of the wall attachment, which does not impede protoplast separation of the cell wall in this stage.

A disadvantage of the plasmometric method is, of course, its limitation to cells of cylindrical or nearly cylindrical shape, but some efforts have been made to extend its application (Härtel, 1963; cf. also Prát, 1923, p. 91f). Also the time needed for the protoplast to reach the higher degree of contraction may be considerably longer than for incipient plasmolysis, which could lead to appreciable alterations of cell sap concentration because of exosmosis or osmoregulatory processes.

Since its introduction, the plasmometric method has become a standard method in cell physiology and proved to be a very useful tool in measuring not only the osmotic ground value O_g, but also, as discussed later, the protoplasmic permeability. The value of O_g found by this method is always in very good agreement with that derived from the same cell material with the method of incipient plasmolysis (cf. Höfler, 1918, p. 147).

Application of the plasmometric method to many cells of the same tissue provides a set of O_g values which may be evaluated statistically. In this way the mean O_g of the tissue is derived, which is comparable with the values found by the method of incipient plasmolysis. When the O_g values and the size of the individual cells are compared for a limited area of the inner epidermis of the bulb scale of *Allium cepa*, no correlation of O_g with cell size is found. However, the standard deviation of the cell size is about 20%, while that for the osmotic ground value (O_g) is only 7%; this indicates that in spite of the variations in cell size a high degree of uniformity of O_g exists, which reflects its physiological significance and the interrelationship of the protoplasm of these cells.

1. EXPERIMENTAL PROCEDURES

The manipulations and precautions here are very similar to those described for the method of incipient plasmolysis, but only one concentration of the test solution is needed. Measurements are performed as soon as the tissue pieces, etc., have been in the test solution long enough to assure that the final volume of the protoplast is reached. This time may be found in preliminary tests, where the same cells are checked after suitable time intervals as long as L reaches a constant value. However, with difficult material, control measurements after suitable time intervals may also be recommended for the main experiment.

For evaluation the tissue pieces, etc., are placed on a slide under a coverglass in a sufficiently large droplet of the test solution. Different measures are possible to avoid appreciable change in the concentration of the test solution, which might cause considerable errors. The easiest precaution is to make measurements so rapidly, that practically no concentration change occurs. However, in most cases this is not practicable, since frequently several cells must be measured in one experiment. Sometimes surrounding the edges of the coverglass with vaseline may help considerably. A frequent perfusion of droplets of new solution placed at one side of the coverglass by applying a piece of filter paper at the other side sometimes is also useful. Of course, the use of a perfusion chamber is the best assurance of constant external concentration (cf. Gahlen, 1934; Stadelmann, 1959).

When evaluation is completed but a later measurement is scheduled, the material is again placed in the bottle with the test solution. A normal eyepiece scale micrometer, with sufficiently accurate subdivisions, is suitable for measuring L, b, and h. No absolute value is needed for the plasmometric evaluation of G or O_g.

2. ACCURACY OF THE DETERMINATION

Accuracy of the determination is reported to be higher than for the method of incipient plasmolysis for suitable cell material and may reach $\pm$ 0.001–0.002 M saccharose solution (Höfler, 1918, p. 130).

3. SOURCES OF ERRORS

The same factors, with the exception of wall attachment, as mentioned for the method of incipient plasmolysis may also interfere in the plasmometric evaluation of a cell. Neglecting the correction factor α for different forms of the protoplast ends usually leads to negligible error for calculation of O_g. In recent measurements on epidermal cells of *Allium cepa* bulb scales, 70% of the maximum curvature was reached at final plasmolysis ($m/b = 0.35$, Stadelmann, 1964, p. 41f). This caused only a minor difference in the resulting values [α for formula (15) 0.667 instead of, correctly, 0.837]. However, effects of geometry, wall attachment, and thickness of the protoplasm envelope on the curvature need further investigation.

Increased hazards of alterations by exosmosis or osmoregulatory processes caused by the longer time interval needed between the beginning of plasmolysis and the end of plasmolysis was mentioned by Ursprung (1939, p. 1238), but considering the experimental results this does not seem to interfere seriously.

4. EXAMPLE OF A TYPICAL EXPERIMENT

a. Cell with spherical protoplast ends or when correction factor α can be neglected. Material: *Zebrina pendula*, longitudinal section of an internode; epidermal cell (Stadelmann, 1952, p. 376, cell No. 2). Solution: 0.5 M glucose (molal concentration). $h = 90.0 - 47.0 = 43.0$ MU (micrometer units); $L = 76.8 - 58.0 = 18.8$ MU; $b = 54.0 - 40.0 = 14.0$ MU. Substituting these values in formula (17) gives

$$O_g = \frac{18.8 - (14/3)}{43.0} \cdot 0.5 = \frac{14.1}{43.0} \cdot 0.5 = 0.164 \ M$$

(molal concentration)

b. *Calculation applying the correction factor* α. Material: *Tradescantia guianensis*, longitudinal section of an internode, cell in the outermost layer of parenchyma (Höfler, 1918, p. 132, cell No. 1). Solution: 0.4 M saccharose (molar concentration). $h = 00.0 - 63.5 = 63.5$ MU (micrometer units); $L = 12.0 - 56.0 = 44.0$ MU; $b = 17.0$ MU, $m = 6.5$ MU, $6.5/17 = 0.38$. For this value interpolated from Table I: $α = 0.80$. Substituting these values in formula (17) gives

$$O_g = \frac{44 - 0.80 \times 6.5}{63.5} \cdot 0.4 = \frac{44.0 - 5.2}{63.5} \cdot 0.4$$

$$= \frac{38.8}{63.5} \cdot 0.4 = 0.244 \ M \quad \text{(molar concentration)}$$

C. Determination of O_g for Whole Tissues

A method evaluating the contraction of tissue pieces was described by Pfeffer (1893, p. 313f). It is based upon the summation of the contraction of the individual cells induced by the water release when the organ or tissue is transferred to an osmotically active solution.

Two clear marks with water-resistant ink are made on the material to be tested, and the distance between these marks is measured with a magnifying glass or low power microscope in a series of increased concentrations of glucose or sucrose, for example. At higher concentrations this distance between the marks becomes constant while decreasing at the beginning of the concentration series. The lowest concentration in which the constant distance is measured represents the average O_g of the tissue.

This method was used for measuring mean values for O_g of the cells of root tips. The material was left for 25–30 minutes in each solution before transferring to a more concentrated one. This time was considered to be sufficient, even for the relatively thick roots of *Vicia faba*, for reaching osmotic equilibrium with the external concentration.

Of course, only an average of the osmotic ground value O_g of the cells composing the organ or tissue is found by this method. The O_g value determined in this way in homogeneous tissue may be a little lower than that calculated from experiments with incipient plasmolysis of the individual cells. This is true because relaxation of the cell wall is the criterion, while for producing incipient plasmolysis the separation of the protoplast from a limited cell wall area may need a slightly higher concentration (cf. Ursprung, 1939, p. 1236).

D. Significance of O_g

The osmotic value for the cell in the turgorless state may be considered a basic reference value to describe the osmotic situation of the cell under a standard condition, which is also useful for comparing different kinds of cells (cf. Höfler, 1917, p. 724). This value gives information on sap concentration and osmotic potentialities of the individual cell and is also often the basis for further calculations of other osmotic quantities. Further, it is needed for determination of the permeability constant.

III. Determination of the Suction Potential of the Cell Contents in Incipient Plasmolysis (Si_g)

The suction potential of the cell contents in incipient plasmolysis (Si_g) is obtained from the osmotic ground value (O_g) from an appropriate table for the osmotic pressure of the concentrations used. For the most frequently applied substances, saccharose and glucose, Tables III and IV may be used. Further tables are published for KNO_3 (Ursprung, 1939, p. 1290), NaCl (Ursprung, 1939, p. 1285), and sea water (Biebl, 1962, p. 112).

TABLE III

OSMOTIC PRESSURE OF SACCHAROSE SOLUTIONS[a]

Saccharose		Saccharose		Saccharose	
Concentration (M)	Osmotic pressure (atm)	Concentration (M)	Osmotic pressure (atm)	Concentration (M)	Osmotic pressure (atm)
0.00	0.0	0.10	2.7	0.20	5.3
0.01	0.3	0.11	2.9	0.21	5.6
0.02	0.5	0.12	3.2	0.22	5.9
0.03	0.8	0.13	3.5	0.23	6.2
0.04	1.1	0.14	3.7	0.24	6.4
0.05	1.3	0.15	4.0	0.25	6.7
0.06	1.6	0.16	4.3	0.26	7.0
0.07	1.9	0.17	4.5	0.27	7.3
0.08	2.1	0.18	4.8	0.28	7.6
0.09	2.4	0.19	5.1	0.29	7.9

[a] Ursprung (1939, p. 1275-1276). Concentrations indicated in M solution (volume-molar).

TABLE III (*Continued*)

Saccharose		Saccharose		Saccharose	
Concen-tration (M)	Osmotic pressure (atm)	Concen-tration (M)	Osmotic pressure (atm)	Concen-tration (M)	Osmotic pressure (atm)
0.30	8.2	0.72	22.5	1.14	42.6
0.31	8.5	0.73	22.9	1.15	43.2
0.32	8.8	0.74	23.4	1.16	43.7
0.33	9.1	0.75	23.8	1.17	44.3
0.34	9.4	0.76	24.2	1.18	44.8
0.35	9.7	0.77	24.6	1.19	45.4
0.36	10.0	0.78	25.0	1.20	46.0
0.37	10.3	0.79	25.4	1.21	46.5
0.38	10.6	0.80	25.9	1.22	47.1
0.39	10.9	0.81	26.3	1.23	47.7
0.40	11.2	0.82	26.7	1.24	48.3
0.41	11.5	0.83	27.2	1.25	48.9
0.42	11.8	0.84	27.6	1.26	49.5
0.43	12.2	0.85	28.1	1.27	50.1
0.44	12.5	0.86	28.5	1.28	50.8
0.45	12.8	0.87	29.0	1.29	51.4
0.46	13.1	0.88	29.4	1.30	52.0
0.47	13.5	0.89	29.9	1.31	52.7
0.48	13.8	0.90	30.4	1.32	53.4
0.49	14.1	0.91	30.8	1.33	54.0
0.50	14.5	0.92	31.3	1.34	54.7
0.51	14.8	0.93	31.8	1.35	55.4
0.52	15.1	0.94	32.3	1.36	56.1
0.53	15.5	0.95	32.7	1.37	56.8
0.54	15.8	0.96	33.2	1.38	57.5
0.55	16.2	0.97	33.7	1.39	58.2
0.56	16.5	0.98	34.2	1.40	58.9
0.57	16.9	0.99	34.7	1.41	59.7
0.58	17.2	1.00	35.2	1.42	60.4
0.59	17.6	1.01	35.7	1.43	61.1
0.60	18.0	1.02	36.2	1.44	61.9
0.61	18.4	1.03	36.7	1.45	62.7
0.62	18.7	1.04	37.3	1.46	63.4
0.63	19.1	1.05	37.8	1.47	64.2
0.64	19.5	1.06	38.3	1.48	65.0
0.65	19.8	1.07	38.8	1.49	65.8
0.66	20.2	1.08	39.3	1.50	66.6
0.67	20.6	1.09	39.9	1.51	67.4
0.68	21.0	1.10	40.4	1.52	68.2
0.69	21.4	1.11	41.0	1.53	69.1
0.70	21.8	1.12	41.5	1.54	69.9
0.71	22.2	1.13	42.0	1.55	70.8

TABLE III (*Continued*)

Saccharose		Saccharose		Saccharose	
Concen-tration (M)	Osmotic pressure (atm)	Concen-tration (M)	Osmotic pressure (atm)	Concen-tration (M)	Osmotic pressure (atm)
1.56	71.6	1.98	114.9	2.40	182.9
1.57	72.5	1.99	116.2	2.41	184.8
1.58	73.3	2.00	117.5	2.42	186.8
1.59	74.2	2.01	118.8	2.43	188.8
1.60	75.1	2.02	120.1	2.44	190.8
1.61	76.0	2.03	121.4	2.45	192.8
1.62	76.9	2.04	122.8	2.46	194.8
1.63	77.8	2.05	124.2	2.47	196.9
1.64	78.7	2.06	125.6	2.48	198.9
1.65	79.7	2.07	127.0	2.49	201.0
1.66	80.6	2.08	128.4	2.50	203.1
1.67	81.5	2.09	129.8	2.51	205.2
1.68	82.5	2.10	131.3	2.52	207.4
1.69	83.5	2.11	132.8	2.53	209.5
1.70	84.4	2.12	134.3	2.54	211.8
1.71	85.4	2.13	135.8	2.55	214.0
1.72	86.4	2.14	137.3	2.56	216.3
1.73	87.4	2.15	138.9	2.57	218.6
1.74	88.4	2.16	140.5	2.58	220.9
1.75	89.4	2.17	142.1	2.59	223.3
1.76	90.4	2.18	143.7	2.60	225.6
1.77	91.4	2.19	145.3	2.61	228.1
1.78	92.5	2.20	146.9	2.62	230.5
1.79	93.5	2.21	148.6	2.63	233.0
1.80	94.5	2.22	150.3	2.64	235.5
1.81	95.6	2.23	152.0	2.65	238.0
1.82	96.7	2.24	153.7	2.66	240.6
1.83	97.7	2.25	155.4	2.67	243.2
1.84	98.8	2.26	157.1	2.68	245.8
1.85	99.9	2.27	158.8	2.69	248.4
1.86	101.0	2.28	160.6	2.70	251.1
1.87	102.1	2.29	162.4	2.71	253.8
1.88	103.2	2.30	164.2	2.72	256.6
1.89	104.3	2.31	166.0	2.73	259.4
1.90	105.5	2.32	167.8	2.74	262.3
1.91	106.6	2.33	169.6	2.75	265.2
1.92	107.7	2.34	171.5	2.76	268.1
1.93	108.9	2.35	173.3	2.77	271.1
1.94	110.1	2.36	175.2	2.78	274.1
1.95	111.3	2.37	177.1	2.79	277.2
1.96	112.5	2.38	179.0	2.80	280.3
1.97	113.7	2.39	180.9	2.81	283.4

E. J. STADELMANN

TABLE IV

OSMOTIC PRESSURE OF GLUCOSE SOLUTIONS[a]

Glucose		Glucose		Glucose		Glucose	
Concentration (M)	Osmotic pressure (atm)	Concentration (M)	Osmotic pressure (atm)	Concentration (M)	Osmotic pressure (atm)	Concentration (M)	Osmotic pressure (atm)
0.1	2.50	0.4	10.31	0.7	18.09	1.06	28.02
0.15	3.80	0.45	11.59	0.75	19.41	1.14	31.80
0.2	5.18	0.5	12.90	0.8	20.80	1.22	34.80
0.25	6.40	0.55	14.19	0.85	22.35	1.30	37.70
0.3	7.70	0.6	15.48	0.9	24.12	1.38	41.20
0.35	9.00	0.65	16.77	1.0	26.19	1.47	45.00

[a] Biebl (1962, p. 95) after Repp (1939, pp. 580, 581). Concentrations indicated in M solution (volume-molar).

IV. Methods for Measurement of the Suction Potential of the Cell Contents in the Normal State (Si_n)

For an understanding of the important problems of water transport and water economy of the plant, a knowledge of the osmotic quantities in the normal state is essential. One of the important factors is the suction potential of the cell contents (Si_n). This value is often used as a significant indicator of the water stress of tissues and whole plants, especially in ecological studies (cf. Walter, 1963, 1965). However, Si_n does not describe completely the water relations of the cell, since the wall pressure (W) is not taken into account.

Several methods are proposed for the determination of Si_n and those most frequently used are described below.

A. Determination of Si_n for a Single Cell from O_g

The basic idea here is to find the ratio

$$\frac{\text{Cell volume in incipient plasmolysis}}{\text{Cell volume in normal state}} = \frac{V_g}{V_n} \qquad (18)$$

and to calculate O_n from the formula

$$O_n = O_g \cdot \frac{V_g}{V_n} \qquad (19)$$

The value of O_n resulting from this equation is used to find the value of Si_n from the appropriate table (for saccharose or glucose, Tables III and IV may be used). Formula (19) cannot be established for Si_g and Si_n itself, since they are not strictly proportional to the inverse volume ratio.

The exact determination of the cell volume is very difficult and often impossible when the cell depth cannot be determined. A good approximation for calculating the ratio is based on the assumption that the cell depth is equal to the cell width, and also that their relative dilatation at turgescence is the same. When the cell width varies over the cell length its mean value may be used. The cell surface is determined by planimetric evaluation of an accurate drawing of the inner cell wall contours (cf. Ursprung, 1939, p. 1349). Ratio of cell length or cell surface in most cases may not be the same as the volume ratio and therefore cannot be used in place of V_g/V_n.

Of course, when the cell wall is only slightly stretchable V_g will differ little from V_n. O_g in this case may be only slightly different from O_n.

1. Experimental Procedure

The first step is the determination of the cell volume in the normal state. The plant material (e.g., part of a leaf blade) is immediately placed in paraffin oil to avoid turgor change by evaporation, and cell volume is determined. After measurement the cell material is transferred to an appropriate sugar solution to produce incipient plasmolysis, and the volume measurement of the cell is repeated in this state.

2. Example for Determination of Si_n

Material (Ursprung, 1939, p. 1305): parenchymatous cell of the cortex of *Vicia faba* root. Area (A_n) of the cell surface measured on an exact drawing of the cell while it was in paraffin oil: 961 mm²; mean cell width (b_n) measured in this drawing: 27.5 mm; osmotic ground value (O_g), i.e., the concentration causing incipient plasmolysis: 0.47 M; area (A_g) of the cell surface measured on an exact drawing of the cell (with same magnification as for A_n) while at incipient plasmolysis: 904 mm²; and mean cell width (b_g) measured in this drawing: 28.0 mm. From these values O_n is calculated:

$$O_n = O_g \cdot \frac{V_g}{V_n} = 0.47 \times \frac{904 \times 28.0}{961 \times 27.5} = 0.47 \times 0.958 = 0.45 \ M$$

For $O_n = 0.45 \ M$, from Table III it follows that

$$Si_n = 12.8 \text{ atm}$$

B. Determination of Si_n of a Single Cell from Isolated Cell Sap

Since the usual methods for determination of the osmotic pressure of a solution require a relatively high minimum volume of the test solution, O_n can be determined in this way only for the giant *Characean* cell. Cell sap is collected after puncturing the cell with a glass capillary. Impurities must be carefully excluded, since their presence might decrease the osmotic value (cf. Wildervanck, 1932, p. 242). Freezing point depression is measured with one of the standard procedures and from its value O_n is calculated. However, some new methods or improvements of old ones were described recently (Ramsay, 1949, p. 48f; Hinzpeter, 1952; Thöni, 1965). The micromethod described by Hargitay et al. (1951) should be applicable for normal-sized plant cells, since only about 10^5–10^6 μ³ of test solution (cell sap) is required.

C. Methods for Mean Values of Si_n of Tissues, Plant Parts, or Plant Organs

1. EVALUATION OF TISSUE STRIPS

When the cells of a tissue are osmotically homogeneous the value of O_n can be derived from the change in length of a narrow and relatively long strip of this tissue. Since changes in thickness and width of the tissue may be neglected, the following approximation may be used to calculate the volume ratio V_g/V_n which is needed to determine O_n from O_g:

$$\frac{V_g}{V_n} \approx \frac{L_g}{L_n} \qquad (20)$$

Here L_n is the length of the tissue strip in the normal state, which should be measured in paraffin oil to avoid water loss during the time needed for the measurement. To measure L_g, the concentration for O_g must first be found. This can be done by Pfeffer's method (e.g., Section II,C) or by microscopic observation of the incipient plasmolysis in the cells of the tissue. When the concentration for O_g is found in this way, the tissue strip is allowed to stay in this solution as long as no further decrease in length is observed. This final length is measured and gives L_g.

2. DETERMINATION OF Si_n FROM EXPRESSED SAP

This method is not based on plasmolysis or turgescence. However, it should be mentioned here since it is widely used in ecological studies (cf. Walter, 1931, 1965; Kozlowski, 1964, p. 44f). Samples of plant parts or of tissues are used for sap extraction (for discussion of technique see Bennet-Clark, 1959, p. 126f). Usually the freezing point depression is measured to find the osmotic concentration of the expressed sap, and this is considered to be about equal to O_n for a suitable tissue and procedure.

Of course, only mean values are obtained with this method. Therefore it is less suitable for detailed studies of water relations inside a plant or a tissue. Details of the technique, limitations to application, and precautions to avoid error are discussed by Walter (1931, p. 27f) and Ursprung (1939, p. 1309f).

When the results of plasmolytic measurements are compared with the values derived from sap expressed from the same material, discrepancies often are reported. These differences result to a large extent from overlooking the effect of the volume change when a turgid cell reaches the turgorless state of incipient plasmolysis. In this case Si_g was compared directly with Si_n and, of course, Si_g was found to be higher. However, even after calculation of Si_n from O_g as described above, a discrepancy often remains (cf. Bennet-Clark, 1959, p. 140, Table III) which has to be attributed to other factors. Some of these are discussed by Mosebach (1932, p. 122f), Kramer (1956, p. 325), Bennet-Clark (1959, p. 140f) and Kozlowski (1964, p. 48f). Furthermore considerable amounts of soluble sugars in the chloroplast (Heber, 1957) and water absorbed in the cell wall (Kreeb, 1965) are considered possible causes for these differences. Lehtorana (1956, p. 11) also reports differences in the composition of the sap obtained by puncture or mechanical pressure from some *Characean* plant cells.

V. Measurement of the Suction Potential of the Cell in the Normal State (Sz_n)

This osmotic quantity (also called diffusion pressure deficit, DPD) is considered most important in characterizing cell water relations, since it measures the ability of a cell to absorb water. It is also important in describing water transport within a tissue. Water moves there along a gradient of Sz_n only. The suction potential of the cell content (Si_n, also called osmotic pressure) does not determine the water transport; Si_n may show a completely different gradient and even decrease while Sz_n increases (cf. Kramer, 1956, p. 320f).

All methods for measurement of Sz_n use basically the same principle: exposure of the cell or the tissue to external conditions where no water uptake or release takes place (equilibrium condition). Quite different means are applied to detect the equilibrium concentration (cf. Crafts *et al.*, 1949, p. 104f; Kozlowski, 1964, p. 40f). The principle of the turgescence methods described here is to observe changes of volume (mostly indicated as changes of area or length under the microscope) and to find the external concentration where the volume of the cell is the same as under the natural conditions of the cell or tissue. When used carefully and with suitable material the turgescence methods give excellent results, as shown in the accurate work of Ursprung and Blum (cf. Blum, 1958).

A. Determination of Sz_n for a Single Cell

The ideal procedure would be to immerse a cell in paraffin oil (to prevent exchange of water with the surrounding medium), to measure its volume, and subsequently to transfer the cell to a series of stepwise increasing concentrations of a nonpermeating solute such as saccharose.

Since a normal plant cell rarely survives such frequent transfer without damage, a modified method must be used. A series of cuttings is made from the same tissue and each cutting is transferred from the paraffin oil to one of the concentrations of the sugar solution. Of course, this method requires tissue pieces with the same osmotic characteristics over an area big enough for making the series of cuttings.

For most cells it is not possible to measure the cell volume easily, thus an indicator of volume changes must be used. Since changes of the cross section area in most cells correspond to the volume changes, they may be used to indicate the volume changes. Sometimes (Hoffmann, 1932, p. 413) the cell length is used for this purpose, but this leads to a higher degree of uncertainty since an increase in this dimension does not always correspond to an increase in cell volume in all cells.

1. EXPERIMENTAL PROCEDURE

The material should be collected in such a way that no loss of water occurs. Immediately after the separation of the tissue or organ (e.g., leaf) from the plant, it should be placed in paraffin oil and remain there for transport to the laboratory. Further preparation of the cuttings for microscopic observation should be made in paraffin oil also.

With some experience the investigator will work fast enough so that in most cases a bath with paraffin oil can be avoided. A sufficiently large droplet of paraffin oil is placed on the area where the cuttings are to be made, so that the new cut surfaces never become exposed to air. A very sharp knife must be used to avoid deformation of the cuttings. The cuttings must be made thick enough so that at least three layers of undamaged cells are obtained.

One cutting is transferred on a slide with paraffin oil to the microscope stage. The coverglass should be supported by other small pieces of coverglass so that no pressure leading to deformation of the cutting is exerted.

At least two or three apparently healthy cells which have sharp contours are selected. Camera lucida drawings of the cross sections of these cells are made. Their location in the cutting is remembered or noted for fast recovery in the later checks. When this is done the cutting is transferred to a tightly closed wide-necked glass bottle containing the first solution of the saccharose concentration series.

A second cutting is then examined in the same way in paraffin oil under the microscope and later transferred to the second saccharose solution. This procedure is continued until a cutting has been immersed in each of the graded saccharose solutions.

The cuttings are left for about 45 minutes in the solutions and thereafter examined in the same sequence in a droplet of the solution in which they were immersed. Cell cross sections are redrawn at exactly the same enlargement as before, and the cuttings are reimmersed in their solution. This check is repeated several times up to about 5–6 hours, depending upon the material. When this evaluation is finished the vitality of the cells is tested by plasmolysis and deplasmolysis. Only those cells which show normal behavior are considered for the final calculation of Sz_n.

This usual procedure for evaluation has the advantage that it is not necessary that osmotic equilibrium be reached. However, the treatment may be modified depending upon the tissue, and it sometimes is suitable to use the same cell for different saccharose solutions. In this case, however, the osmotic equilibrium in each of these solutions must be reached before evaluation of the cell.

From the drawings of the cell circumference the area of the cell cross section is determined by planimetry and the difference: (area in paraffin oil) — (area in saccharose solution) is expressed in percent of the area in paraffin oil. With these data a table is established with each line corresponding to the time of the measurements (see Table V). For the calculation of Sz_n each cell showing irregularities in these values is eliminated. Sz_n is calculated from two concentrations, where an increase and a decrease in area took place, respectively. Linear interpolation is used to calculate Sz_n from the relation

$$\frac{P_2 - P_1}{P_2 - Sz_n} = \frac{D_1 - D_2}{-D_2} \tag{21}$$

by the formula

$$Sz_n = (P_2 - P_1) \cdot \left(\frac{D_2}{D_1 - D_2}\right) + P_2 \tag{22}$$

where P_1 and P_2 are the suction potentials of the two closest sucrose solutions, and D_1 and D_2 are the respective area changes expressed in percent of the area in paraffin oil, with D_1 or D_2 being negative.

2. Sources of Errors

Obviously many factors must be taken carefully into consideration in such a delicate procedure as the determination of Sz_n of individual cells, to make it successful. Some of the pitfalls have led to heavy criticism, which has been overgeneralized without sufficient justification (Ernest, 1935; Oppenheimer, 1932).

Certain factors should be considered for possible error in the determination of Sz_n: (1) The cell walls must have a sufficiently high degree of *elasticity* (cf. Blum, 1958, p. 15; Walter, 1963, p. 63). Only those cells with a wall allowing a minimum of 3% of area change of the cross section are suitable for this method. For cells with less elasticity only Si_g can be measured, and this must always be greater than Sz_n. (2) Cells under *tissue tension* (cf. Ursprung, 1939, p. 1362) must be excluded from this method since it is almost impossible to determine the contribution of tissue tension to the osmotic condition of the cell. (3) *Swelling* of the cell wall or the protoplasmic envelope also produces incorrect results. Sometimes changing to another plasmolyticum may correct this problem. (4) *Tension from water cohesion* in the vessels may cause errors. When this tension is released by the cutting process the parenchyma cells may take up some additional water. This can be avoided when cutting is made in such a way that vessels are excluded.

Two important criteria for the reliability of an experiment are found in evaluating the area differences (expressed in percent) with regard to the concentration corresponding to Sz_n: (a) The differences must all have the same sign on one side of Sz_n; for all cells in the concentrations lower than that for Sz_n these differences must be positive and for all cells in a higher concentration they must be negative. (b) The differences must increase more and more from one concentration to the next as one considers concentrations farther away from that corresponding to Sz_n. No decrease of this difference is allowed and only for two immediately following concentrations may this difference have the same value. Any experiment which does not comply with these two criteria should be rejected.

3. EXAMPLE FOR THE DETERMINATION OF Sz_n

Material (Ursprung, 1939, p. 1360): epidermal cell of the lower surface of the leaf of *Sambucus ebulus* over the middle rib.

a. Results of the area measurements. From Table V it is seen that Sz_n lies between 5.3 and 6.7 atm so that the values found in the concentrations 0.20 M and 0.25 M should be used for calculation of Sz_n with formula (22).

b. Calculation of Sz_n with formula (22). The values of cells 4 and 5 for $t = 45$ minutes are used here. However, other combinations are also possible as long as the values are taken from the same line in the table (i.e., treated for the same length of time), and it can be shown that such other values give essentially the same results (cf. Ursprung, 1939, p. 1360, Table 64, 2nd part):

$$Sz_n = (6.7 - 5.3) \times \frac{-2.9}{18.5 + 2.9} + 6.7$$

$$= 1.4 \times (-0.135) + 6.7 = 6.4 \text{ atm}$$

TABLE V

DETERMINATION OF AREA CHANGES OF INDIVIDUAL CELLS WHEN TAKEN FROM
PARAFFIN OIL TO SACCHAROSE SOLUTIONS OF DIFFERENT CONCENTRATIONS[a]

Time in saccharose (min)	Changes in the cross section area of the cells[b] in cutting number:							
	1[c]		2[d]		3[e]		4[f]	
	1	2	3	4	5	6	7	8[g]
45	+ 4.9	+17.0	+19.5	+18.5	−2.9	−7.9	− 6.6	−1.1
90	+10.5	+23.1	+25.3	+22.1	−2.2	−6.1	−14.3	0.0
135	+10.2	+23.7	+26.9	+22.5	−9.9	−9.0	−16.5	−1.5
315	+11.4	+23.7	+27.8	+23.7	−9.7	−7.9	−16.1	−1.5
375	+11.0	+24.4	+28.2	+22.7	−8.7	−7.6	−14.7	0.0

[a] From Ursprung (1939, p. 1360).
[b] Expressed in per cent of the area found in paraffin oil after transfer to the saccharose solution.
[c] Saccharose concentration, 0.15 M; osmotic pressure, 4.0 atm.
[d] Saccharose concentration, 0.20 M; osmotic pressure, 5.3 atm.
[e] Saccharose concentration, 0.25 M; osmotic pressure, 6.7 atm.
[f] Saccharose concentration, 0.30 M; osmotic pressure, 8.2 atm.
[g] Cell number.

B. Determination of the Mean Value of Sz_n for Tissues, Plant Parts, etc.

While measurement of Sz_n of an individual cell may be important for detailed studies of water relations, it is often sufficient to know just the average value, even when cells belonging to different tissues are tested together. Less experimental work is also involved. Therefore in many experiments only this mean value is determined. Several methods have been published, some of them specialized for certain tissues (e.g., hard leaf method; Ursprung and Blum, 1927). The most widely used of the turgescence methods is the *tissue strip method* from Ursprung (1923; 1939, p. 1369f). The change in length of the strip serves as the indicator of the volume change of the cells in this method. Therefore time-consuming measurements of surface area are avoided. Of course, the principles and the procedures for this method are essentially the same as described above.

1. EXPERIMENTAL PROCEDURES

The plant material (e.g., leaf) is put on a small cork plate in a Petri dish sufficiently filled with paraffin oil. Strips about 2×20 mm are cut with a sharp knife in such a way that they remain completely immersed in paraffin oil. To get two sharp points for the measurements, the strips should have a slightly trapezoid form. Thicker vascular bundles or any region with thick-walled cells should be excluded from the strips. If this is not possible the strips must be cut at a right angle to such vascular bundles, etc., to minimize their effect on the volume change of the strips when they are later transferred to saccharose solutions.

Meanwhile, a series of graded solutions of saccharose has been prepared, with concentrations chosen in such a way that the expected suction potential (Sz_n) of the tissue strip lies inside the range of the suction potential of these saccharose solutions.

Each strip is placed in a drop of paraffin oil on a slide provided with calibration marks of 0.5 mm over about 20 mm, and the length is measured carefully with a microscope. Then the bulk of the paraffin oil is quickly removed from the strip with filter paper, and the strip is placed immediately in a wide-necked glass bottle with about 10 ml of the least concentrated saccharose solution. In this way, one strip is transferred to each of the saccharose concentrations.

After about 45 minutes (for some material even 5–10 minutes is sufficient) the length of each strip is measured again on the calibrated slide in a large drop of the solution in which the tissue was immersed.

From these two series of measurements the difference in length of the strip is calculated in mm. With these values a table is established (see Table VI) and the two nearest concentrations are used to interpolate Sz_n with formula (22), where for D, the absolute difference in mm, is introduced.

2. Sources of Errors

In general, sources of error are the same as those for the determination of Sz_n for a single cell. Only the effect of tissue tension may be reduced; it will be almost the same in a tissue strip as in the intact material itself. Additional errors may be introduced by changing frequencies of the different cell types from one strip to another. Diluting effects of the water vapor of the intercellular space on the external concentration applied certainly may be neglected. They might be important as a source of errors for the gravimetric method (Kreeb, 1960, p. 274). In any case the two criteria for reliability mentioned above must be verified in these experiments also.

3. Example for the Determination of Sz_n with the Tissue Strip Method

Material (Ursprung, 1939, p. 1372): petal of *Rosa* sp. Results of the length measurements are given in Table VI. This experiment can be used for the evaluation of Sz_n since the two criteria mentioned in Section V,A, 2 are fulfilled: on the one side of Sz_n all differences are positive while on the other side all are negative. The difference increases when starting from Sz_n in both directions. Only for the two concentrations 0.35 M and 0.40 M does it remain the same.

TABLE VI

CHANGE IN LENGTH OF A STRIP OF A PETAL OF *Rosa* SP. WHEN TRANSFERRED FROM PARAFFIN OIL TO GRADED SACCHAROSE CONCENTRATIONS

Suction potential (atm) of the saccharose solution	Concentration (M)	Length of the tissue strip (mm)		
		In paraffin oil	After 15 min in the sucrose solution	Difference
5.3	0.20	11.3	11.6	+0.3
6.7	0.25	13.2	11.35	+0.15
8.2	0.30	9.9	9.95	+0.05
9.7	0.35	12.1	12.05	−0.05
11.2	0.40	13.55	13.50	−0.05
12.8	0.45	13.55	13.40	−0.15

The value of Sz_n must be between 8.2 and 9.7 atm, since in the first concentration there is a dilatation, and in the latter a contraction, of the strip. Sz_n is calculated by linear interpolation with formula (22):

$$Sz_n = (9.7 - 8.2) \times \frac{-0.05}{0.05 + 0.05} + 9.7$$

$$= 1.5 \times (-0.5) + 9.7 = 8.9 \, \text{atm}$$

C. Significance of the Suction Potential of the Cell in the Normal State (Sz_n)

The accurate measurement of Sz_n for cells in different locations in the higher plant provides one of the principal keys to understanding of the mechanism of water transport in intact plants. Further measurements on the same material under different ecological conditions have increased our knowledge about the availability of water to plants under natural or controlled environmental conditions.

The general rule can be established for higher plants that the suction potential in the normal state increases more and more, with distance of the cell from the point of water entrance, i.e., the root epidermis and root hairs. Hence, an increase of Sz_n is found in the absorption zone of the root from the epidermis inward until the endodermis is reached. Here the suction potential drops drastically (an effect related to the peculiar physiological activity of the endodermis). It rises again inside the central cylinder and becomes gradually greater in the higher regions of the stem. From there it increases toward the leaves.

TABLE VII

THE SUCTION POTENTIAL OF THE CELL IN ITS NORMAL STATE (Sz_n) AND IN
INCIPIENT PLASMOLYSIS (Sz_g) IN THE ABSORPTION ZONE (1 CM FROM
THE ROOT TIP) OF A LATERAL ROOT OF *Phaseolus vulgaris*[a]

Tissue and/or cell layer	Sz_n (atm)	$Sz_g = Si_g$ (atm)
Epidermis	0.9	8.5
Cortex layer		
1 (outermost)	1.3	8.2
2	1.7	8.8
3	2.0	9.4
4	2.6	9.7
5	3.2	9.1
6	3.6	8.8
7	4.2	8.8
Endodermis	1.3	8.8
Pericycle	0.9	8.5
Parenchyma of vascular bundle	0.8	8.5

[a] From Ursprung and Blum (1921, p. 72).

TABLE VIII

DISTRIBUTION OF THE SUCTION POTENTIAL IN THE NORMAL STATE
(Sz_n) IN *Hedera helix*[a]

Plant part	Pith	Cell near vessel	Cortex Inner	Cortex Outer	Epidermis or phellogen
Root tip	—	—	—	1.6 ←——— 1.0	
Root (18 cm behind the tip)	—	2.1 ————————→ 2.4 ———→ 3.2			
Stem					
Lower part (35 cm above the soil)	2.4 ←—— 2.1 ——→ 2.9 ——→ 3.4 ——→ 3.7				
Upper part (225 cm above the soil)	4.8 ←—— 4.2 ——→ 5.0 ——→ 7.3 ——→ 7.4				
Petiole	8.4 ←—— 8.1 ——→ 8.7 ——→ 9.0 ——→ 9.3				
Blade					
Lower epidermis	7.3				
Spongy parenchyma	{ 10.5 / 10.1				
Site of the vascular bundle	—				
Pallisade cells	{ 11.9 / 12.2 / 12.5				
Upper epidermis	8.0				

[a] Blum (1958, p. 58); Ursprung and Blum (1918, p. 617). Arrows indicate direction of water flow.

Table VII gives the detailed values of Sz_n for the layers of the root cortex and other important regions. Also, the value of Sz_g ($=Si_g$) is noted, and this stays nearly constant or shows only slight variations without relation to the position of the cell layer. This clearly indicates that Sz_g is insufficient for explaining water transport in the plant.

An example of the distribution of the value of Sz_n in a whole plant is given in Table VIII, using data from a *Hedera helix* plant. While the absorption zone of the root shows the above mentioned distribution, in the older regions of the root the outer parts have a higher suction potential. This, of course, reflects the changed conditions of the water supply: in the absorption zone the outer layers have easier access to the external water, but in the older root the outer zones obtain their water from the vessels and the farther they are from them, the higher Sz_n must be.

The distribution of the suction potential in the stem shows the expected pattern. The same holds true for the leaf with the exception of the epidermis, which exhibits a definitely smaller value of Sz_n than that for the cell adjacent to the vessels. No sufficient explanation can be given as yet for this.

The effect of changes in external conditions on the value of Sz_n in different parts of a plant has not yet been thoroughly investigated. Some work has dealt with the experimental restriction of the soil water supply.

TABLE IX

INCREASE OF Sz_n IN THE LEAF BLADE OF SOME SPECIES KEPT IN POTS WITH GARDEN SOIL IN THE GREENHOUSE UNDER EQUAL ENVIRONMENTAL CONDITIONS (1) WITH READILY AVAILABLE WATER AND (2) AFTER WATERING IS STOPPED AND WILTING TAKES PLACE[a]

	Suction potential Sz_n (atm)	
Plant	Under field capacity conditions	Last measurable value during wilting
Heliophytes		
Astrantia maior	14.3	34.6
Helianthus annuus	8.1	29.8
Datura stramonium	9.5	17.7
Sciophytes		
Convallaria majalis	12.3	23.4
Sanicula europaea	20.3	39.8
Impatiens parviflora	6.7	34.6
Impatiens noli tangere	8.1	34.6
Circaea lutetiana	8.1	34.6
Asarum europaeum	11.1	25.5
Viola silvatica	12.6	34.6

[a] From Hauck (1929, p. 479).

As expected, this leads to an increase of Sz_n. However, the amount varies considerably with the plant species investigated (see Table IX). Sciophytes have in the state of beginning wilting a Sz_n value about 3 times as high as the one found under no water stress conditions.

Some studies dealt with the effect of rain, humidity, and temperature on Sz_n. A change in air and culture-medium temperature was found to have about the same effect on Sz_n as a change in root temperature alone (Molz, 1926, p. 454). However, these results are fragmentary and much work is needed on such interrelationships.

VI. Determination of the Passive Permeability of the Plant Cell with the Plasmolytic Method

A. General Considerations

Since the very beginning of physiological experimentation with plant cells the passability of the protoplasm envelope of the fully developed plant has been one of the main objects of interest. It was recognized early that the protoplasmic envelope is the barrier to the exchange of material between vacuole and external solution. The early investigations were possible since the organization of the plant cell and the partial permeability of the protoplasmic layer allowed the application of nondestructive methods based mainly on microscopic observations and osmosis. Early investigators attempted to explain the uptake of all substances in the cell by osmotic action, but it soon became clear that the uptake of nutrients (ions) followed a quite different mechanism (cf. Overton, 1899, p. 102). However, the penetration of harmless nonelectrolytes from the external solution into the cell sap is still a diffusion process which leads to a concentration equilibrium of the permeating substance between external solution and cell sap (Collander and Bärlund, 1933, p. 19). Results of such permeability experiments are of great importance since they give information on the function, the structure, and the composition of these parts of the protoplasmic envelope which are the sites of resistance to permeation.

The basis for the experimental setup in the work on passive permeability is the following: The cell material is placed in a solution containing in an appropriate concentration the substance to be tested. After a specified time the amount of the permeated substance in the vacuole of the cell is measured. Different criteria are used to estimate this amount (cf. Stadelmann, 1956b, p. 693f). When such experiments are done with a series of different substances, some kind of relative permeability measure of the protoplasm is obtained.

The plasmolytic methods use a hypertonic concentration as the external solution. The amount or the concentration of the permeating substance is inferred from the *volume* or *volume change* of the plasmolyzed protoplast. When the protoplast has a geometrically simple shape it is also possible to determine the volume and the surface of the protoplast, which are needed for the calculation of a more precise measure of permeability.

Different formulas for calculation of permeability have arisen as a result of the specific methods and materials of various authors (for review, see Stadelmann, 1956c). To allow comparison of the permeability of different cells the *absolute permeability constant* must be calculated, which is derived from a simplified Fick's Equation:

$$\frac{dn}{dt} = - K \cdot A \cdot (C - k) \tag{23}$$

where $n =$ the amount of the penetrating substance, $K =$ the permeability constant, $A =$ surface area of the protoplasm layer, $C =$ external concentration of the permeating substance, $k =$ actual internal concentration of the permeating substance, and $t =$ time. When the amount (n) is indicated in moles and the CGS system is used, the dimension for K is

$$\frac{\text{moles}}{\text{sec}} = K \times \text{cm}^2 \times \frac{\text{moles}}{\text{cm}^3}; K = \text{cm sec}^{-1}$$

The absolute permeability constant indicates the amount of the permeating substance which penetrates during the unit of time and the unit of surface area when the concentration difference between both sides of the membrane is unity.

Integration of the differential equation (23) is possible for cylindrical cells. Then a formula for K can be derived allowing one to calculate the absolute permeability constant from experimental data (Stadelmann, 1951, 1952). For cells of irregular shape, however, it is almost impossible to calculate K and only approximations can be obtained.

1. SOURCES OF ERROR

To measure permeability reliably, different sources of error must be considered. Errors normally are not serious in measuring permeability reliably when the usual experimental conditions are established and the cell material exhibits no peculiarities.

Below are listed the requirements which must be fulfilled in the experiments (cf. Stadelmann, 1956c, p. 141f); the first items [(i) to (xii)] are valid for every kind of permeability experiment while the later items concern either the osmotic methods [(*xiii*) to (*xvi*)] in general or the plasmolytic methods [(*xvii*) to (*xix*)].

a. *Requirements Concerning the General Conditions of the Experiment.* (*i*) The migration of the permeating substance does not require a supply of energy: diffusion alone is involved, a process without energy conversion.

(*ii*) In analogy with Dalton's law for partial pressures, the total osmotic pressure of a solution is always equal to the sum of all osmotic partial pressures.

(*iii*) According to the law of van't Hoff and the law of Boyle-Mariotte, the osmotic pressure of a solution is inversely proportional to the volume of the solution, when the number of the dissolved particles remains constant. Any further correction according to van der Waal's equation should not be necessary (cf. Resühr, 1935, p. 339; 1936, p. 437).

(*iv*) Once an osmotic equilibrium is established between external solution and total concentration of the cell sap, this equilibrium will be retained as long as no energy supply occurs or no elastic forces from the cell wall elongation interfere (cf. Jacobs and Stewart, 1932, p. 72; Stadelmann, 1951, p. 768).

b. *Requirements Concerning the Experimental Setup.* (*v*) The concentration of the external solution must remain the same over the entire length of the experiment; this should hold true particularly for those parts of the solution in immediate contact with the protoplasmic envelope with respect to the other regions of the solution (cf. Bärlund, 1929, p. 49; Jacobs and Stewart, 1932, p. 73; Resühr, 1936, p. 437; Dainty, 1963, p. 300).

(*vi*) An eventual effect of light and temperature should be considered. Experiments must be carried out with the same light and temperature conditions so that the data will be comparable (cf. Stadelmann, 1951, p. 768).

(*vii*) Equimolar solutions of nonelectrolytes are assumed to be exactly isosmotic (Bärlund, 1929, p. 49; for isotonic coefficient, see Fitting, 1920).

c. *Requirements Concerning the Cell Material.* (*viii*) The external solution applied in the experiments must have no harmful effect on the material and the material must not change its permeability during the experiment (cf. Bärlund, 1929, p. 15).

(*ix*) The permeability of the protoplasmic envelope for water must be considerably higher than for the substances to be tested.

(*x*) There must be none or only negligible resistance of the cell wall against the migration of the external solution through it (Resühr, 1935, p. 339).

(*xi*) The resistance against the migration of the solvents must be localized mainly in a very thin layer of the protoplasmic envelope (Resühr, 1935, p. 339; 1936, p. 347; cf. Höfler, 1965, p. 155).

(*xii*) The thickness of this thin layer must not change during the experiment (cf. Resühr, 1935, p. 339; 1936, p. 347; Jacobs and Stewart, 1932, p. 74).

(*xiii*) The amount (m_u) of the original osmotically active substance in the vacuole must not change, i.e., no catatonose, anatonose, or exosmosis may occur, so that the volume-concentration relationship (reciprocity law)

$$m_u = V_1 \cdot c_1 = V_2 \cdot c_2 = V_n \cdot c_n$$

can be applied.

(*xiv*) The osmotic activity of the solute molecules of the external solution must not be changed when they are inside the vacuole (cf. Bärlund, 1929, p. 49).

(*xv*) When the final equilibrium is reached the concentration of the permeating substance inside the vacuole must be equal to the concentration in the external solution (cf. Collander and Bärlund, 1933, p. 19).

(*xvi*) The resistance against diffusion of the permeating substance must be very small inside the vacuole so that virtually at every place in the vacuole the same concentration of the permeating substance is present at the same moment (cf. Jacobs and Stewart, 1932, p. 74; Huber, 1943, p. 442f; Höfler, 1949, p. 117).

(*xvii*) The viscosity and wall adhesion of the protoplast must be small enough so that a volume change as induced by osmotic release or uptake of water is not delayed (cf. Bärlund, 1929, p. 49).

(*xviii*) No change in the degree of the swelling of the protoplasm may occur during the experiment, and the volume of the protoplasm must present only a small fraction of the total volume of the protoplast (no "protoplasm correction"; cf. Höfler, 1918, p. 113; 1920, p. 292; Höfler and Weber, 1926, p. 652).

(*xix*) There must be no difference in the permeability of those parts of the protoplasmic envelope separated from the cell wall and those lining the cell wall (cf. Scarth, 1939, p. 132).

2. Analysis of a Typical Permeability Experiment

Let us consider a tissue in its normal state with cells each containing a large vacuole with cell sap, exhibiting a certain concentration of osmotically active material, and the thin protoplasmic envelope lining the inside of the cell wall. The protoplasmic envelope which is essentially nonpermeable to the material inside the vacuole acts as a semipermeable[2] membrane. Hence, a suction potential develops inside the vacuole. This causes an intake of water until equilibrium is attained between the elastic wall pressure and the suction potential of the cell sap. The water intake leads to a certain degree of turgescence of this tissue.

When such a tissue is transferred to an osmotically active solution this original equilibrium between wall pressure and suction potential of the cell sap will be disturbed by the suction potential of the external solution, which acts against the suction potential of the cell sap.

How fast the equilibrium of the suction potential is established depends mainly on the permeability of the protoplasmic envelope to water, on protoplasmic viscosity, and on wall attachment. The attainment of a concentration equilibrium is controlled by solute permeability through the protoplasmic envelope. Since only substances which permeate much slower than water are dealt with here, the equilibrium of the suction potentials will always be established much earlier than the equilibrium of the permeating substance. Therefore the first part of such an experiment is always similar to the plasmolysis experiment in a nonpermeating solution.

To establish the equilibrium of the suction potentials after this transfer the vacuole must lose water until the new equilibrium is reached according to the formula

$$W = - (Si - E) \qquad (24)$$

where W = the wall pressure in atm, Si = the suction potential of the cell content (= vacuolar sap) in atm, and E = the suction potential of the external concentration in atm.

However, withdrawal of water from the vacuole involves, of course, a decrease in the vacuolar volume and therefore a decrease of elastic tension of the cell wall, which in turn leads to a considerable decrease of the wall pressure (P). Since the use of hypertonic solutions is postulated, the osmotic equilibrium will not have been attained by the time the vacuolar volume becomes small enough so that no further wall pressure is developed (state of incipient plasmolysis).

Since the equilibrium of the suction potential is not yet reached at this time, water withdrawal will continue, therefore leading to a further separation of the protoplasmic envelope from the cell wall, i.e., to plasmolysis. Equilibrium is now established between the suction potentials of the external solution and the vacuolar sap.

Withdrawal of water from the vacuole will continue until the concen-

[2] Also called differentially permeable, a more descriptive term for a membrane which is impermeable or virtually impermeable to some substances, while others pass through them readily (Meyer and Anderson, 1947, p. 118).

trations in the vacuole and the external solution are equal. However, the composition of the vacuolar contents has altered: beginning with the transfer of the tissue to the permeating solution a certain amount of solute has already permeated through the protoplasmic envelope into the vacuole. Thus a concentration (k) of the permeating substance is developed in the vacuole. This partial concentration, of course, is quite low at the beginning of the experiment. The formula for the equilibrium of the suction potential therefore becomes

$$Sl = Si_p = Sv_p + Se_p \qquad (25)$$

where $Sl =$ suction potential of the external solution in atmospheres, $Si_p =$ suction potential of the cell contents ($=$ vacuolar sap) in the state of plasmolysis of the cell, in atmospheres, and $Se_p =$ partial suction potential brought about by the amount of solvent which permeated into the vacuole, in the state of plasmolysis of the cell, in atmospheres.

Since the suction potential is proportional to the concentration, formula (25) may be rewritten:

$$C = c + k \qquad (26)$$

where $C =$ the concentration of the external solution in moles per liter, $c =$ the partial concentration of the cell sap brought about by the osmotically active material originally present in the vacuolar sap, in the state of plasmolysis of the cell, in moles per liter, and $k =$ the partial concentration of the cell sap brought about by the amount of solvent permeated into the vacuole, in the state of plasmolysis of the cell, in moles per liter.

Once this osmotic equilibrium is attained it will be maintained as is required from thermodynamics, since no energy is supplied to the system. However, the concentration equilibrium for the permeating solute between cell sap and external concentration is not yet reached, as it follows from formula (26) that

$$C > k$$

Thus, permeation still continues from the external solution into the vacuole. But the permeating solute takes with it an amount of water into the vacuole, adequate to avoid a disturbance of the equilibrium of the suction potential. In this way the total concentration of the cell sap will not be changed. The intake of this accompanying amount of water into the vacuole causes an increase in its volume and the cell deplasmolyzes. During this period the partial concentration (c) of the originally present osmotically active material decreases, since the volume of the vacuole decreases. In the same degree the partial concentration of the permeating substance (k) increases so that the total concentration inside the vacuole ($c + k$) remains constant and stays in equilibrium with the external concentration (C).

The vacuole now increases in volume so much that the protoplasmic envelope completely lines the inner side of the cell wall, and the state of incipient plasmolysis is reached again. However, at the moment of its attainment the partial concentration of the permeating substance (k) is still smaller than the external concentration (C). Thus the concentration equilibrium for the permeating substance is not reached and its permeation into the vacuole continues. Thereafter a further increase of the vacuolar volume will stretch the cell wall as an internal pressure begins to develop again. Water uptake into the vacuole no longer will correspond to the amount of the permeated solute, but will be much less. Hence, an excess suction potential is developed, which is equilibrated by the wall pressure resulting from the elastic stretching of the cell wall.

The diffusion of the permeating substance into the vacuole ends when the concentration inside the vacuole is equal to that of the external solution. In this *final equilibrium* the wall pressure is the same as that developed when the cell was placed in pure water. But the external concentration is here C and the total concentration inside the vacuole is also increased by this amount.

Theoretically the final equilibrium will be reached only for $t = \infty$, but is approached quite closely soon after the end of deplasmolysis. Recently it was possible to prove this increase in turgidity experimentally (Wattendorff, 1964).

The whole sequence of changes in cell and protoplast volume described above may be divided into four phases (Stadelmann, 1951, p. 770):

1. *Initial phase,* begins at the start of the experiment when the cell is transferred from pure water to a hypertonic solution of the permeating substance. The initial phase ends when the cell has attained the state of incipient plasmolysis.

2. *Phase of contraction,* begins at incipient plasmolysis and ends when the protoplast attains maximum contraction.

3. *Phase of dilatation,* begins with the maximum contraction of the protoplast and includes the deplasmolysis until the protoplast completely lines the inner side of the cell wall, and the state of incipient plasmolysis is reached again.

4. *End phase,* begins when deplasmolysis is completed and ends when the concentration of the permeating substance reaches the concentration of the external solution. When this occurs the cell has regained the original size it had at the beginning of the experiment in pure water.

The change in vacuolar volume as described above is shown diagrammatically in Fig. 4. The length of the phase of dilatation and of the end phase is much reduced in time. Also the relative volume change of the protoplast during the initial phase and at the end phase is not as high as presented in the diagram.

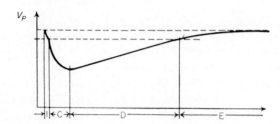

Fig. 4.   Volume changes of the vacuole during a permeability experiment with the plasmometric method (volume vs. time diagram). I = initial phase; C = phase of contraction; D = phase of dilatation; E = end phase; x axis: time; y axis: volume of the vacuole.

The most important phase in the evaluation of such an experiment is the phase of dilatation, since, using formula (12) as a basis, the absolute permeability constant can be calculated from the time course of deplasmolysis.

B. Plasmolysis Frequency Method

This method was developed by Fitting (1915, 1920) to find some relative measure of cell permeability. However, when more advanced calculations are applied some approximate values for the absolute permeability constant of the tissue can also be derived.

The main advantage of this method is its applicability for cells of any shape because no regular cell forms are necessary. Thus most tissues are suitable for this method. However, wide application of the method requires a large population of cells of highly uniform osmotic values and permeability. Quite frequently the available tissues do not provide a sufficiently large number of uniform cuttings, rendering the method unusable. A suitable material for this method is the lower epidermis of the midrib of the leaf of *Rhoeo discolor*, which has been frequently used in such experiments.

The basic idea is to measure the frequency of occurrence of plasmolysis and to infer from this value the partial concentration of the permeated substance inside the vacuole at a given time. Here any degree of plasmolysis, including incipient plasmolysis, is considered in the same way.

A series of graded hypertonic concentrations of the substance to be tested for permeation is prepared, each one in a small glass bottle. A tissue cutting is placed in each of the graded solutions at constant time intervals. These intervals should be sufficient to allow for the later microscopic examination of one cutting (5–10 minutes).

After the cutting has remained in the solution for a constant time interval (15–30 minutes may be suitable) each cutting is examined on a microscope slide in a drop of the solution in which it was immersed. The frequency of plasmolysis is estimated or determined microscopically by counting the number of plasmolyzed cells. Cuttings are then replaced into the bottles in which they were immersed. To compare the frequencies of plasmolysis, the time interval from immersion of the cutting to examination must be the same for each cutting.

TABLE X

EXAMPLE OF AN EXPERIMENT USING THE PLASMOLYSIS FREQUENCY METHOD[a]

Time (min)	Frequency of plasmolysis found (%) in cutting number[b]:							
	1	2	3	4	5	6	7	8
15	0	5	50	75	100	100	*100*	100
30		0	35	50	100	100		
45		0	0	0	50	100		
60					0	75	100	100
75					0	35	75	100

[a] Modified, from Fitting (1920, pp. 9, 10). The permeating substance was KNO_3. The time indicates the interval between the transfer of the cuttings to the solution and the determination of the plasmolysis frequency. Material: epidermal cells of the midrib of the lower side of a leaf from *Rhoeo discolor*. The values in italics are used in the example discussed in the text.

[b] Concentration of external solution of cutting: $1 = 0.1900\ M$; $2 = 0.1925\ M$; $3 = 0.1950\ M$; $4 = 0.1975\ M$; $5 = 0.2000\ M$; $6 = 0.2025\ M$; $7 = 0.2050\ M$; $8 = 0.2075\ M$.

Such determinations of the plasmolysis frequencies are repeated every 15 or 30 minutes and a table is established from the data (see Table X) with each line showing the results of one of these determinations.

Since the substance from the external solution permeates into the vacuole, deplasmolysis takes place and the frequency of plasmolysis decreases with time. But permeation takes place only slowly so that the amount permeated at the time of the first series of measurements may be neglected. Therefore these values for the plasmolysis frequencies are considered to reflect conditions like those in a nonpermeating solution. Since osmotic equilibrium is reached at the time of the first measurements, each of these values of the plasmolysis frequency characterizes the state of the cell material when the internal concentration in the vacuoles is in equilibrium with the respective external concentration. In this way the plasmolysis frequencies of the first measurements become a reference for the concentration of the vacuolar sap. Thus, the difference in plas-molysis frequency found in the first measurements between two of the cuttings is a measure of the difference in concentration of the two re-spective solutions in which these cuttings were placed.

This relation between plasmolysis frequency and cell sap concentration is used to determine the increase in partial concentration of the permeat-ing substance in the vacuole during deplasmolysis: it is assumed that the difference in plasmolysis frequency in a cutting, from the first measure-ment to a later one (when deplasmolysis is noticeable), indicates the buildup of a partial concentration of the same magnitude as the concen-tration difference which corresponds to the identical changes in plas-molysis frequency in the first measurement. An example of this inference is shown in Table X: cutting No. 7 has, in the first measurement in 0.2050 M, a plasmolysis frequency of 100%. Since deplasmolysis progresses the plasmolysis frequency decreases and 1 hour later $(75 - 15$ minutes) is 75%. The same plasmolysis frequency was found in the first measurements for cutting No. 4.

The difference in the concentrations wherein cuttings Nos. 7 and 4 were placed is $0.2050 - 0.1975 = 0.0075$ M. Therefore it can be concluded that the permeating substance entered the vacuole in such quantity that a partial concentration of 0.0075 M was built up in the cell sap during the 75 minutes that cutting No. 7 was in the concentration of 0.2050 M. This increase in concentration of the permeating substance can be related to the time interval used. Hence, the concentration increase per unit time can be calculated and some relative measure of permeability obtained. This measure may be adequate for comparing similar results with dif-ferent substances on the same kind of cell material. For the example cited above the concentration increase was 7.5 mM during 75 minutes, there-fore an average concentration increase of 0.1 mM per minute results.

An approximation of the absolute permeability constant can be calcu-lated from these experiments when the average surface (A), the average volume (V_p) of the protoplasts for the time t_1 of the first measurement (A_1, V_{p1}), and for the time t_2 of the second measurement (A_2, V_{p2}) can be estimated.

For this approximation, formula (23) is applied without integration and the differentials are replaced by the differences:

$$\frac{n_2 - n_1}{t_2 - t_1} = - K \cdot A \cdot (C - k) \tag{27}$$

The amount of the permeated substance (n_2) at the time t_2 is found from the partial concentration k_2 and the average volume for the protoplast V_{p2}: $n_2 = k_2 \times V_{p2}$. If at t_1 only a negligible amount of the permeating substance entered the vacuole, $n_1 = 0$.

The surface of the protoplast (A) is calculated as a mean value from A_1 and A_2:

$$A = \frac{A_1 + A_2}{2}$$

The external concentration of the permeating substance C is known from the experiment, and for k the mean value between k_1 and k_2 (determined from the plasmolysis frequencies as shown above) is substituted.

When length is measured in centimeters, the concentration in moles/cm^3, and the time in seconds, the value for K has the unit of cm sec^{-1}. K can be compared with any value of the absolute permeability constant found in other material and is a fairly good approximation for the cells tested.

C. Plasmometric Method

For cylindrical (or nearly cylindrical) cells the permeability constant can be calculated accurately for the individual cell. This is based on the plasmometric measurements introduced by Höfler (1918) and opens the way for a more detailed study of cell permeability. In a cylindrical cell the protoplast takes the form of a cylinder with two hemispheres at its ends during the phase of dilatation (plasmometric shape; see p. 157 and Fig. 3). Under such conditions the surface and volume of the protoplast can easily be calculated and it is possible to substitute a single variable, the length of the protoplast (L), in formula (23) for all dependent variables. Formula (23) may then be integrated (Stadelmann, 1951). From the resulting formula it is possible to calculate the absolute permeability constant directly from the experimental data:

$$K = \frac{b}{4} \frac{[L_2 - L_1 - (b/3) \cdot \ln (L_2/L_1)]}{[L_0 - (b/3)] \cdot (t_2 - t_1)} \tag{28}$$

where K the absolute permeability constant is indicated in cm sec^{-1}; b the cell diameter in centimeters; L_0 the protoplast length in the state of complete plasmolysis in a sugar solution of the same concentration as the solution of the permeating substance to be tested, in centimeters; L_1 and L_2 the lengths of the protoplast at the times t_1 and t_2 in centimeters; and t_1 and t_2 the times of the measurements, both times being in the phase of dilatation, in seconds.

Since special care must be taken to maintain a constant external concentration, the use of a perfusion chamber is highly recommended. Spe-

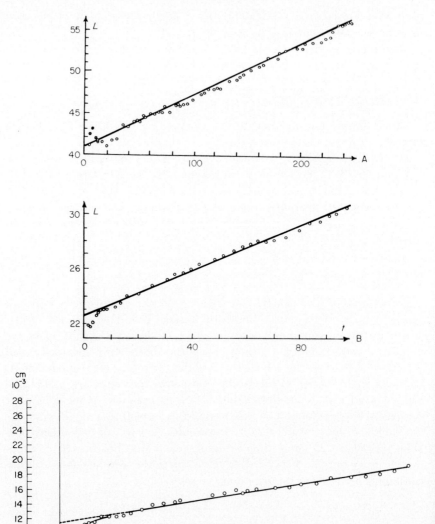

FIG. 5. Time course of the deplasmolyses of cylindrical cells in a hypertonic solution of a permeating nonelectrolyte; x axis: time; y axis: length of the protoplast. A: *Zebrina pendula*. Parenchyma cell of the stem; $h = 75.5$ micrometer units (MU); $b = 23.0$ MU. Solution: 1 M urea. 1 MU $= 2.78\mu$ (Stadelmann, 1951, p. 379, Fig. 1a). B: *Sonchus laciniatus*. Base of the leaf. Cell of the upper epidermis over the midrib; $h = 36.0$ MU; $b = 7.5$ MU. Solution: 1 M urea. 1 MU $= 2.78\mu$ (Stadelmann, 1951, p. 379, Fig. 1f). Time indicated in minutes. C: *Allium cepa*. Bulb scale. Cell of the upper epidermis; $h = 20.7 \times 10^{-3}$ cm; $b = 8.3 \times 10^{-3}$ cm. Solution: 1 M urea (Stadelmann and Wattendorff, 1966, Fig. 11f). Time indicated in hours.

cially suited are the perfusion chambers described by Stadelmann (1959) and Werth (1961).

Only two measurements of L at the times t_1 and t_2 are needed for calculating formula (28). However, to be sure that the cell tested is in healthy condition, several measurements of L over a considerable time of the dilatation phase should be made. In fact, if possible, L and t should be closely recorded for the entire dilatation phase so that a graph of length of the protoplast over time can be made. Since L increases proportionally with time in an intact and healthy cell, a straight line relationship results (Fig. 5). For some cells this line is broken, indicating an abrupt change in permeability (cf. Stadelmann, 1964, p. 32, Fig. 3). Two values of L and t, preferably at the end of the first and second thirds of the straight part of the dilatation curve, are chosen to be used for formula (28).

The value L_0 may be extrapolated from the graph of protoplast length over time of the dilatation phase by taking the value of L for the time $t = 0$, i.e., when the tissue was placed in the permeating solution. Therefore a prior plasmolysis in the sugar solution can be avoided. This is preferable because in some cells a longer stay in a hypertonic sugar solution may lead to a hardening of the external layers of the protoplasm, which would cause the formation of hernial protuberances (cf. Stadelmann, 1964, p. 54).

1. EXPERIMENTAL PROCEDURES

When a perfusion chamber is used, the necessary manipulations are simple. At first the tissue is placed in a drop of spring water in the open perfusion chamber. After the chamber is sealed, spring water is perfused for a few moments in order to check the proper functioning of the perfusion. Then water is replaced by the permeating solution (e.g., 1 M urea) and a suitable cell is selected for the measurement. Such a cell must have almost perfectly parallel lateral walls. Even slight deviations from parallelism result in a change in the slope of the protoplast length over time diagram and therefore cause errors in the value of the permeability constant. As soon as the protoplast ends are rounded, if possible before deplasmolysis begins, the length of the protoplast is recorded at suitable time intervals with an eyepiece micrometer, which is also used to measure cell length and width. From the recorded data the protoplast length vs. time diagram (Fig. 5) is established. The values for L are selected, located at the end of the first and second thirds of the straight part of the dilatation curve. If needed, L_0 is extrapolated, using this part of the diagram.

Since the permeability constant of the individual cells may show a high statistical scatter for the same material, it is desirable to take measurements on as many cells as possible for calculating the mean value. Since direct measurement of the protoplast length as described above is time-consuming, only about 2–3 cells can be measured as frequently as needed for establishing the exact time course of the deplasmolysis. To overcome this limitation, time lapse photographs at low magnification have been used. The measurements have been made on the frames on the film or on enlarged prints. In this way every cell with cylindrical shape and sufficiently sharp contours on the film may be evaluated.

2. Example for the Calculation of K

Material (Stadelmann, 1951, p. 383f): leaf of *Sonchus laciniatus*. Cell of the upper epidermis over the midrib. Tested solution: 1 M urea. Inner cell length $h = 24.0$ MU[3]; inner call width $b = 5.5$ MU; 1 MU = 2.78μ; L_0 (extrapolated for the time of the transfer to the urea solution) = 11.7 MU. Measurements of L: for $t_1 = 19$ min = 1140 sec, $L_1 = 12.4$ MU; for $t_2 = 55$ min = 3300 sec, $L_2 = 14.1$ MU. From these data are calculated:

$$\frac{b}{3} = 1.8 \text{ MU} \qquad \frac{b}{4} = 1.4 \text{ MU} \qquad L_0 - \frac{b}{3} = 9.9 \text{ MU}$$

$$t_2 - t_1 = 2160 \text{ sec} \qquad L_2 - L_1 = 1.7 \text{ MU} \qquad \frac{L_2}{L_1} = 1.14$$

$$\log \frac{L_2}{L_1} = 0.0567 \qquad \ln \frac{L_2}{L_1} = 0.131 \qquad \frac{b}{3} \cdot \ln \frac{L_2}{L_1} = 0.24 \text{ MU}$$

When these values are introduced into formula (28):

$$K = 1.4 \times \frac{1.7 - 0.24}{9.9 \times 2160} = 0.96 \times 10^{-4} \text{ MU/sec} = 2.7 \times 10^{-8} \text{ cm sec}^{-1}$$

D. Measurement of Permeability to Water with the Plasmometric Method

The passability of the living protoplasm to water is an important factor in many physiological processes. The methods of measuring protoplasmic permeability have varied with the experimental conditions and tissues used (cf. Stadelmann, 1956c, p. 172f). The evaluation of plasmolysis or deplasmolysis permits the calculation of the *absolute permeability constant for water*. This can be directly compared with the absolute permeability constant for solutes (Stadelmann, 1963).

The experimental setup and the procedure for determining permea-

---

[3] MU = micrometer units.

bility to water are the same as described above. The basic idea here is to follow as closely as possible the relatively rapid changes in the protoplast length L, when the cell is transferred suddenly from one nonpermeating solution to another with a markedly different concentration of the same solute. The time course for establishing osmotic equilibrium between cell sap and the new external concentration is used for calculating the water permeability constant.

However, this time course also depends on other factors such as protoplasmic viscosity or resistance against diffusion inside the vacuole. Since volume changes under these conditions occur very rapidly, such factors may interfere more significantly than in the experiments dealing with permeating solutes. In spite of these increased chances for complications, experiments with the plasmometric method give good results when suitable material is used. The values for the absolute water permeability constants derived with this method are of the same order of magnitude (about 1–20×10^{-4} cm sec^{-1}; cf. Stadelmann, 1963, pp. 706, 707, Fig. 2a and b) as found with diffusion method on plant material and as found for most animal cells with a variety of methods (cf. Dick, 1959, p. 433f).

For better evaluation of the experiments, time lapse photography should be applied in establishing accurately the relationship between the length of the protoplast and time. Either a higher or a lower concentration can be used in the second treatment solution. In the first case a partial plasmolysis occurs and in the latter a deplasmolysis. In both cases three measurements of L are made: two during the main phase of the change of the protoplast length (L_1 and L_2) and the third when the final equilibrium in the second concentration is reached (L_0).

The absolute permeability constant is calculated from these data with formula (29) or (30):

$$K_{w0} = 32.0 \cdot \frac{b}{C \cdot L_0 \cdot (t_2 - t_1)} \cdot \left[\left(L_0 - \frac{b}{3} \right) \right.$$
$$\left. \cdot \log \frac{L_1 - L_0}{L_2 - L_0} + \frac{b}{3} \cdot \log \frac{L_1}{L_2} \right] \quad (29)$$

$$K_{w0} = 13.9 \cdot \frac{b}{C} \cdot \frac{L_2 - L_1}{t_2 - t_1} \cdot \frac{1}{\overline{L} - L_0} \cdot \frac{\overline{L} - (b/3)}{\overline{L}} \quad (30)$$

where K_{w0} is the absolute water permeability constant, which can be compared directly with the absolute permeability constant for solutes, in units of cm sec^{-1}, b the inner cell width in centimeters, c the concentration of the nonpermeating substance (e.g., saccharose) used as the second concentration in the experiment, in moles per liter, L_1 and L_2 the lengths of the protoplast at the times t_1 and t_2, in centimeters, t_1 and t_2 the times of the measurements made during the period of protoplasmic contraction or dilatation, in seconds, L_0 the length of the protoplast when final equilibrium in the second concentration is almost reached, in centimeters:

$$L_0 = \frac{L_1 + L_2}{2}$$

and 32.0 and 13.9 are numerical factors with the unit moles per liter.

Formula (29) is derived by integrating the appropriate differential equation, while formula (30) gives only an approximation but requires less calculating. The error will stay within the limits -25% and $+16\%$ of the correct value from formula (29) when cells are used with b between $0.4 \times L_0$ and $1.0 \times L_0$ and the difference $(L_2 - L_0)$ is at least as high as 10% of L_0 (cf. Stadelmann, 1966b).

EXAMPLE FOR THE CALCULATION OF K_{w0}

Material (Hofmeister, personal communication, 1965): *Lamium purpureum*, epidermal cells of the stem. First solution 1.0 M glucose, second solution 0.6 M glucose (deplasmolysis was used for testing the permeability to water). Inner cell length $h = 75.6\mu$; inner cell width $b = 34.0\mu$. $L_1 = 40.6\mu$, $L_2 = 53.2\mu$, $L_0 = 58.8\mu$; $t_1 = 43$ minutes 55 seconds, $t_2 = 44$ minutes 56 seconds, $t_0 = 49$ minutes 15 seconds. From these data are calculated:

$$\frac{b}{3} = 11.3\mu \qquad L_2 - L_1 = 12.6\mu \qquad \frac{L_1 + L_2}{2} = \bar{L} = 46.9\mu$$

$$L_0 - \frac{b}{3} = 47.5\mu \qquad t_2 - t_1 = 61 \text{ sec} \qquad \bar{L} - \frac{b}{3} = 35.6\mu$$

$$L_1 - L_0 = -18.2\mu \qquad L_2 - L_0 = -5.6\mu \qquad \frac{L_1}{L_2} = 0.763$$

a. *Calculation with formula* (29).

$$K_{w0} = \left[32.0 \times \frac{34}{0.6 \times 58.8 \times 61} \left(47.5 \times \log \frac{-18.2}{-5.6} \right. \right.$$

$$\left. \left. + 11.3 \times \log 0.763 \right) \right] \times 10^{-4} \text{ cm sec}^{-1}$$

$$= \left[32.0 \times \frac{34.0}{2152} (47.5 \times 0.512 - 11.3 \times 0.117) \right] \times 10^{-4} \text{ cm sec}^{-1}$$

$$= [32.0 \times 0.0158 (24.3 - 1.33)] \times 10^{-4} \text{ cm sec}^{-1}$$

$$= 32.0 \times 0.343 \times 10^{-4} = 11.6 \times 10^{-4} \text{ cm sec}^{-1}$$

b. *Calculation with formula* (30).

$$K_{w0} = \left(13.9 \times \frac{34.0}{0.6} \times \frac{12.6}{61} \times \frac{1}{46.9 - 58.8} \times \frac{35.6}{46.9} \right) \times 10^{-4} \text{ cm sec}^{-1}$$

$$= 13.9 \times 56.7 \times 0.207 \times 0.0840 \times 0.758 \times 10^{-4} \text{ cm sec}^{-1}$$

$$= 13.9 \times 0.983 \times 0.758 \times 10^{-4} = 10.4 \times 10^{-4} \text{ cm sec}^{-1}$$

The difference in the K_{w0} values calculated with formulas (29) and (30) is only about 10% and therefore stays inside the limits postulated above.

VII. Evaluation of Wall Attachment (Adhesion) and Protoplasmic Viscosity

A. General Considerations

During plasmolysis the protoplast not only decreases in volume but, in most cases, also changes in shape. These changes are often very characteristic and lead to a sequence of different protoplast forms until the final form is reached. This final form often appears sometime after the protoplast has attained its final volume. Many forms of plasmolysis have been described. They vary with cell species, cell shape, and external and internal conditions (cf. Stadelmann, 1956a, p. 77f). Also the time interval from the beginning of plasmolysis to the appearance of the final plasmolysis form (here called *plasmolysis time*) is highly variable.

Plasmolysis form and plasmolysis time are the results of interplay of several factors (cf. Schaefer, 1955, p. 423f), the most important of which are as follows.

1. CELL SHAPE, SIZE, AND MECHANICAL QUALITIES OF THE PROTOPLASM

The separation of protoplasm from the cell wall and its contraction require a longer time when the protoplasmic envelope is thick (cf. Ursprung, 1939, p. 1202). The chromatophore may determine to a considerable degree the plasmolysis form (Cholnoky, 1928, p. 459; 1931, p. 322; Eibl, 1939, p. 534) and may even lead to a peculiar shape of the plasmolyzed protoplast, when it has a sufficiently high mechanical strength ("screw plasmolysis", cf. Scarth, 1924, p. 103; Weber, 1925a). Sometimes crystals (carotene or vital stains) deposited inside the cytoplasm may also influence the plasmolysis form (Küster, 1929a, p. 193f; Beckerowa, 1935, p. 386).

2. CONSISTENCY OF THE VACUOLE

When the cell sap is rich in colloids and therefore has a high low fluidity the plasmolyzed protoplast of such a cell may retain the contours of the cell on a small scale (cf. Kenda and Weber, 1952, p. 461).

3. STRENGTH AND QUALITY OF THE EXTERNAL SOLUTION (PLASMOLYTICUM)

An increase in concentration of the solution in which plant cells were immersed either increased (cf. el Derry, 1930, p. 6; Takamine, 1940, p. 309) or decreased (Huber and Höfler, 1930, p. 357; Prud'homme van Reine, 1935, p. 485f) plasmolysis time, depending upon the material and other external conditions. The plasmolysis form also may depend upon the concentration strength (Prud'homme van Reine, 1935, p. 481) and the direction from which the plasmolyticum has access to the cell (Eibl, 1939, p. 532f).

Of course, any effect on protoplasmic viscosity or wall attachment of the plasmolyticum or other substance supplied in addition will also alter the time and form of the plasmolysis.

4. PROTOPLASMIC VISCOSITY AND WALL ATTACHMENT

These factors are most important for plasmolysis form and time. High viscosity and/or wall attachment[4] produces concave or angular plasmolysis and an increase in plasmolysis time. Low viscosity and wall attachment have the opposite effect and lead in a short time to well-rounded protoplasts, with a shape corresponding to the minimum surface for the given protoplast volume as limited by the cell walls. The ratio between the mechanical resistance of the protoplasm to separation from the wall and the resistance against increase of surface in the already detached areas will finally determine if concave or convex plasmolysis takes place (Höfler, 1932a, p. 194).

Local differences in wall attachment, which may be related to structural peculiarities of the cell wall, determine separation or attachment of the protoplasmic envelope in the very first moment of the beginning plasmolysis (*positive* and *negative loci of plasmolysis*, where the protoplast either does not line or lines the cell wall, Weber, 1929c, p. 583). Also differences in the physical or chemical qualities of the protoplasmic envelope may interfere here.

---

[4] In earlier work often referred to as "adhesion." This term, however, describes a very specific and well-defined physical phenomenon, while in the living plant cell a complex of several factors is involved in the cell wall-protoplasm contact.

The localization of loci of plasmolysis on the surface of the cell wall is very characteristic for certain cells and leads to peculiar plasmolysis forms (e.g., X-plasmolysis in *Oscillatoria tenuis;* Kuchar, 1950, p. 215; and band plasmolysis of endodermal cells; Gravis, 1898; cf. Weber, 1929c, p. 593). In certain cells of the cotyledon of *Soja hispida* and in the prothallium of *Pteris*, positive loci of plasmolysis develop only under such parts of the protoplasmic envelope which lack plastids (Hellweger, 1935, p. 235; Reuter, 1948, p. 393).

The strength of the wall attachment may depend mainly upon (1) the submicroscopic ramifications of the protoplasm into the macrofibril skeleton of the cell wall, (2) the resistance against tearing of the external part of the protoplasmic envelope and its ramifications, and (3) the occurrence of electrical potentials between cell wall and external protoplasmic surface, which may cause more difficult separation even in such areas where only little or no protoplasmic ramification into the cell wall occurs (Schaefer, 1958, p. 414).

Variations of plasmolysis time are also found to be correlated with the season of the year (cf. Küster, 1942a, p. 135; Pirson and Göllner, 1953, p. 493f).

5. ENDOGENOUS FACTORS

The time and form of plasmolysis in stomatal cells vary with the physiological state and depend upon the degree of stomatal opening (Weber, 1925b). Other changes in plasmolysis time or form occur when cells are in the state of cell division, of copulation, or ready to undergo cell division (Weber, 1924, p. 264; 1925c, p. 148, 154; Cholnoky, 1930, p. 290f). In *Mougeotia* filaments, plasmolysis form was correlated with the position of the cell (cf. Cholnoky, 1931, p. 327). Of course, the immediate causes for these relationships are differences in the strength of wall attachment and viscosity, brought about by the physiological state of these cells.

6. EXPERIMENTAL CONDITIONS

Some external factors also influence plasmolysis form and time: the pH value of the milieu (Prát, 1926, p. 250; el Derry, 1930, p. 17f; Takamine, 1940, p. 313) and the temperature affect these qualities and for a certain temperature the plasmolysis time becomes minimum (cf. Kolkwitz, 1896, p. 224; el Derry, 1930, p. 14; Prud'homme van Reine, 1935, p. 488f). However, for *Nitella flexilis* a maximum viscosity was reported near 21°C, since at this temperature concave plasmolysis forms occur (Romijn, 1931, p. 295). Application of alkali salts in the external solution causes a decrease in viscosity and more or less limited swelling leading to *cap plasmolysis* (see p. 205; cf. Stadelmann, 1956a, pp. 89, 92f). Increase of viscosity and even irreversible hardening of the protoplasm has been observed after treatment with salts of alkaline earth metals. Also aluminum salts may change viscosity. Strong wall attachment results from pretreatment of the cells in weak solutions of Al, La, Li, or Th salts and sometimes no separation of the protoplast from the cell wall is possible, leading to cytorrhysis (Stadelmann, 1956a, p. 73, 82; Höfler, 1958, p. 250). Other factors influencing wall attachment are light conditions and the atmosphere to which the tissue was exposed before being plasmolyzed (cf. Weber, 1929b, p. 257; Hellweger, 1935, p. 232; Takamine, 1940, p. 313f; Schaefer, 1955, p. 433; 1956). Here too the direct action of these factors concerns protoplasmic viscosity and sometimes wall attachment also. Their alteration finally causes the observed changes in plasmolysis time and form.

B. Determination of Viscosity and Wall Attachment

Although these factors which have been discussed above affect plasmolysis form and time, one can still evaluate viscosity and wall attachment from plasmolysis time and form. When standard cell material with the usually high fluidity of the vacuolar sap is chosen for these experiments, the effect of the consistency of the vacuole will be negligible. Also, uniformity of cell material assures the same cell size and shape, the same mechanical qualities of the protoplasm, and the same endogenous factors. Under these conditions, any alteration in plasmolysis form and time resulting from different external conditions will indicate changes in protoplasmic viscosity and wall attachment.

The relative contribution of each of these two factors can be estimated from the time course of the changes of the plasmolysis form: When the protoplast is completely separated from the cell wall, a low wall attachment may be inferred. The protoplast form and its further change after separation from the wall are determined only by the protoplasmic viscosity and may serve as a reliable relative measure of it.

However, in the majority of all plasmolysis experiments, the protoplast envelope remains in contact with the cell wall in some areas. Convex plasmolysis indicates low viscosity combined with low wall attachment in such cells. When concave plasmolysis forms appear they are caused by high viscosity or strong wall attachment. If high viscosity is the cause, the shape of the protoplast will change very slowly. However, if strong wall attachment is the governing factor, the concave forms will disappear as soon as the separation of the protoplast from the cell wall is completed (Stadelmann and Wattendorff, 1966).

First estimations of protoplasmic viscosity from plasmolysis experiments were made by Weber (1925c). He evaluated the plasmolysis form of different *Spirogyra* material (copulating filaments and filaments in the resting state). The protoplast in these cells separates completely from the cell wall. The plasmolysis forms of these two stages were examined at the same time after the beginning of plasmolysis. Samples of the same materials were subjected to centrifugation, and displacement of the chromatophore was determined after equal centrifugation time. Here the degree of the displacement depends almost exclusively upon the viscosity of the protoplasm. A smaller displacement occurred in the same material where the plasmolysis forms were angular; thus, high viscosity will be indicated by angular plasmolysis. Heavy displacement of the chromatophores corresponds to round plasmolysis forms, i.e., low protoplasmic viscosity.

Later the *plasmolysis time* [defined by Weber (1929a, p. 637) as the time interval from the beginning of plasmolysis to the appearance of convex forms] was also introduced as the first numerical measure of protoplasmic viscosity from evaluation of plasmolysis. The plasmolysis time increases with viscosity and becomes ∞ in cells in which the concave plasmolysis forms which developed at the beginning of the experiment remain unchanged. Accurate measurement of the time when convex forms are reached often is difficult. To overcome this inconvenience, Schmidt *et al.* (1940, p. 571) introduced as plasmolysis time the time interval needed to produce convex plasmolysis in 50% of the cells of uniform material. This time is interpolated from two estimates of the frequency of convex plasmolysis forms, which are close to the 50% value. Of course, this plasmolysis time cannot be applied for single cells or small tissues where the number of uniform cells is inadequate for statistical evaluation.

Quantitative evaluation of plasmolysis form was first attempted by Cholodny and Sankewitsch (1934, p. 69). The lengths of the contours of the plasmolyzed protoplast were determined for those portions where it is separated from the cell wall. Otherwise the corresponding sections of the cell wall contours which are freed from the protoplast are measured and the ratio is determined. However, this ratio concerns linear magnitudes and, therefore, does not correspond properly to the spatial relationship and the changes of the protoplast surface when the plasmolysis form develops.

1. The Rounding Coefficient α

A numerical value indicating the viscosity can be derived from the changes of the protoplast surface with the time, when the protoplast is completely separated from the cell wall (Schaefer, 1955). When this occurs the protoplast is assumed to be in osmotic equilibrium with the external solution. To calculate α, the ratio (r) of the actual surface (A_a) to the minimum surface (A_{min}) of the protoplast for the same

TABLE XI

SURFACE RATIO v FOR THE DIFFERENT PLASMOLYSIS FORMS
CALCULATED FOR CELLS OF THE SECOND SUBEPIDERMAL
LAYER OF THE ROOT OF *Lemna minor*[a]

Plasmolysis form		Surface ratio $r = \dfrac{A_a}{A_{min}}$		
Diagrammatic presentation	Designation	Mean	Minimum	Maximum
	Extreme concave III ("Cramped" plasmolysis)	3.5	3.0	4.0
	Extreme concave II	2.5	2.0	3.0
	Extreme concave I	1.8	1.6	2.0
	Concave III	1.55	1.5	1.6
	Concave II	1.45	1.4	1.5
	Concave I	1.35	1.3	1.4
	Angular	1.25	1.2	1.3
	Irregular convex	1.15	1.1	1.2
	Convex	1.0	1.0	1.05

[a] The maximum and minimum values correspond to those plasmolysis forms which are at the limit between each consecutive pair of the forms illustrated above. After Schaeffer (1955, p. 426, Table I).

volume must be determined ($r = A_a/A_{min}$). This calculation is made for a series of different surfaces, A_a corresponding to the various degrees of concave, angular, and convex plasmolysis for a given cell form (Table XI). The surface area A_a and the actual volume of the protoplast are found by approximation, through subdividing the actual protoplast shape into simple stereometric bodies.

In this way the qualitative notion of the plasmolysis form can be transformed to a fairly reliable numerical value, for the cell form and size for which it was calculated. Since the changes of the plasmolysis form during plasmolysis are directed toward a decrease in protoplasm surface, the value of r becomes smaller with the time, A_a approaching more and more the minimum value A_{min}. It can be assumed that the decrease in protoplast surface progresses exponentially with time and this was also proven by experimental results (Schaefer, 1955, p. 427, Fig. 1). Thus, ($\ln r$) shows inverse proportionality with time, starting with the maximum value ($\ln r_0$), when (at the time t_0) the protoplast just becomes separated from all sides of the cell wall, and becoming zero for the plasmolysis time (t_e), when minimum surface is reached. Next the rounding coefficient α is calculated as the change of ($\ln r$) per unit time for the interval $t_e - t_0$:

$$\alpha = \frac{\ln r_0}{t_e - t_0} \qquad (31)$$

where α is the rounding coefficient in units of min^{-1}, $r_0 =$ the maximum surface ratio taken from Table (XI) from the plasmolysis form, $t_0 =$ the time (in minutes) when during plasmolysis the protoplast just becomes separated completely from all parts of the cell wall, and $t_e =$ the time (in minutes) when final plasmolysis form with minimum surface is reached. Since the value of α is small, Schaefer (1955, p. 428) multiplied it by 100 to avoid high decimal fractions. From the maximum and minimum value of r from Table XI the error for the rounding coefficient α can be calculated (cf. Schaefer, 1955, p. 428).

Example for calculation of α: Material (Schaefer, 1955, p. 430): *Lemna minor*, root, second subepidermal cell layer, 2.2 mm from the root tip. Plasmolysis form when protoplast is completely separated from the cell wall: concave I (see Table XI), which gives $r = 1.35$; time when protoplast is completely separated from the wall: 5 minutes after start of the plasmolysis; time when final plasmolysis form with minimum surface is reached: 30 minutes after the plasmolysis begins. Therefore, α becomes

$$\alpha = \frac{\ln 1.35}{30 - 5} = \frac{0.30}{25} = 0.012 \ min^{-1}$$

From the extreme values for r (1.3 and 1.4) the error of α can be calculated (see Schaefer, 1955, p. 428f) as $\pm$ 0.0017, so that

$$\alpha = 0.012 \pm 0.0017 \, \text{min}^{-1}$$

The rounding coefficient is the most advanced measurement for protoplasmic viscosity that can be obtained by observation of plasmolysis.
However, the rounding coefficient is dependent not only upon protoplasmic viscosity, but also upon the surface tension of the protoplast and
the thickness of the protoplasmic layer. This relation may be expressed
by the equation:

$$\alpha = f \left[\frac{\text{surface tension}}{\text{viscosity} \times \text{plasma thickness}} \right]$$

where f stands for "function of," indicating that the exact relationship to
α is not yet known. Hence the value of α is a reliable measure of protoplasmic viscosity when the effect of surface tension and protoplasm
thickness on the rounding process of the protoplast is relatively small.
However, α is only a relative measure and its relation to the absolute
measure which indicates viscosity in poises (dyne $\times$ sec $\times$ cm^{-2}) must
still be established.

Determination of absolute viscosity of the living protoplasm is possible
for suitable material and several methods were developed for this purpose (cf. Heilbrunn, 1958, p. 13f). One of the most appropriate of these
evaluates the displacement caused by Brownian movement of small particles. This method can be used when protoplasm is not streaming, the
protoplasmic envelope exceeds a certain minimum thickness, and the
granules in the protoplasm are not too close to each other (Pekarek, 1930,
1932).

2. Measurement of the Wall Attachment

While the viscosity can be measured when the effect of wall attachment is eliminated, it is not possible to measure wall attachment of the
protoplast in such a way that viscosity does not interfere.

An indirect measure of the wall attachment is derived from measurements of the osmotic ground value (O_g) with the plasmometric method
(see Section II,B, p. 157) and the method of incipient plasmolysis (see
Section II,A, p. 152). Since wall attachment makes the separation of the
protoplast difficult, a higher concentration of the plasmolyticum must be
applied than that which corresponds to the cell sap, in order to produce
incipient plasmolysis. When the plasmometric method is used, the effect
of wall attachment is almost completely eliminated. Thus, for the same
material the value of O_g determined by the plasmometric method will be

higher than that from the method of incipient plasmolysis (cf. Höfler, 1918, p. 147f). The difference O_g (incipient plasmolysis) $- O_g$ (plasmometric) can be used as a measure of wall attachment. Further elaboration of this method appears to be promising since it might be possible to develop a measure which expresses the wall attachment in pressure units (force per area). Such a method would be more useful than others (cf. Schaefer, 1958, p. 419, Fig. 2; p. 420).

Another approach is based upon the evaluation of the time course of the change in length of the protoplast contours (as seen in the microscopic picture) and in the area included by these contours (Masuda and Takada, 1957). It is assumed that the protoplast contours can be considered as representative of the protoplast surface area and that the protoplast area (m) as seen in the microscope can be taken for the degree of plasmolysis. A ratio (s) is calculated from the actual length of the contours (f_a), and its value (f_{min}) when final minimum surface of the protoplast is reached:

$$s = \frac{f_a}{f_{min}}$$

When the values of m and s are measured for different times during plasmolytic contraction, proportionality can be found between $(\log s)$ and m. A higher value of $(\log s)$ corresponds to a higher value of m (Masuda and Takada, 1957, p. 652, Fig. 3). The quotient A (cf. Schaefer, 1958, p. 420) is considered to be a measure of the wall attachment:

$$A = \frac{\log s}{m - m_{min}}$$

A was found to be lower for the second plasmolysis of a cell than for the first one, i.e., when after the first plasmolysis and deplasmolysis the cell was again subjected to the plasmolyzing solution. This decrease of A confirms many earlier observations, where easier separation of the protoplasmic envelope from the cell wall occurs during the second (and third) plasmolysis, resulting in smoother plasmolysis form and shorter plasmolysis time (cf. Masuda and Takada, 1957, p. 651, Fig. 1).

A third way to evaluate wall attachment is based upon the determination of the plasmolysis time of cells which were tested with hypotonic solutions close to the concentration for incipient plasmolysis (Schaefer, 1958). These solutions appreciably decrease the wall attachment, most probably by causing retraction of the plasmodesmata from the external layers of the cell wall. This retraction seems to be a physiological response to the stimulus resulting from the external solution rather than merely an osmotic effect. Comparison is made with the plasmolysis time

of a nontreated portion of the same material, and the difference in plasmolysis time is considered to be a measure of wall attachment.

VIII. Cytomorphological Alterations Generated by Plasmolysis

One of the most interesting aspects of the plasmolysis experiment is
its application to cause and observe cytomorphological alterations of
the living protoplasm. Some of these alterations are pathological and the
degree of reversibility depends upon their intensity. When irreversible
they lead sooner or later to the death of the cell. Studies of cytomorphological changes, often neglected, are a valuable tool for verifying any
effect of a specific external condition on the living protoplasm.

Shortly after beginning of plasmolysis the protoplast exhibits a series
of changes in morphology and structure. Protoplasmic strands frequently
develop on the surface of the protoplasmic envelope or cross the vacuole.
In some cells, lamellae of protoplasmic material are formed which may
more or less completely subdivide the cell sap, or filamentous protrusions
may emerge from the cytoplasm into the vacuole. Most of these forma-
tions develop from dislocations of cytoplasmic material by protoplasmic
streaming. These dislocations are the primary factors for the genesis of
these morphological alterations.

Simultaneously structural changes (especially swelling or de-swelling)
in the protoplasm may occur leading to the separation of the cytoplasmic
material in different zones, vacuolization, or segregation of swollen cytoplasm (cf. Germ und Kubelka, 1965, p. 392f).

Depending upon their origin these alterations can be classified into
two groups:

1. Changes mainly involving the colloidal consistency of the protoplasm and considered to be a more immediate consequence of the applied treatment concerning merely physicochemical effects.

2. Changes showing more clearly the feature of a response to a
stimulus. Here the applied condition should be considered as the stimulus
and a more complicated reaction chain must be postulated than is assumed for the first type.

A more detailed classification scheme using only topographical and
optical criteria was also established (Stadelmann, 1957, p. 480f). The
high diversity of these morphological changes in such experiments (cf.
Küster, 1929b, p. 4f) results from the differences in cell material and experimental conditions.

Only the two most typical alterations will be discussed here: cap plasmolysis as the example of the first type of alteration, and systrophe of the protoplasm which illustrates the second type. A comprehensive description of protoplasmic alterations by plasmolysis is given by Küster (1929b, 1956, p. 17f) and Stadelmann (1956a, p. 85f).

A. Cap Plasmolysis

Cells placed in a sufficiently high hypertonic solution of an alkali salt (e.g., potassium nitrate) at first plasmolyze as usual. When left in such a solution for several hours (sometimes much less) the protoplast swells and caps of the swollen protoplasm develop at the protoplast ends

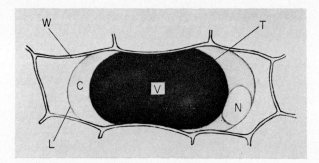

FIG. 6. Diagrammatic presentation of a cell in the state of cap plasmolysis. C = caps of swollen protoplasm; N = swollen nucleus; V = vacuole; L = plasmalemma; T = tonoplast; W = cell wall. [After Höfler, 1928, p. (74).]

[Höfler, 1928, p. (74); see Fig. 6]. The increase in protoplasm volume can be considerable, sometimes up to 50–100 times the original volume.

The swelling is a direct consequence of the applied salt solution: the alkali ions increase the plasmalemma permeability when the cell is placed in the external solution for such a period. Because permeability is increased the alkali ions are able to migrate into the protoplasm more rapidly and to build up there a high concentration, which causes the enormous water uptake [high "intrability" (= passability of the plasmalemma only)]. However, no appreciable permeation into the vacuole occurs, hence the vacuolar volume remains constant until the cells become damaged by such treatment. It appears that in cells in which the tonoplast is at first slightly permeable to alkali ions, this permeability decreases as a result of the prolonged action of the potassium salt. In this way cap plasmolysis gives striking evidence of the remarkable difference in permeability of the plasmalemma and the tonoplast.

The viscosity of the swollen protoplasm is decreased as a consequence of the high increase in water content (Höfler, 1939, p. 548). This is shown by the formation of the completely rounded caps and the confinement of the swollen plasma to the protoplast ends only, indicating the low mechanical resistance of the swollen plasma against the surface tension of the tonoplast. Brownian movement is extensive in the caps, while protoplasmic streaming always stops. The nucleus increases in volume, becoming well-rounded without further visible internal structures, and the nuclear membrane is better contrasted than under normal condition.

The swelling often is reversible for a certain length of time after the beginning of the treatment. The caps recede when the cell material is transferred to solutions of Ca or Sr salts or into a mixture of K and Ca salts (Höfler, 1940, p. 299). After extended application of the alkaline salt solution, segregation in the swollen cytoplasm occurs, leading to zonation and finally to disintegration and death of the protoplasm (Kaiserlehner, 1939, p. 593).

In general, the salts of the alkaline earth metals have an antagonistic effect on the alkaline salts. An appropriately mixed solution of potassium and calcium salts, for example, will cause no change of the original degree of swelling of the protoplasm. The prohibitive effect of calcium occurs at a very low partial concentration of the Ca salt. A mixture of 1 part equimolar Ca salt solution to 25 parts of K salt solution may even inhibit cap plasmolysis.

Occasionally cap plasmolysis is also found in a hypertonic solution of urea after α or ultraviolet irradiation or some other treatment (cf. Stadelmann, 1956a, p. 89), but reliable occurrence of cap plasmolysis is confined to the application of alkali salt solutions.

B. Systrophe

This is the *local balling of cytoplasm or plastids* (Schimper, 1885, p. 221; for terminology, see Höfler, 1963, p. 40) mainly accumulating around the nucleus, but sometimes in other parts of the cell (Küster, 1910, p. 274; Germ, 1932, p. 614). Systrophe is a typical and *reversible* response reaction of the living protoplasm to a variety of stimuli. It is also observed in naked protoplasts or partial protoplasts without the nucleus (Höfler and Weber, 1926, p. 698), and very frequently involves a limited increase in protoplasmic volume by swelling (Höfler, 1940, p. 293).

One can distinguish between *plastid* and *protoplasm systrophe* when plastids or protoplasm accumulate around the nucleus or on another location. In both cases dislocation is performed by protoplasmic streaming, but in plastid systrophe this does not lead to an accumulation of the cytoplasm itself.

Systrophe also occurs in the normal life of a cell *in situ*. Systrophe may be brought about experimentally by high light intensity (Küster, 1929b, p. 72f), treatment with acids (Weber, 1932, p. 289), application of Te salts (Moder, 1932, p. 19f), vital staining (Beckerowa, 1935, p. 388f; cf. Küster, 1942b, p. 252), and plasmolysis.

When systrophe is caused by plasmolysis the stimulus most probably is the separation of the protoplast from the cell wall rather than the osmotic withdrawal of water (Germ, 1932, p. 569). There is a certain concentration of the external solution which is optimal. In low concentrations of solutions only the first stages of systrophe appear (Germ, 1933a, pp. 511f, 528f), and later complete re-establishment of the previous protoplasm distribution takes place.

The steps leading to the final state of systrophe, when caused by plasmolysis, depend upon the viscosity of the cytoplasm and the concentration of the external solution applied. Two different sequences can be distinguished:

1. The nucleus is moved along the cell wall to its final location, and later the chloroplasts begin dislocation and accumulate around the nucleus.

2. The nucleus is displaced at first into the center of the cell, moving along diagonal protoplasmic strands. The plastids and the cytoplasm start to migrate to a specific point along the cell wall where they accumulate and also move toward the nucleus. Finally, the nucleus also moves to this specific location along the lateral wall. This kind of development of systrophe is shown diagrammatically in Fig. 7.

The second type of systrophe always occurs in cells with highly viscous cytoplasm and slightly hypertonic solutions. The first type occurs when high concentrations of a plasmolyticum are used or when low concentrations are applied to cells with low protoplasmic viscosity (Germ, 1932, pp. 577, 583f).

Plasmolysis produces protoplasm and/or plastid systrophe in cells of flowering plants, ferns, and the $2n$-generation of mosses. In the n-generation of mosses, plasmolysis stimulates plastid systrophe only and protoplasm systrophe never has been observed (Germ, 1933b, p. 271f; exception: *Plagiothecium;* Prát, 1930, p. 429).

The balling of plastids and cytoplasm at first *extends* into the vacuole (cf. Fig. 7D–F). When it is positioned at the protoplast ends it later prolapses into the space between protoplast end and transverse cell wall, indicating that the surface tension of the tonoplast is overcoming that of the plasmalemma (cf. Küster, 1929b, p. 74; Lanz, 1942, p. 389). In fairly long cells the balling may form a closed band of cytoplasm and plastids, located around the middle portion of the plasmolyzed protoplast.

The *systrophe time* (i.e., the time interval from the beginning of the plasmolytic treatment to the stage of final systrophe; Germ, 1933a, p. 532) decreases as the temperature increases, probably as a consequence of lower protoplasmic viscosity (Germ, 1933a, pp. 541, 547).

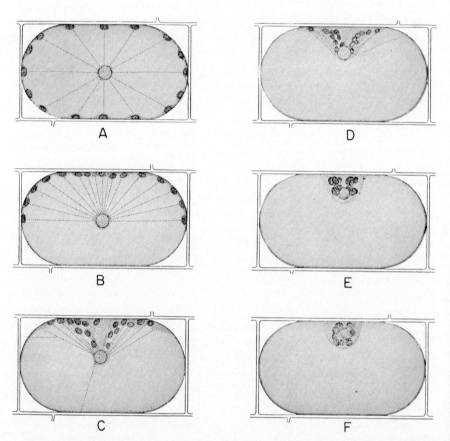

Fɪɢ. 7. Diagrammatic presentation of the development of protoplasm and plastid systrophe with temporary "centration" of the nucleus in the cell (type 2 of the text). A: "Centration" of the nucleus and symmetrical dispersal of plastids. B: Asymmetrical dispersal of the plastids. C: Grouping. D: Balling. E: Perfect systrophe. F: Final systrophe. (Germ, 1932, p. 575.)

When the cells are deplasmolyzed the systrophe becomes resolved and the distribution of cytoplasm and plastids is re-established as it was prior to the beginning of the plasmolysis. In most cells resolution of the accumulated cytoplasm occurs at first, with plastids later returning to their original positions.

IX. Plasmolysis and Turgidity as Test Method for the Living State of the Plant Cell

The differential permeability of the protoplasmic envelope is strongly dependent on the intact organization of the living cell. Most conditions leading to pathological situations and lethal damage of the cell at first cause the breakdown of the permeability barriers of the protoplasm. The important function of a protective shield was recognized for the plasmalemma primarily from studies on the damaging effect of certain hypertonic salt solutions (cf. Höfler, 1951, p. 452). Significantly, such a pathological or premortal increase in permeability very often first affects the permeability of the protoplast to the surrounding hypertonic solution rather than to the original osmotically active material in the vacuole. This can be concluded from the fact that, in most cells, when a plasmolyzed protoplast becomes heavily damaged, deplasmolysis occurs, often with increasing speed. Such very rapid deplasmolysis may cause bursting of the protoplast. The occurence of deplasmolysis instead of shrinkage of the damaged protoplast indicates that the osmotically active substances inside the vacuole are unable to exosmose, or at least that their exosmosis is at a lower rate than the permeation of the external solute into the vacuole. Bursting of the protoplast occurs when there is an insufficient supply of new material for the increase of the protoplast surface. Data on this phenomenon were first found for the upper epidermis of the bulb scale of *Allium cepa*. In that study, a surface increase rate of $23\mu^2$ per second was observed without causing immediate bursting (Stadelmann, 1964, p. 41).

The partial permeability of the protoplasmic envelope leads, in all intact cells under normal conditions, to plasmolysis when a hypertonic solution is applied as the external milieu. Therefore the occurrence of plasmolysis is a reliable indicator of the living state, and failure of plasmolysis is proof of cell death. Although in isolated cases plasmolysis-like contractions of dead protoplasts may occur (Küster, 1924, p. 976; Schneider, 1924, p. 38f), the trained experimenter can easily recognize these artifacts.

The reliability of plasmolysis for testing the state of life was early recognized and is widely known. In fact it is the only application of the plasmolysis experiment mentioned in many textbooks. The other and more important evaluations of osmotic quantities and permeability based upon plasmolysis are often completely overlooked.

Determination of cell turgidity usually is too cumbersome for use as an indicator of the living state of the cell. In most cells it would require an accurate measurement of the cell dimensions and their changes during certain experimental conditions. However, Wattendorff (1964) has recently developed a method of evaluating turgor without making special measurements.

The Wattendorff method is based upon the optical effect of the outward bending of the external cell wall of the individual turgid cell in a monolayer such as the epidermis. This outward bending is caused by the water saturation of the intact cell in the state of turgescence, resulting from the tendency of the cell to reach the maximum volume for the given cell surface.

For the experimental evaluation an epidermal layer is mounted in a perfusion chamber in such a way that a portion of the outer surface is uncovered by the lamella and exposed to the surrounding air. When the cells in this layer are turgescent and the external wall bent outward, the light will be refracted at the peripheral portion of each cell in the direction away from the objective, so that it does not reach the front lens. Therefore the peripheral part of each turgescent cell produces a completely dark region in the microscopic field (see Fig. 8). Dead cells which, of course, cannot develop turgidity, show no dark edges, and can easily be distinguished from the living cells by using this criterion.

Since plasmolyzed cells have zero turgidity, they also show no darkening. When deplasmolyzing, however, the cells become turgescent again as soon as deplasmolysis is completed and gradually the dark regions on the cell wall appear. Later, when some of the deplasmolyzed cells die the black edges disappear again. The time needed for the complete disappearance of the dark regions gives a good indication of the speed of the breakdown of the permeability barrier in the cell. Therefore, it may also be used as a measure of the period required for the death of the cell. In most instances this breakdown requires several minutes, but may be considerably longer in some individual cells.

A special coverglass, provided with a hole of about 4-mm diameter, is

---

Fig. 8. Time lapse photographs of the end of deplasmolysis and the reappearance of turgor pressure in cells of the upper epidermis of the bulb scale of *Allium cepa*. The epidermis was mounted under a hole of the coverglass, the cuticular (external) surface being in direct contact with air. Treatment of the epidermis: 66 minutes in 0.5 M glucose, 71 minutes in 0.65 M glucose, and 111 minutes in 0.8 M glucose; thereafter transferred to 1 M urea. Times of the photomicrographs: *upper left*, 135 minutes in urea, *upper right*, 185 minutes in urea, *middle left*, 265 minutes in urea, *middle right*, 715 minutes in urea, and *lower middle*, 1015 minutes in urea. Between 175 minutes 50 seconds and 178 minutes 5 seconds in urea the epidermis was exposed to α-irradiation. (This example was taken from an experiment for testing the effects of α-irradiation on epidermal cells. (From Wattendorf, private communication.)

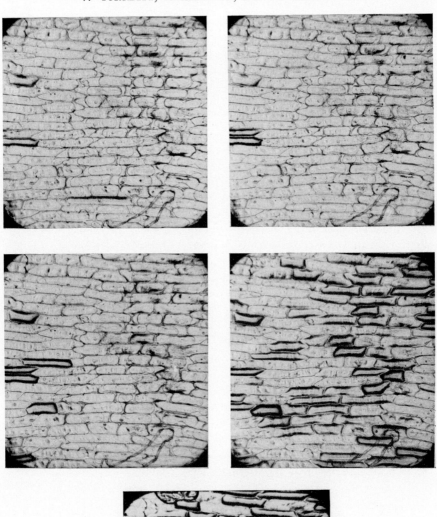

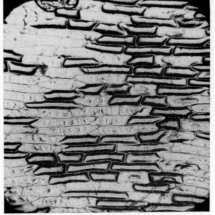

used for mounting the epidermis in such a way that its outer side is free. The epidermis of the upper side of the bulb scale of *Allium cepa* was chosen for these experiments and was sealed with vaseline, with its cuticular side to the coverglass. The coverglass itself is placed on the perfusion chamber and sealed with vaseline (cf. Wattendorff, 1964, p. 194), which provides a continuous flux of the external solution assuring constant concentration.

Time lapse photographs of a portion of this epidermis which is on the cuticular side in direct contact with air are seen in Fig. 8. The fate of the individual cells can easily be followed on the photomicrographs. For most cells turgescence is achieved between 12 and 17 hours after the transfer to the urea solution. Some cells become turgescent earlier. In this latter case often the turgescence disappears after a short period, indicating the death of this cell. An increased speed of deplasmolysis in these cells must be considered as pathological, resulting from the partial damage of the permeability barrier. This leads also to the subsequent, early death of these cells. Undamaged cells, however, normally remain turgescent for several days at least, which indicates that plasmolysis and deplasmolysis usually do not cause damage.

While this method is of great interest for the evaluation of the living state of individual cells, and also for the study of the effect of radiation on the cell layer, it must be tested to determine the additional limitations to its use. It can be expected that exposure of the surface of the material to air will result in harmful effects for some cell material. However, epidermal layers with a more or less protected external surface should be most suitable. This was proven by the experiments with *Allium* epidermis, since no difference in physiological reactions was found between cells covered by the glass lamella or exposed at the cuticular side to air in such an experiment using a lamella with a hole.

REFERENCES

Bärlund, H. (1929). *Acta Botan. Fennica* **5**, 1.
Beckerowa, Z. (1935). *Protoplasma* **23**, 384.
Bennet-Clark, T. A. (1959). *In* "Plant Physiology" (F. C. Steward, ed.), Vol. II, p. 105. Academic Press, New York.
Biebl, R. (1954). *Protoplasma* **44**, 73.
Biebl, R. (1962). *Protoplasmatologia* **12**, Part I.
Blum, G. (1958). *Protoplasmatologia* **2C**, Part 7a.
Boyer, J. S. (1965). *Plant Physiol.* **40**, 229.
Braun, N. (1852). *Ber. Verhandl. K. Preuss. Akad. Wiss. Berlin* p. 220.
Cholnoky, B. V. (1928). *Intern. Rev. Hydrobiol.* **19**, 452.
Cholnoky, B. V. (1930). *Protoplasma* **11**, 278.
Cholnoky, B. V. (1931). *Protoplasma* **12**, 321.

Cholodny, N., and Sankewitsch, E. (1934). *Protoplasma* **20**, 57.

Collander, R., and Bärlund, H. (1933). *Acta Botan. Fennica* **11**, 1.

Crafts, A. S., Currier, H. B., and Stocking, C. R. (1949). "Water in the Physiology of Plants." Chronica Botanica, Waltham, Massachusetts.

Dainty, J. (1963). *Advan. Botan. Res.* **1**, 279.

De Vries, H. (1877). "Untersuchungen über die mechanischen Ursachen der Zellstreckung ausgehend von der Einwirkung von Salzlösungen auf den Turgor wachsender Pflanzenzellen." Engelmann, Leipzig.

De Vries, H. (1889). *Botan. Ztg.* p. 309.

Dick, D. A. (1959). *Intern. Rev. Cytol.* **8**, 388.

Eibl, K. (1939). *Protoplasma* **33**, 531.

el Derry, B. H. (1930). *Protoplasma* **8**, 1.

Ernest, E. C. M. (1935). *Plant Physiol.* **10**, 553.

Eyster, H. C. (1943). *Botan. Rev.* **9**, 311.

Fitting, H. (1915). *Jahrb. Wiss. Botan.* **56**, 1.

Fitting, H. (1920). *Jahrb. Wiss. Botan.* **59**, 1.

Gahlen, K. (1934). *Protoplasma* **22**, 337.

Germ, H. (1932). *Protoplasma* **14**, 566.

Germ, H. (1933a). *Protoplasma* **17**, 509.

Germ, H. (1933b). *Protoplasma* **18**, 260.

Germ, H., and Kubelka, E. (1965). *Protoplasma* **59**, 392.

Gravis, A. (1898 to 1899). *Mém. Cour. Mém. Sav. Acad. Roy. Sci. Belg.* **57**, 1.

Härtel, O. (1963). *Protoplasma* **57** (Höfler Jubilee Vol.), 354.

Hargitay, B., Kuhn, W., and Wirz, H. (1951). *Experientia* **7**, 276.

Hauck, L. (1929). *Botan. Arch.* **24**, 458.

Heber, U. (1957). *Ber. Deut. Bot. Ges.* **70**, 371.

Heilbrunn, L. V. (1958). *Protoplasmatologia* **2c**, Part 1.

Hellweger, H. (1935). *Protoplasma* **23**, 221.

Hinzpeter, A. (1952). *Z. Elektrochem.* **56**, 683.

Höfler, K. (1917). *Ber. Deut. Botan. Ges.* **35**, 706.

Höfler, K. (1918). *Denkschr. Akad. Wiss. Wien., Math.-Naturw. Kl.* **95**, 99.

Höfler, K. (1920). *Ber. Deut. Botan. Ges.* **38**, 288.

Höfler, K. (1928). *Ber. Deut. Botan. Ges.* **46**, (73).

Höfler, K. (1932a). *Protoplasma* **16**, 189.

Höfler, K. (1932b). *Ber. Deut. Botan. Ges.* **50**, (53).

Höfler, K. (1939). *Protoplasma* **33**, 545.

Höfler, K. (1940). *Ber. Deut. Botan. Ges.* **58**, 292.

Höfler, K. (1942). *Ber. Deut. Botan. Ges.* **60**, 179.

Höfler, K. (1949). *Phyton* (*Ann. Rei Botan.*) **1**, 105.

Höfler, K. (1951). *Protoplasma* **40**, 426.

Höfler, K. (1958). *Protoplasma* **49**, 248.

Höfler, K. (1959). *Ber. Deut. Botan. Ges.* **72**, 236.

Höfler, K. (1963). *Protoplasma* **56**, 1.

Höfler, K. (1965). *Protoplasma* **60**, 150.

Höfler, K., and Höfler, L. (1963). *Publ. Staz. Zool. Napoli* **33**, 315.

Höfler, K., and Weber, F. (1926). *Jahrb. Wiss. Botan.* **65**, 643.

Hoffmann, C. (1932). *Planta* **16**, 413.

Hoffmann, C. (1935). *Protoplasma* **24**, 286.

Hofmeister, L. (1938). *Jahrb. Wiss. Botan.* **86**, 401.

Huber, B. (1943). *Protoplasma* **37**, 439.

Huber, B., and Höfler, K. (1930). *Jahrb. Wiss. Botan.* **73**, 351.

Jacobs, M. H., and Stewart, D. R. (1932). A simple method for the quantitative measurement of cell permeability. *J. Cellular Comp. Physiol.* **1**, 71.

Janse, J. M. (1887). *Versl. Mededel. Koninkl. Ned. Akad. Wetenschap. Amsterdam, Afd. Natuurk., 3e Reeks* **4**, 332.

Kaiserlehner, E. (1939). *Protoplasma* **33**, 579.

Kenda, G., and Weber, F. (1952). *Protoplasma* **41**, 458.

Kolkwitz, R. (1896). *Beitr. Wiss. Botan. Abt. II* **1**, 221.

Kozlowski, T. (1964). "Water Metabolism in Plants." Harper, New York.

Kramer, P. J. (1956). *In* Handbuch der Pflanzenphysiologie—Encyclopedia of Plant Physiology" (W. Ruhland, ed.), Vol. II, p. 316. Springer, Berlin.

Kreeb, K. (1960). *Planta* **55**, 274.

Kreeb, K. (1965). *Ber. Deut. Botan. Ges.* **78**, 159.

Kuchar, K. (1950). *Phyton (Ann. Rei Botan.)* **2**, 213.

Küster, E. (1910). *Flora (Jena)* **100**, 267.

Küster, E. (1924). *In* "Handbuch der biologischen Arbeitsmethoden" (E. Abderhalden, ed.), Abt. 11, Teil 1, p. 961. Urban & Schwarzenberg, Berlin.

Küster, E. (1929a). *Protoplasma* **5**, 191.

Küster, E. (1929b). *Protoplasma-Monogr.* **3**.

Küster, E. (1942a). *Protoplasma* **36**, 134.

Küster, E. (1942b). *Z. Wiss. Mikroskop.* **58**, 245.

Küster, E. (1956). "Die Pflanzenzelle," 3rd ed. Fischer, Jena.

Lanz, I. (1942). *Protoplasma* **36**, 381.

Lehtoranta, L. (1956). *Ann. Bot. Soc. Zool. Bot. Fenn. Vanamo* **29**, (1) 1.

Lepeschkin, W. W. (1909). *Ber. Deut. Botan. Ges.* **27**, 129.

Levitt, J. (1951). *Science* **113**, 228.

Macfarlane, M. G. (1964). *Conf. Metabol. Physiol. Significance lipids* pp. 399–412. Wiley, New York.

Masuda, Y., and Takada, H. (1957). *Physiol. Plantarum* **10**, 649.

Meyer, B. S. (1938). *Botan. Rev.* **4**, 531.

Meyer, B. S., and Anderson, D. B. (1947). "Plant Physiology." Van Nostrand, Princeton, New Jersey.

Mitchell, P., and Moyle, J. (1951). *J. Gen. Microbiol.* **5**, 981.

Moder, A. (1932). *Protoplasma* **16**, 1.

Molz, F. C. (1926). *Amer. J. Bot.* **13**, 433.

Mosebach, G. (1932). *Beitr. Biol. Pfl.* **24**, 113.

Nägeli, C. (1855). *In* "Pflanzenphysiologische Untersuchungen" (C. Nägeli and C. Cramer, eds.), p. 1. Schulthess, Zürich.

Oppenheimer, H. R. (1932). *Planta* **16**, 467.

Overton, E. (1895). *Vierteljahresschr. Naturforsch. Ges. Zürich* **40**, 159.

Overton, E. (1899). *Vierteljahresschr. Naturforsch. Ges. Zürich* **44**, 88.

Pekarek, J. (1930). *Protoplasma* **10**, 510.

Pekarek, J. (1932). *Protoplasma* **17**, 1.

Pfeffer, W. (1877). "Osmotische Untersuchungen. Studien zur Zellmechanik." Engelmann, Leipzig.

Pfeffer, W. (1893). *Abhandl. K. Sächs. Ges. Wiss., Math.-Phys. Kl.* **20**, 235.

Pirson, A., and Göllner, E. (1953). *Flora (Jena)* **140**, 485.

Pirson, A., and Schaefer, G. (1957). *Protoplasma* **48**, 215.

Prát, S. (1923). *Preslia (Prague)* **2**, 90.

Prát, S. (1926). *Kolloid-Z.* **40**, 248.

Prát, S. (1930). *Proc. 5th Intern. Botan. Congr., Cambridge, 1930* p. 428. Cambridge Univ. Press, London and New York.

Pringsheim, N. (1854). "Untersuchungen über den Bau und die Bildung der Pflanzenzelle," Abt. 1: Grundlinien einer Theorie der Pflanzenzelle. Hirschwald, Berlin.

Prud'homme van Reine, W. J., Jr. (1935). *Rec. Trav. Botan. Neerl.* **32**, 467.

Ramsay, I. A. (1949). *J. Exptl. Biol.* **26**, 54.

Repp, G. (1939). *Jahrb. Wiss. Botan.* **88**, 554.

Resühr, B. (1935). *Protoplasma* **23**, 337.

Resühr, B. (1936). *Protoplasma* **25**, 435.

Reuter, L. (1948). *Oesterr. Botan. Z.* **95**, 373.

Romijn, C. (1931). *Proc. Sect. Sci. Koninkl. Ned. Akad. Wetenschap. Amsterdam, Afd. Natuurk.* **34**, 289.

Scarth, G. W. (1924). *Quart. J. Exptl. Physiol.* **14**, 99.

Scarth, G. W. (1939). *Plant Physiol.* **14**, 129.

Schaefer, G. (1955). *Protoplasma* **44**, 422.

Schaefer, G. (1956). *Flora (Jena)* **143**, 327.

Schaefer, G. (1958). *Planta* **51**, 414.

Schimper, A. F. W. (1885). *Jahrb. Wiss. Botan.* **16**, 1.

Schmidt, H., Diwald, K., and Stocker, O. (1940). *Planta* **31**, 559.

Schneider, E. (1924). *Z. Wiss. Mikroskopie* **42**, 32.

Sitte, P. (1963). *Protoplasma* **57**, 304.

Spanner, D. C. (1952). *Ann. Botany (London)* [N.S.] **16**, 379.

Stadelmann, E. (1951). *Sitzber. Oesterr. Akad. Wiss., Math-Naturw. Kl., Abt. I* **160**, 761.

Stadelmann, E. (1952). *Sitzber. Oesterr. Akad. Wiss., Math-Naturw. Kl., Abt. I* **161**, 375.

Stadelmann, E. (1955). *Bull. Soc. Fribourg. Sci. Nat.* **46**, 76.

Stadelmann, E. (1956a). *In* "Handbuch der Pflanzenphysiologie—Encyclopedia of Plant Physiology" (W. Ruhland, ed.), Vol. II, p. 71. Springer, Berlin.

Stadelmann, E. (1956b). *Protoplasma* **46**, 692.

Stadelmann, E. (1956c). *In* "Handbuch der Pflanzenphysiologie—Encyclopedia of Plant Physiology" (W. Ruhland, ed.), Vol. II, p. 139. Springer, Berlin.

Stadelmann, E. (1957). *Protoplasma* **48**, 452.

Stadelmann, E. (1959). *Z. Wiss. Mikroskopie* **64**, 286.

Stadelmann, E. (1961). *In* "Recent Advances in Botany" Vol. II, p. 1182. Univ. of Toronto Press, Toronto.

Stadelmann, E. (1963). *Protoplasma* **57**, 660.

Stadelmann, E. (1964). *Protoplasma* **59**, 14.

Stadelmann, E. (1966a). *In* "Handbook on Environmental Biology" (P. L. Altman, ed.). Federation Am. Soc. Exptl. Biol., Washington, D.C. (in press).

Stadelmann, E. (1966b). *Protoplasma* **61**, 387.

Stadelmann, E., and Wattendorff, J. (1966). *Protoplasma* (in press).

Takamine, N. (1940). *Cytologia* **10**, 302.

Tamiya, H. (1937). *Cytologia* **8**, 542.

Taylor, S. A., and Slatyer, R. O. (1961). *Arid Zone Res.* **16**, 339.

Thöni, H. (1965). *Experientia* **21**, 112.

Unger, F. (1855). "Anatomie und Physiologie der Pflanzen." C. A. Hartleben, Leipzig.

Ursprung, A. (1923). *Ber. Deut. Botan. Ges.* **41**, 338.

Ursprung, A. (1935). *Plant Physiol.* **10**, 115.

216   E. J. STADELMANN

Ursprung, A. (1939). *In* "Handbuch der biologischen Arbeitsmethoden" (E. Abderhalden, ed.), Abt. 11, Teil 4/2, p. 1109. Urban & Schwarzenberg, Berlin.

Ursprung, A., and Blum, G. (1916). *Ber. Deut. Botan. Ges.* **34**, 525.

Ursprung, A., and Blum, G. (1918). *Ber. Deut. Botan. Ges.* **36**, 599.

Ursprung, A., and Blum, G. (1921). *Ber. Deut. Botan. Ges.* **39**, 70.

Ursprung, A., and Blum, G. (1927). *Jahrb. Wiss. Botan.* **67**, 334.

Walter, H. (1931). "Die Hydratur der Pflanze und ihre physiologisch-ökologische Bedeutung (Untersuchungen über den osmotischen Wert)." Fischer, Jena.

Walter, H. (1952). *Planta* **40**, 550.

Walter, H. (1963). *Ber. Deut. Botan. Ges.* **76**, 40.

Walter, H. (1965). *Ber. Deut. Botan. Ges.* **78**, 104.

Wattendorff, J. (1964). *Protoplasma* **59**, 193.

Weber, F. (1921). *Oesterr. Botan. Z.* **73**, 261.

Weber, F. (1924). *Oesterr. Botan. Z.* **73**, 261.

Weber, F. (1925a). *Ber. Deut. Botan. Ges.* **43**, 217.

Weber, F. (1925b). *Jahrb. Wiss. Botan.* **64**, 687.

Weber, F. (1925c). *Z. Wiss. Mikroskopie* **42**, 146.

Weber, F. (1929a). *Protoplasma* **5**, 622.

Weber, F. (1929b). *Protoplasma* **7**, 256.

Weber, F. (1929c). *Protoplasma* **7**, 583.

Weber, F. (1932). *Protoplasma* **16**, 287.

Werth, W. (1961). *Protoplasma* **53**, 457.

Whaley, W. G., Kephart, J. E., and Mollenhauer, H. H. (1964). *In* "Cellular Membranes in Development," 22nd Symp. Soc. Study Develop. Growth (M. Locke, ed.), p. 135. Academic Press, New York.

Wildervanck, L. S. (1932). *Rec. Trav. Botan. Néerl.* **29**, 227.

Chapter 8

Culture Media for Euglena gracilis[1]

S. H. HUTNER, A. C. ZAHALSKY, S. AARONSON,
HERMAN BAKER, AND OSCAR FRANK

*Haskins Laboratories, New York, New York, and The New Jersey College of
Medicine and Dentistry, Jersey City, New Jersey*

I. Introduction

Euglena gracilis group II strains, which include the Z and T strains (Pringsheim and Pringsheim, 1952), formally belong to var. *saccharophila* (Pringsheim, 1955), which may also include "var. *bacillaris*." All have an intense photosynthesis of the green-plant type and in the dark, given various simple substrates, grow as colorless cells with equal or even superior vigor. They re-green in a few hours on restoration to light. Moreover, permanently bleached ("apochlorotic") strains are easily induced (heat, streptomycin, ultraviolet, some antihistamines, and certain other

[1] Recent experiments by the authors were aided by NIH grant RG-9103, NSF grant GB-285, and grant DR 6-827 from the Damon Runyon Memorial Fund for Cancer Research.

agents). Many bleached strains grow as vigorously as the original green strains. In addition, naturally occurring vigorous colorless counterparts of *E. gracilis, Astasia longa,* are available, and also pure cultures in nearly defined media of the unmistakably euglenoid, colorless, voracious *Peranema* (Allen *et al.,* 1966).

This array of organisms simplifies decision as to whether a substance is localized in the chloroplast; examination of light- vs. dark-grown cultures generally settles the question. As such inquiries increasingly revolve around trace constituents, and large quantities of cells are often needed, this chapter emphasizes high-yield media and how to compound media as dry mixes, which may be prepared cheaply in bulk and have a long shelf life. Our work in recent years has centered on the Z strain, but probably applies to *bacillaris*. The use of *Euglena* for vitamin B_{12} assay (Baker and Sobotka, 1962) enhances the desirability of high-yield media which permit a linear response through a wide range of vitamin B_{12} concentrations along with rapid growth.

II. Maintenance

Disasters originate from inattention to maintenance of photosynthetic vigor in stock cultures. Even brief exposure to bleaching temperatures is hazardous. The critical temperature is $\sim33°C$ (Anderson, 1964). The curved glass of a culture vessel may focus heat and by the greenhouse effect raise the temperature. Also many stocks have a high—1–2%—spontaneous rate of appearance of yellow or white mutants, as well as "petit" greens (Gibor and Granick, 1962). In rich media many of the bleached strains have a slight but definite advantage: they eventually outgrow the green strains. Therefore, sooner or later, if kept in rich media, the culture will become obviously impaired photosynthetically. To impose selection pressure for photosynthetic efficiency, maintenance media must be substrate-poor. Slant growth on 0.2% peptone solidified with agar 0.4–0.8%, pH 6.2–6.8, is good. A duplicate subline may be kept as a shallow stab in semisolid media (agar 0.2–0.3%). The cultures are grown in light, then stored in dim light between 10° and 20°C. Such cultures may keep more than a year. Cultures tend to die out rather quickly in darkness at refrigerator temperature (4–6°). The traditional north window in a cool basement is especially safe; air conditioners often fail. Virtually any animal peptone is suitable; yeast extract or soy peptones need supplementation with vitamin B_{12}. If faster, heavier growth is desired, the medium may be supplemented with Na acetate·$3H_2O$ 0.01–0.04%. If less carry-over of organic material is desired, one may use Baker and Sobotka's (1962) maintenance medium solidified with agar. A peptone medium eases detection of contaminants. Thanks to mucus secretion, *Euglena* grows well on moist agar surfaces. Use of liquid rather than solid media increases risk of contamination. Screw-cap borosilicate culture tubes, 75×16 mm, are sturdy, convenient, and generally survive mailing. The liner of the cap should be inspected for good sealing. Cotton plugs are obsolete: tubes so plugged soon dry out, and cotton is susceptible to molds and mites. Addition of sterile water to overdried maintenance media generally makes tubes usable by the next day as the agar rehydrates.

III. Autotrophic Media

The medium of Cramer and Myers (1952) has been widely used; an acetate-supplemented version is given by Padilla and James (1964). Although the Cramer-Myers medium contains citrate, the high pH—6.8— practically eliminates the possibility of citrate being metabolized. In many experiments on light and dark growth, with some media at pH 3.0 or lower to favor penetration of un-ionized forms of acids, we have not detected substrate activity of citrate, alone or as a sparker for other substrates. It seems safe therefore to regard the Cramer-Myers medium as autotrophic.

A useful low-pH autotrophic medium is given in Table I; it was intended to (a) be reasonably adequate in trace elements, (b) be readily compounded as a dry mix, (c) be heavily buffered and requiring little or no pH adjustment, and (d) permit a sensitive growth response to substrates.

A. Comment on Table I

The ingredients were chosen for constancy of composition at the temperature of a domestic freezer ($\sim -20°C$) or dry laboratory air. Thus $MgSO_4 \cdot 3H_2O$ was selected, since $MgSO_4 \cdot 7H_2O$ effloresce at the low humidity of a freezer or in dry winter air at room temperature. HEDTA was chosen as auxiliary chelator over EDTA, since its Ca and Mg complexes are fairly soluble. Histidine serves as another auxiliary chelator; it seems not to be used as a substrate; it is probably taken up. The dry ingredients are mixed in a ball mill or mortar, then kept frozen. For dry mixes, thiamine is dispensed as a 1:1000 triturate in pentaerythritol, hence 0.1 gm would be weighed out instead of 0.1 mg, permitting use of a sturdy 2-mg-sensitivity torsion balance instead of a microbalance; this also reduces the risk of seriously uneven distribution of the vitamin in the dry mix. Pentaerythritol, aside from its osmotic pressure, is metabolically inert for all microorganisms tested. If enough dry mix is compounded for, say, 20 liters and the total weight of mix comes out to 200 gm, one weighs out 200 gm $\div$ 200 $=$ 1.0 gm/100 ml of final medium. For use, the

TABLE I

Low-pH "Autotrophic" Growth Medium

Compound	Concentration (mg/100 ml)
$MgCO_3$	30.0
$MgSO_4 \cdot 3H_2O$	20.0
$CaCO_3$	2.0
HEDTA[a]	20.0
K_3 citrate $\cdot H_2O$	40.0
Citric acid $\cdot H_2O$	400.0
KH_2PO_4	15.0
L-Histidine HCl $\cdot H_2O$	100.0
"Metals 60A" (see Table IV)	18.0
NH_4HCO_3	50.0
Thiamine HCl	0.1
Vitamin B_{12}[b]	0.002
pH	3.2–3.5

[a] Hydroxyethylethylenediaminetriacetic acid (Geigy Chemical Corp., Ardsley, New York; Aldrich Chemical Co., 2371 North 30th St., Milwaukee, Wisconsin 53210.)

[b] Conveniently dispensed as commercial 1:1000 triturate in mannitol. Therefore for 0.002 mg, weigh out 2.0 gm of the triturate. *Euglena* does not utilize mannitol.

mix powder is suspended in water to 0.5–0.75 final volume, and brought to a near boil on the water bath. Mechanical stirring helps. At this low pH, retention of CO_2 is negligible in heated media and so the pH shift upward on autoclaving is slight. Growth in the dark is negligible. C.P. anhydrous citric acid is now available and may be preferable to the monohydrate.

B. Comment on Trace Elements and Inorganics

Euglena has exceptional resistance to trace-metal toxicity, hence there have been few difficulties reported in the literature despite obvious latitude in adapting published directions to chemicals available on the shelf. Thus, while our papers generally specify Ca prepared by dissolving $CaCO_3$ in minimal HCl, it is common, even in marine laboratories, to see some such salt as $CaCl_2 \cdot 2H_2O$ specified, without assurance that the concentration of solutions of this very hygroscopic salt is checked by analysis for Ca or Cl$^-$. Since there may be less Ca than specified in the original recipe, the effective concentrations of the trace elements are lower, as there is less Ca to compete for the chelating agent. With some organisms this could result in trace-element deficiency. Conversely, when a salt such as $CaCl_2 \cdot 2H_2O$ or $Ca(NO_3)_2 \cdot 2H_2O$ is specified, the medium in other hands may precipitate badly, evidently because the salts in the original recipe were more hydrated than realized; and, as another consequence, the trace elements may be at toxic levels. The following salts should be shunned because of proneness to deliquescence or efflorescence: $MgCl_2$ and its hydrates, $FeCl_3 \cdot 6H_2O$, $FeSO_4 \cdot 7H_2O$, $CuSO_4 \cdot 5H_2O$, K_2HPO_4, and chlorides of Mn, Zn, Co, Cu, Ni, etc.

Below pH 4.0 or so metal requirements rise steeply, rather than diminishing as expected were requirements governed entirely by increased dissociation of metal-ligand (chelate) complexes; unexplored properties of cell membranes must come into play.

Pentaerythritol is added as an inert diluent to minimize salts reacting with each other, and again to help bring the amount of mix to be weighed out within the range of a torsion balance. The trace-element mix can be stored at room temperature. If left exposed, the $CuSO_4$ becomes partly hydrated, deepening the color of the mix. Therefore the metals mix is kept well-stoppered or frozen. If it is desired to titrate the trace-element mix, a 1% solution may be prepared in 0.05% sulfosalicylic acid and then diluted as desired with water. These solutions darken on standing but seem to be stable indefinitely. Fe solutions may be prepared in the same way with sulfosalicylic acid.

Direct evidence is lacking that *Euglena* requires V, B, and Co (apart from vitamin B_{12}), but V and B requirements are virtually certain and hence V and B are included as insurance. Future inorganic mixes will include Ni [required for autotrophic growth of *Hydrogenomonas* (Bartha and Ordal, 1965)] and NH_4VO_3, a fairly stable salt (Sandell, 1959), instead of $NaVO_4 \cdot 16H_2O$. Se and I are extremely toxic in acid media and appear to be amply present as impurities. While it is desirable to freeze the composition of media, this is best done by buying key chemicals in large batches, then testing each new batch. The trace elements supplied most organisms had to be increased some years ago when we began using glass-distilled water (see Allen *et al.*, 1966, on all-glass stills) throughout Unpublished experiments with other organisms indicate that biologically unfamiliar elements may emerge as essential in very acid media compounded of ordinary C.P. chemicals; when these are identified, the inorganic components of acidic media may be lowered without lowering yield. The chloride supplied in thiamine HCl probably satisfies any trace chloride requirement.

IV. Heterotrophic Acidic Medium

A. General Comment

The medium in Table II was designed for compounding in batches enough for 50 or 100 liters of medium, using cheap, reasonably pure, commercial chemicals. It supports dense growth in light in well-illu-

TABLE II

HETEROTROPHIC ACIDIC MEDIUM

Compound	Weight per 100 ml of final medium
KH_2PO_4	0.04 gm
$MgSO_4 \cdot 3H_2O$	0.01 gm
$MgCO_3$	0.04 gm
$CaCO_3$	0.01 gm
DL-Malic acid	0.5 gm
L-Glutamic acid	0.5 gm
Glucose (anhydrous)	1.0 gm
Urea	0.04 gm
Na_2 succinate $\cdot 6H_2O$	0.01 gm
Glycine	0.25 gm
DL-Aspartic acid	0.2 gm
"Metals 60A" (see Table IV)	9.0 mg
Thiamine HCl	0.06 mg
Vitamin B_{12}	0.05 μg
pH	3.1–3.4

minated cultures by a balanced-aquarium effect: oxidation of substrates by photosynthetically produced O_2 yields CO_2, which is used in photosynthesis. Photoassimilation of substrates perhaps contributes to the heavy growth; this possibility has not been adequately explored. For use as a basal medium for vitamin B_{12} assay, vitamin B_{12} of course is omitted.

B. Substrates

Malic, aspartic, and glutamic acids are only used substantially at pH's below 6.0. As acetate concentrations must be lowered in media more acid than pH 6 or so, not enough acetate can be tolerated to sustain dense growth, hence it is omitted. Aspartic acid is preferred to asparagine because it does not make the pH rise as much (Hutner *et al.*, 1956). Earlier media contained sucrose; because of the acidity of the medium, autoclaving hydrolyzed it to glucose + fructose. The availability of cheap, pure anhydrous glucose made it practical to replace sucrose with glucose even for large-scale work. As utilization of glucose is adaptive (App and Jagendorf, 1963), organisms passaged once or twice in the heterotrophic medium should be used as inocula; cultures in large screw-cap tubes or screw-cap micro-Fernbach flasks (Bellco) are convenient. The danger of employing substrate-rich media for subcultures serving as inocula has been emphasized under "Maintenance" (Section II). Repression of chlorophyll synthesis by substrates (App and Jagendorf, 1963) is evidenced by the cultures being initially light green, then becoming fully green during later growth. In the early development of sugar-containing media an intense acidification (an aerobic fermentation?) was occasionally observed in the form of a sharp pH drop. Now that glucose utilization is recognized as adaptive, these experiments warrant repetition. Likewise it is not known how well *Euglena* can live fermentatively at low O_2 tensions. This might be investigated with permanently bleached strains in semisolid media containing redox indicator dyes. Sugar utilization in media > pH 4.0 or so has been erratic, but again adaptation and effects of glycine were inadequately taken into account.

The high malic and glutamic acid concentrations come from a recent follow-up of findings by Robbins *et al.* (1953), who recommended a pH 3.0 medium containing 0.8% each of glutamic and DL-malic acids. Earlier media had lower concentrations, e.g., one extensively employed by Greenblatt and Schiff (1959), and a similar medium employed by Zahalsky *et al.* (1963). Malic acid acts both as chelator and Krebs-cycle substrate; presumably, as the L-isomer is consumed, metals are released in better adjustment to demand than permitted by a nonmetabolizable chelator. An earlier high-substrate medium (Wolken, 1961) contains commercial (DL) lactic acid, a substrate used appreciably only at pH 4.0 and below. Lactic acid or another favorite (and pH-insensitive) substrate, ethanol (Eshleman and Danforth, 1964), perhaps will permit a further growth increment; they have not yet been so tested. The mode of glycine utilization is unknown; it is taken up and favors utilization of glucose (Hurlbert and Rittenberg, 1962). Glycolic acid was not used alone or as a sparker, even in very acid (pH 2.6) media in our experiments.

An ethanol medium was used to study Zn (Price and Vallee, 1962) and Fe (Carell and Price, 1965) requirements.

By employing Mg and Ca as carbonates, these cations can be supplied in stable form and suitable for dry mixes, yet the high concentration of very acid chelating agents, especially malic acid, permits quick solution on heating and avoids ballasting the medium with chloride or sulfate. As a general rule, dry mixes do not dissolve well, especially those containing carbonates, unless the pH is below 6.0 or so. Alkali should be added separately after solution is effected if a less acid medium is desired. Tris is recommended for this purpose (see next section). How close the present medium comes to exhausting Euglena's osmotic tolerance has not been determined. If the osmotic margin of safety becomes too narrow, it may become desirable (a) to replace DL-aspartic acid by L-aspartic acid or L-asparagine; (b) to find a wholly utilizable substitute for DL-malic acid (L-malic acid is about 20 times more expensive than DL-malic acid, i.e., it is 10 times more expensive on an L-malic acid basis). One may return to sucrose, especially now that media lower then 3.0 are feasible and thus hydrolysis of sucrose on autoclaving is more nearly complete. In these very acid media decomposition of glucose on autoclaving is negligible.

The rather high Ca may help overcome inhibition by hemoglobin in vitamin B_{12} assays (Coelho and Rege, 1963).

C. Other Ingredients

Na_2 succinate·$6H_2O$ serves as a stable Na salt to meet a possible Na requirement; succinic acid is utilized as a substrate. Urea perhaps serves directly as an N source, as it does for many green algae; at the low pH of the medium it is probably largely hydrolyzed on autoclaving.

Euglena gracilis, from many unpublished experiments on antimetabolites and reversing factors, and from scattered reports in the literature on the uptake of minor isotopically labeled nutrients, seems to share much the same qualitatively restricted permeability pattern as higher green plants and such fungi as yeasts. This resemblance was overlooked because of its obviously high capacity for metabolizing acetate, ethanol, and other low-molecular aliphatic acids and alcohols and, in acid media, a few Krebs-cycle intermediates. There is no obvious growth increment when amino acid mixtures, peptones, and the like are added to the present heterotrophic medium. Nevertheless it is likely that a combination of compounds known to penetrate and be metabolically active, such as α- and γ-aminobutyric acids, leucine, methionine, tryptophan, tyrosine, adenine, guanine, uracil, thymine, p-aminobenzoic and nicotinic acids, biotin, pyridoxine, and inositol, could well increase growth collectively even though, tried singly, growth stimulations are hardly beyond experimental error. These experiments are in progress. A comprehensive survey of substrates for Euglena is warranted although the odds are that no advantageous, unfamiliar ones will be found.

V. Neutral and Alkaline Media

As a general rule, below pH 6.0 or so it is unsafe to autoclave agar media: even a slight hydrolysis of the agar liberates acid, which accelerates hydrolysis, which liberates more acid. A trivial change in autoclaving routine can thus lead to hydrolysis of the agar. One occasionally used way around this difficulty is to autoclave the agar and the liquid base separately (both at double strength), mixing them aseptically while hot. Near-neutral or alkaline media are also desirable to favor the penetration of bases such as streptomycin. Older media contained acetate and tended to become so alkaline as to brake growth. The pH shift is less with butyrate or caproate, but even as Na salts they have objectionable odors; vapors of the free acids dissolve in body oils, and so entail a thorough scrubdown on the part of the worker after a day's work to avoid social ostracism. Ethanol, perhaps the best substrate of all, gives rise to a slight acidification, and must be added aseptically because of its volatility; besides, especially for large-scale work, it entails extra paper work because of the legal restrictions attending nonbeverage, tax-free use.

In an attempt to get around these difficulties, an acetate–*n*-butanol medium was recommended which permitted growth at an initial pH of 8.0. A published version (Wolken, 1961) was somewhat imprecise in respect to trace elements, and butanol is variably volatilized on autoclaving despite its high boiling point (117.5°C). Use of hexanol has not been adequately explored. A workable solution to these problems was found by using ethyl esters of amino acids (medium in Table III), although

TABLE III
pH 6.8 MEDIUM[a]

Compound	Weight per 100 ml of final medium
K_3 citrate·H_2O	0.1 gm
Na acetate·$3H_2O$	0.1 gm
Glycine ethyl ester HCl	0.1 gm
L-Glutamic acid γ-ethyl ester HCl	0.1 gm
L-Asparagine·H_2O	0.15 gm
Na_2 glycerophosphate·$5H_2O$	0.05 gm
$MgSO_4$·$7H_2O$	0.05 gm
$(NH_4)_2SO_4$	0.002 gm
Thiamine HCl	1.0 mg
Vitamin B_{12}	0.4 μg

[a] Trace elements (mg): Fe 1.0, Mn 0.8, Zn 0.5, Mo 0.05, Cu 0.05, Co 0.05, B 0.01, V 0.005, I 0.004, Se 0.002; pH adjusted to 6.8.

growth was inferior compared with that supported by the low-pH heterotrophic medium. Probably a trace-element mixture such as 60A (Table
IV) or the like would serve, but this mixture has not been titrated for
optimal levels and proportions of metals.

TABLE IV
TRACE ELEMENT SUPPLEMENT ("METALS 60A")[a]

Compound	For 1000 liters of final medium (gm)	Metal in 100 ml of final medium (mg)
$Fe(NH_4)_2(SO_4)_2 \cdot 6H_2O$	42.0	Fe 0.6
$MnSO_4 \cdot H_2O$	15.5	Mn 0.5
$ZnSO_4 \cdot 7H_2O$	22.0	Zn 0.5
$(NH_4)_6Mo_7O_{24} \cdot 4H_2O$	3.6	Mo 0.2
$CuSO_4$ (anhydrous)	1.0	Cu 0.04
$Na_3VO_4 \cdot 16H_2O$	3.7	V 0.04
$CoSO_4 \cdot 7H_2O$	0.48	Co 0.01
H_3BO_3	0.57	B 0.01
Pentaerythritol	88.8	

[a] To obtain the indicated trace-element concentration in the final medium, metals
60A is used at 18.0 mg/100 ml of final medium. Thus metals 60A is "full strength"
in Table I and "½ strength" in Table II.

This medium served well for drug studies but was never agarized for
plating work—it was too expensive. Originally the pH was adjusted with
"Quadrol," a liquid, dibasic, noncorrosive alkali having a bulky, presumably nonpenetrating molecule (Packer *et al.*, 1961). The substantial
lowering in the price of Tris buffer and absence of toxic effects make Tris
preferable, especially since Tris is a stable powder. Once the required
Tris concentration is known, the appropriate amount can be weighed out
directly for new batches. The glycerophosphate is the commercially
mixed α- and β-isomers and is both buffer and phosphate source. Use of
amino acid esters, asparagine, and glycerophosphate enabled the precipi-
tate-forming inorganic phosphate to be eliminated, and the concentration
of the potentially toxic NH_4^+ ion to be lowered.

VI. General Comments

The scanty array of substrates for *Euglena* constricts the design of
media. *The* badly missing medium for *E. gracilis* is the one for specific
enrichment (Hutner, 1964). Perhaps the critical selective feature will
exploit resistance to, and utilization of, higher acids and alcohols, perhaps

combined with exploitation of *Euglena*'s remarkable resistance to high concentrations of PO_4, Fe, Mn, and Cu, especially at low pH's; perhaps too in the presence of glucose. There has been no serious reported effort to cross-test many physiologically similar strains in the hope of detecting a phage. The scarcity of nutritional markers (as with *Chlamydomonas reinhardi*) hinders attempts to detect sexuality via recombination. Hence a thorough survey of substrates for *Euglena* and of its tolerances to heavy metals could be helpful. Also, as *Euglena gracilis* withstands O_2-poor environments such as oxidation ponds, its ability to grow in O_2-poor media should be explored.

Another aid to research would be an antibiotic that would repress the growth of acid-tolerant sooty molds, e.g., *Pullularia* and *Cladosporium*, but without inhibiting *Euglena*. The low pH of the standard medium appears to eliminate bacterial contamination altogether. The only known acid-tolerant bacteria are chemoautotrophs such as *Thiobacillus thioöxidans* and acid-bog forms, all probably unable to withstand concentrated media such as the heterotrophic medium. Hence an antifungal antibiotic might ease the use of unautoclaved media, and so eliminate a frequent bottleneck in large-scale work: few biochemists have easy access to hospital- or industrial-size autoclaves.

REFERENCES

Allen, J. R., Lee, J. J., Hutner, S. H., and Storm, J. (1966). *J. Protozool.* 13, 103-108.
Anderson, B. B. (1964). *J. Clin. Pathol.* 17, 14–26.
App, A. A., and Jagendorf, A. T. (1963). *J. Protozool.* 10, 340-343.
Bach, M. K. (1960). *J. Protozool.* 7, 50-52.
Baker, H., and Sobotka, H. (1962). *Advan. Clin. Chem.* 5, 173-235.
Bartha, R., and Ordal, E. J. (1965). *J. Bacteriol.* 89, 1015-1019.
Carell, E. F., and Price, C. A. (1965). *Plant Physiol.* 40, 1-6.
Coelho, J., and Rege, D. V. (1963). *J. Protozool.* 10, 473-477.
Cramer, M., and Myers, J. (1952). *Arch. Mikrobiol.* 17, 384-402.
Eshleman, J. N., and Danforth, W. F. (1964). *J. Protozool.* 11, 394-399.
Gibor, A., and Granick, S. (1962). *J. Protozool.* 9, 327-334.
Greenblatt, C. L., and Schiff, J. A. (1959). *J. Protozool.* 6, 23-28.
Hurlbert, R. E., and Rittenberg, S. C. (1962). *J. Protozool.* 9, 170-182.
Hutner, S. H. (1964). *In* "Biochemistry and Physiology of Protozoa" (S. H. Hutner, ed.), Vol. 3, pp. 1-7. Academic Press, New York.
Hutner, S. H., Bach, M. K., and Ross, G. I. M. (1956). *J. Protozool.* 3, 101-112.
Packer, E. L., Hutner, S. H., Cox, D., Mendelow, M. N., Baker, H., Frank, O., and Amsterdam, D. (1961). *Ann. N.Y. Acad. Sci.* 92, 486-490.
Padilla, G. M., and James, T. W. (1964). *In* "Methods in Cell Physiology" (D. M. Prescott, ed.), Vol. I, pp. 141-157. Academic Press, New York.
Price, C. A., and Vallee, B. L. (1962). *Plant Physiol.* 37, 428-433.
Pringsheim, E. G. (1955). *Arch. Microbiol.* 21, 414-419.
Pringsheim, E. G., and Pringsheim, O. (1952). *New Phytologist* 51, 65-76.

228   S. H. HUTNER *et al.*

Robbins, W. J., Hervey, A., and Stebbins, M. E. (1953). *Ann. N.Y. Acad. Sci.* **53**, 818-830.
Sandell, E. B. (1959). "Colorimetric Determination of Traces of Metals," 3rd ed. Academic Press, New York.
Wolken, J. J. (1961). "Euglena. An Experimental Organism for Biochemical and Biophysical Studies." Rutgers Univ. Press, New Brunswick, New Jersey.
Zahalsky, A. C., Hutner, S. H., Keane, M. M., and Burger, R. M. (1962). *Arch. Mikrobiol.* **42**, 46-55.
Zahalsky, A. C., Keane, M. M., Hutner, S. H., Lubart, K. J., and Amsterdam, D. (1963). *J. Protozool.* **10**, 421-428.

Chapter 9

General Area of Autoradiography at the Electron Microscope Level

MIRIAM M. SALPETER

Laboratory of Electron Microscopy, Department of Engineering Physics,
Cornell University, Ithaca, New York

I. Introduction

Within the last decade, autoradiography has become a useful tool for the electron microscopist. Caro reviews the early literature (Vol. I, Chapter 16 of this series), discusses some theoretical considerations, and describes technical steps for preparing specimens for electron microscope autoradiography.

An optimum specimen for electron microscope autoradiography consists of a high-contrast thin section coated with a uniform monolayer of an emulsion which has silver halide crystals of the minimum size compatible with adequate sensitivity. Of primary importance for resolution, sensitivity, and quantitation is the formation of the emulsion layer. In this chapter the factors involved in emulsion coating are considered and criteria established for evaluating thickness and uniformity of the emulsion layers over the autoradiographic specimen. Problems of contrast, sensitivity, resolution, and quantitation are also discussed and all applied to a recommended specimen preparation procedure. Special emphasis is placed on the commonly used Ilford L4 emulsion (silver halide crystal size ~1200 Å) and the newer Kodak NTE emulsion (silver halide crystal ~500 Å).

The techniques, experimental results, and theoretical discussions presented in this chapter are primarily the result of a collaboratory effort by Dr. Luis Bachmann and the author, and most of the data has already been published elsewhere (Bachmann and Salpeter, 1964, 1965; Salpeter and Bachmann, 1964, 1965).

II. Factors in Autoradiographic Technique

No two experimenters can exactly reproduce each other's procedures. It is therefore essential that, for any crucial step of a given technique, criteria be established for evaluating the results, thus allowing a degree of freedom in the detailed manipulations.

A. Emulsion Layer

1. Coating

The formation of an optimum emulsion layer is a critical step for obtaining reproducibility and reliability in electron microscope autoradiography. Numerous methods have been proposed for producing adequate layers. These include dipping specimens in diluted liquid emulsions or dropping the emulsion in this form onto the specimens (Granboulan, 1963; Hay and Revel, 1963; Koehler et al., 1963; Young and Kopriwa, 1964; Salpeter and Bachmann, 1964); centrifuging the emulsion directly onto specimens grids (Dohlman et al., 1964); and preforming thin layers of emulsion in metal loops (Caro and Van Tubergen, 1962; Moses, 1964; Revel and Hay, 1961) or on agar (Caro and Van Tubergen, 1962) before applying them to the section. As Caro (1964) points out, emulsions which are applied in liquid form outline the surface irregularities of the specimens and thus cannot be distributed in uniform layers. This is especially a problem if the sections are placed on metal grids before the emulsion is applied, because then the silver halide crystals tend to cling to the grid wires. The difficulty can be overcome either by using a fully pre-formed emulsion layer (e.g., as in the agar method of Caro and Van Tubergen, 1962) or by providing a substrate of uniform physical properties on which to place the liquid emulsion (e.g., as in the procedure recommended here, in which the emulsion layer is formed over a thin layer of carbon or silicone monoxide that has been evaporated on the biological sections mounted on collodion-filmed glass slides) (Salpeter and Bachmann, 1964; see also Pelc *et. al.*, 1961). Caro and Van Tubergen's (1962) procedure for forming pre-gelled emulsion layers in metal loops is widely used. Whether such layers are truly pre-gelled, however, depends greatly on the skill of the technician. While the technique may be mastered successfully with the Ilford L4, the new emulsions with smaller silver halide crystals, such as the Gevaert 307 (Granboulan, 1963; Young and Kopriwa, 1964) and the Kodak NTE (Salpeter and Bachmann, 1964), do not easily form uniform pre-gelled layers when handled in this way.

2. Optimum Emulsion Layers

An optimum emulsion layer for maximum resolution consists of a sheet of closely packed silver halide crystals with little overlap (i.e., a monolayer). This type of layer precludes the presence of much gelatin. While a gain in sensitivity may be obtained with a multilayered emulsion, such a gain is at the expense of resolution. When tritium is the radioactive source, this effect on sensitivity and resolution is less evident with the Ilford L4 emulsion than with the finer-grained emulsions, since a β-particle from tritium is already heavily scattered within *one* silver halide crystal the size of those of Ilford L4 (i.e., it is scattered through 1 radian in 700 Å of silver hallide) (Bachmann and Salpeter, 1965). The probability that this particle will reach the upper layers of the emulsion is thus markedly decreased. With higher energy β-emitters, or very fine-grained emulsions, however, scattering within the emulsion is decreased and the range increased, thus a multilayer of crystals does result in increased sensitivity. In such cases, offsetting sensitivity against resolution becomes an experimental variable to be manipulated.

It should be emphasized that both sensitivity and resolution are reduced by using layers in which the silver halide crystals are not tightly packed. The effect on sensitivity is obvious for if a β-particle hits gelatin rather than silver halide it will not be recorded. The two effects of a loosely packed layer on resolution are less apparent. Caro (1962) called the first factor the shielding of one crystal by those in front of it. In

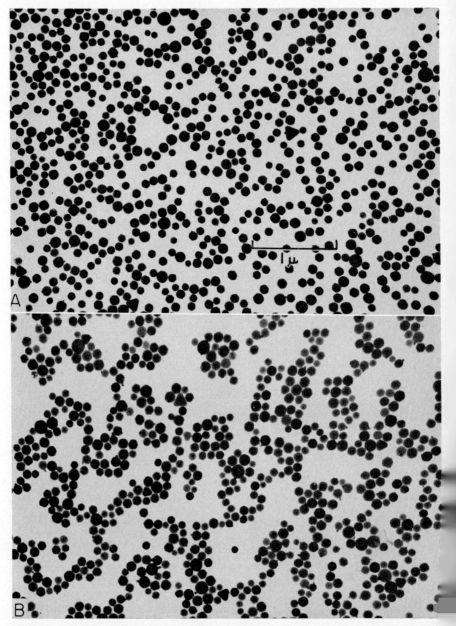

FIG. 1. Layer of Ilford L4. A: Silver halide evenly distributed but not closely packed. B: Similar layer after immersion in water for 1 minute. The over-all grain density in both cases is about the same. (From Bachmann and Salpeter, 1965).

effect, a radioactive particle will be scattered more heavily in silver halide than in gelatin. Thus, a silver halide crystal at some distance from the source is less likely to be hit by a radioactive particle which is zig-zagging through silver halide than by one which is pursuing a relatively straight-line course through gelatin. The silver halide diminishes the spread of electrons from the source and thus a close-packed layer of silver halide enhances resolution.

The second effect of a loosely packed layer on resolution was pointed out by Bachmann and Salpeter (1965) and is illustrated in Fig. 1. If a uniform but loosely packed layer of Ilford L4 (Fig. 1A) is dipped for a few seconds in water, the silver halide crystals tend to reorganize and clump (Fig. 1B) with some crystals moving several thousand angstroms. Interferometric measurements show that the layer itself is decreased in thickness by 20%, an indication of a loss of water-soluble components from the gelatin. The loss in resolution that can result from such a re-arrangement of crystals during development of an exposed emulsion is obvious. This rearrangement does not occur in a closely packed layer of Ilford L4. In the Kodak NTE emulsion, where water-soluble components are mostly removed during centrifugation, such a rearrangement of crystals has not been observed.

3. CRITERIA FOR EVALUATING UNIFORMITY AND THICKNESS OF EMULSION LAYERS

Thin emulsion layers on glass slides show, in reflected white light, interference colors which depend on their thickness. These interference colors

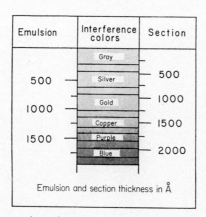

FIG. 2. Interference color—thickness scale for sections and emulsion. Interference colors of section on water and emulsion on glass slide. (From Bachmann and Salpeter, 1965.)

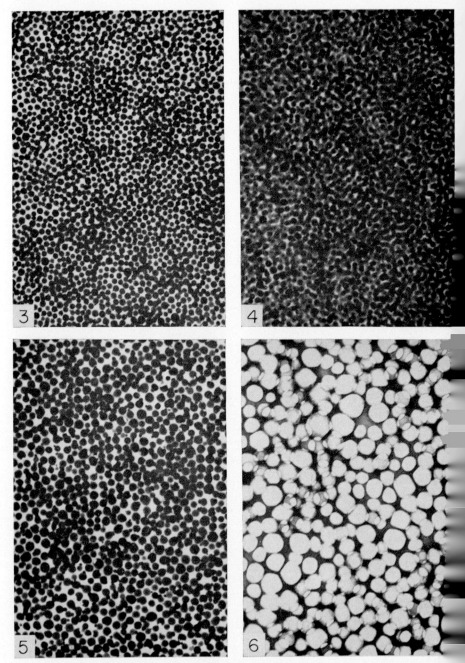

Fig. 3. Monolayer of centrifuged Kodak NTE emulsion. Interference color—silver to pale gold; measured thickness—600 Å.

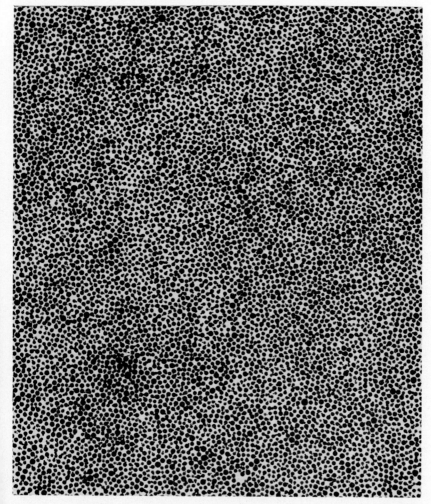

FIG. 7. Monolayer of Kodak NTE (silver interference color) to show uniformity of layer over large area. × 22,500.

FIG. 4. Slightly overlapped layer of centrifuged Kodak NTE emulsion. Interference color—gold; measured thickness—1000 Å.

FIG. 5. Monolayer of Gevaert nuclear 307 emulsion. Interference color—silver-to-gold.

FIG. 6. Monolayer of Ilford L4 emulsion. Interference color—purple; measured thickness—1500 Å (The emulsion gelatin was stained with phosphotungstic acid and the emulsion layer was then fixed. The images of the silver halide crystals are thus negatively stained ghosts. A close-packed monolayer of Ilford L4 is otherwise difficult to photograph without grossly disturbing the silver halide crystals due to the high beam intensity necessary for adequate illumination.) (From Salpeter and Bachmann, 1965).

are not influenced by an underlying collodion film or a section because the refractive indices of collodion and embedding plastic are very similar to that of glass. In all practical cases, the interferences colors are also insensitive to air humidity as long as the specimen is dry. The thicknesses of layers of different interference colors were measured with a Normarski interferometer (Salpeter and Bachmann, 1964, 1965). Figure 2 correlates the interference color with thickness. These data have been validated for both the Ilford L4 and centrifuged Kodak NTE emulsions. It was further determined by interferometry that the thickness remains constant on a light microscope level as long as the interference color remains uniform to the naked eye, that a spotty interference color indicates an uneven layer, and that a much more uniform layer is obtained if the section is first coated with a layer of carbon. When emulsion layers of different interference colors (i.e., thicknesses) were examined with the electron microscope, it was found that a silver-to-pale gold layer of centrifuged Kodak NTE (600 Å) is a monolayer (Figs. 3 and 7) while a gold layer (1000 Å) already shows some overlap (Fig. 4), and that a purple layer of Ilford L4 (1300–1500 Å) is a monolayer of that emulsion (Fig. 6). Again on the electron microscope level, uniformity of interference color reflected uniformity of emulsion layer. The uniformity of interference color can, therefore, be a criterion for an adequate coating procedure and the interference color itself a criterion for emulsion thickness. The values in Fig. 2 adequately apply to any emulsion if its gelatin content is low enough to allow the formation of a closely packed monolayer. For the Gevaert 307 emulsion a silver-to-gold layer is a monolayer (Fig. 5). We have, however, been less successful in consistently obtaining uniform emulsion layers with the Gevaert 307 than with the centrifuged Kodak NTE emulsion.

For quantitative work the exact thickness of an emulsion layer over a particular ribbon of sections must be known. Interference colors over the sections can be viewed in the darkroom (yellow safe light, Wratten filter OA) where they are seen as density differences. An even more accurate way to make this determination is available if developers such as Dektol or Elon-ascorbic acid (see Section III, F,2), which do not attack the silver halide crystals, are used. In such cases the developing procedure can be interrupted after the specimen has been developed and stopped, but before it is finally fixed. After the stop bath, the specimen can be rinsed and air-dried and the interference colors then safely viewed in white light without introducing background. Exact information can thus be obtained regarding the uniformity and actual thickness of the emulsion over any given ribbon of sections. Fixation then follows in the usual manner.

B. Sensitivity

Sensitivity of an emulsion layer can be defined as the ratio of the number of developed grains over the number of electrons hitting a given area of this layer. In other words, sensitivity gives the probability that the radiation hitting the emulsion will be registered. Two processes are involved: an electron emitted from the radioactive source has to form a latent image somewhere in the emulsion, and this latent image has to be transformed into a visible silver grain during development. The probability of forming a latent image depends on the energy of the particle and its path length in the silver halide. With electrons of the energy 5 keV or more (average energy of tritium β-electrons is 6 keV), self-absorption in a plastic embedded section of the thickness used in electron microscope autoradiography (300–1000 Å) need not be of concern. Thus it can be assumed that almost all the electrons emitted into the upper hemisphere (2π geometry) will hit the emulsion. (Tritium electrons emitted at very grazing angles will, of course, not reach the emulsion within their range, yet these electrons constitute less than 10% of the total.)

The probability of developing a latent image depends on the size of this latent image and on the developer. For any emulsion-developer combination there are unique threshold conditions for development.

Sensitivity may be adversely affected by oxidation of the latent image during exposure, caused either by the atmosphere or by an interaction with the biological specimen. When the sensitivities of the Ilford L4 and Kodak NTE emulsions were compared, using various developing and exposing procedures, it was generally found that the smaller the silver halide crystals or the finer grained the development, the more carefully the latent image has to be protected from oxidation. In some instances it is even necessary to enhance the latent image before development. In earlier publications (Salpeter and Bachmann, 1964; Bachmann and Salpeter, 1965) we reported that, when layers of Ilford L4 and Kodak NTE were placed on carbon-coated collodion slides, irradiated with beams of 10 keV electrons of measured intensity (to simulate irradiation from tritium), and developed (the Ilford L4 emulsion with Microdol-X for 3 minutes, and the Kodak NTE with Dektol for 1 minute), the sensitivity values for monolayers of the two emulsions were the same. Repeating these observations recently, we found that the sensitivity of currently supplied commercial batches of Kodak NTE emulsion is approximately a factor of 2 lower, and the enhancement of sensitivity due to gold latensification

(Section III,F,2,b) is less. All the relative changes in sensitivity due to external factors, however, remained the same.

An important external factor effecting sensitivity is the atmosphere during exposure. If the emulsions were stored in air for a 2-month period after irradiation but before development, the sensitivity of the Kodak emulsion dropped approximately 60% due to latent image fading. Storing the Kodak layers in helium for this period eliminated this drop in sensitivity. The sensitivity of the Ilford L4 emulsion was unaffected by storage in air for 2 months, nor did storage in helium produce any marked improvement in the sensitivity.

Sensitivity as measured by 10 keV irradiation could be expected to be similar but not identical to that in autoradiography. Recent studies on the sensitivity of Ilford L4 and Kodak NTE emulsion monolayers under conditions of electron microscope autoradiography with both tritium and S^{35} (Bachmann and Salpeter, unpublished observations, 1966) confirm indeed that the currently available Kodak NTE is less sensitive than the Ilford L4 emulsion. Specimens consisting of uniformly dispersed H^3 or S^{35} sources in gelatin layers of measured thickness were used for these determinations. The number of radioactive decays in the specimen area was then correlated with the number of developed grains in the overlying emulsion obtained after different developing procedures. We found that for H^3 the average sensitivity (ratio of developed grains to half the radioactive decays, i.e., those emitted toward the emulsion) was 1/5 for a monolayer of Ilford L4 (1300 Å) developed with Microdol-X for 3 minutes (24°C). For a monolayer of Kodak NTE (650 Å) developed at 24°C with Dektol for 1 minute it was 1/18; with Dektol for 2 minutes it was 1/10; and with gold latensification–Elon-ascorbic acid it was 1/5. For S^{35} the sensitivity was about a factor of 2 lower in both emulsions.

The higher sensitivity obtained with the H^3-labeled specimen than with the 10 keV irradiation is consistent with theoretical expectation: the probability of forming a latent image is proportional to the path length in the emulsion of a charged particle, which is at a minimum in the case of the 10 keV irradiations since the electrons all hit the emulsion at normal incidence. Greater backscattering from the glass substrate is expected in the H^3-labeled gelatin specimen than with the irradiations, and the average energy of the emitted particles is somewhat lower.

In actual autoradiography, when the emulsion overlies a biological section, oxidation by the tissue becomes another factor in decreasing sensitivity, a factor which may be magnified by prestaining of the section and by the fixative used. If a thin carbon layer (50–100 Å) is evaporated over the section before it is coated with emulsion, however, the effects of tissue section on emulsion sensitivity are eliminated or greatly reduced.

Direct contact between section and emulsion affects sensitivity most markedly in the Kodak NTE emulsion, but is also a factor with the Ilford L4 emulsion (Bachmann and Salpeter, 1964) when it is developed with p-phenylenediamine as recommended by Caro and Van Tubergen (1962). See also Caro (1964). Adjacent ribbons of thymidine-H^3 labeled nuclei were coated with Ilford L4 and developed with either Microdol-X for 3 minutes or with p-phenylenediamine for 1 minute. The average grain count over 20 nuclei of the p-phenylenediamine-developed sections was one half that of those developed with Microdol-X. A third ribbon, coated with 50 Å of carbon before the emulsion was applied and also developed with p-phenylenediamine after an equivalent exposure time, had a grain count almost equivalent (80–90%) to that of the Microdol-X-developed sections. Storage in helium was not preferable to storage in air (plus Drierite) in the refrigerator. It is likely that an intermediate layer other than carbon, for instance one of collodion or Formvar, would have a similar protective effect. This type of layer is automatically deposited over the biological section in the agar method of emulsion coating suggested by Caro and Van Tubergen (1962); such a layer is, however, usually thicker than the 50 Å of carbon found to be effective, and thus decreases resolution by a greater amount. Furthermore, emulsion layers over collodion show distinctly mottled interference colors signifying that the layers are not of uniform thickness. This unevenness in thickness of emulsion layers over collodion was confirmed interferometrically.

Finally, sensitivity can be increased by controlled manipulation of the developing procedure, a technique which presumably transforms subthreshold latent images to images capable of development. An instance of such a manipulation is given by Salpeter and Bachmann (1964) in the use of gold latensification to enhance sensitivity in fine-grained development of the Kodak NTE emulsion (for detailed procedure, see Section III,F,2,b). If this emulsion is treated with gold thiocyanate (James, 1948) before development, exceedingly small developed grains (100–1000 Å, depending on the choice of developing time) can be obtained with an Elon-ascorbic acid developer (Hamilton and Brady, 1959), and with sensitivity 2–3 times greater than that obtained in a similarly exposed Kodak NTE layer developed with Dektol. The choice of developed grain size depends on the factors that affect resolution (see Section I,C). From the discussion of such factors, it becomes clear that relatively little resolution is gained by making the size of the developed grain much smaller than the size of the silver halide crystal of the emulsion. Therefore, since small developed grains are hard both to see during rapid scanning and to differentiate from extraneous dirt on the tissue, a good developed grain size for the Kodak NTE emulsion is approximately 500 Å.

C. Resolution

A detailed discussion of the factors affecting resolution is beyond the scope of this chapter. Several authors have considered these problems (Pelc, 1963; Caro, 1962; Moses, 1964; Bachmann and Salpeter, 1965). Although their analyses do not coincide precisely, these authors are agreed that the specimen thickness, the size of the developed grain, and the size of the silver halide crystal all contribute to a final error in localizing the site of the radioactive decay and thus in limiting resolution. The relative extent to which these individual factors influence resolution is discussed by Bachmann and Salpeter (1965). They point out that these factors limit resolution independently in a statistically varying manner. The final error of the technique is therefore not a simple sum of the individual resolution-limiting errors but is more like a root mean square. For calculating a "total error"[1] they propose the formula

$$E_t = (E_g + E_x + E_{dg}^2)^{1/2}$$

where E_g is the error caused by geometric factors, i.e., specimen and emulsion thickness and degree of scattering; E_x is the error caused by the size of the silver halide crystal; and E_{dg} is the error caused by the size of the developed grain. From the above formula it is obvious that, if any one of these three factors that limit resolution is significantly larger than the others, it predominates in determining the final value, and little can be gained by decreasing the other two. In consequence, a simple calculation shows that, with the currently available emulsions and fine-grained developing procedures (see below), the resolving power of electron microscope autoradiography is now primarily limited by geometric factors and will continue to be so limited until a method is found in which the distance between the radioactive decay and the detector is significantly reduced.

[1] In this discussion E_t is considered as the radius of a circular zone around the midpoint of a developed grain. This circle represents the zone within which the radioactive source has a given probability of being located. The actual value of E_t depends on the chosen value for the probability. Because of the relatively small number of developed grains in electron microscope autoradiography, this consideration of resolution is more useful than the one inherited from light optics which deals in density gradients around a radioactive source.

III. Specimen Preparation

A recommended specimen preparation procedure is illustrated in Fig. 8. The following requirements guided the choice of this procedure: (a) to use the objective criteria of emulsion uniformity and thickness as presented above; (b) to form a uniform surface of consistent physical properties on which to form the emulsion layer; (c) to prepare and expose the specimen in such a way as to minimize latent image fading; and (d) to obtain a clean specimen with contrast adequate for the electron microscope, without risking a disturbance of the developed grains.

A. Section Mounting

As was mentioned previously, mounting sections on metal grids may cause trouble, expecially when liquid emulsions are used, since the silver halide crystals tend to collect at grid bars. If, however, specimens are first placed on a flat surface, a uniform close-packed layer can be obtained when using either pre-gelled or liquid emulsions.

Ribbons of thin sections from tissue embedded in either methacrylate or epoxy resin are sectioned in the conventional manner. Only ribbons of uniform thickness with no imperfections should be used. They are transferred with a sharpened orange stick or glass rod to a drop of distilled water or 10% acetone on a collodion-coated slide (Fig. 8A). Care must be taken not to tear the collodion film. If dirt is a problem it helps to transfer the sections from the sectioning boat to a beaker of clean distilled water before finally transferring them to the slide. The drop of water is then drawn away from the ribbon with the orange stick and removed with a thin strip of filter paper. The location of the sections should be indicated on the back of the slide by a circle inscribed with a diamond pencil. The use of frosted end slides facilitates specimen handling in the darkroom.

B. Staining

Specimens prepared for electron microscope autoradiography have very low contrast unless some special treatment is introduced. Figure 9 shows a typical autoradiogram of a specimen in which the contrast has not been enhanced. Most procedures for improving contrast involve specimen manipulation, either staining or gelatin removal or both, after the photographic process is complete (for review, see Moses, 1964).

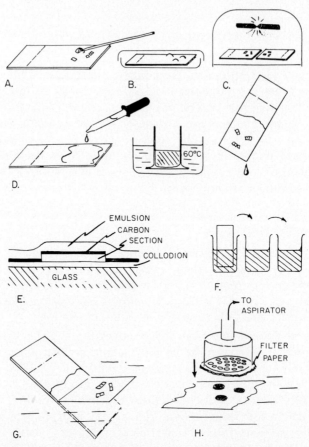

FIG. 8. Specimen preparation procedure. A: Ribbons of sections are placed on collodion-coated slides and allowed to dry. B: Sections are stained in a tightly closed Petri dish with drops of stain placed over the individual ribbons. After the staining period the stain is flushed off with distilled water. C: The stained sections are vacuum coated with a thin layer of carbon (50 Å). D: Liquid emulsion is dropped over the section and then drained and dried in a vertical position. Specimen may equally well be dipped into diluted liquid emulsion. E: The final specimen sandwich. F: Slides are developed in a series of beakers, with distilled water rinses in distilled water. G: Specimen sandwich is stripped onto a water surface; grids 3% acetic acid stop bath for 10 seconds, distilled water rinse, fixer for 1 minute, three rinses in distilled water. G: Specimen sandwich is stripped onto a water surface, grids are then placed over the tissue sections. H: A procedure for picking up the specimen from the water surface by suction onto a moist filter paper applied over a filter plate. (From Salpeter and Bachmann, 1965.)

These methods are not always successful; they involve a risk of removing or disturbing developed grains and can introduce dirt. To eliminate these possibilities, tissue sections can be stained before coating with emulsion. High-contrast specimens are obtained provided a thin layer of carbon (~50 Å) is evaporated over the stained section, being interposed between the stained section and the emulsion (Section III,C; Figs. 10–14). (Koehler *et al.*, 1963; Salpeter and Bachmann, 1964).

Procedure for staining: The slide to be stained is placed in a Petri dish, a few drops of stain are placed over the sections, and the dish is covered. Only clean, precipitate-free drops of stain should be used and they should be prevented from evaporating during the staining period. After staining, the drops of stain should be flushed off quickly with an ample stream of distilled water.

Words of caution on staining are necessary. Staining may extract radioactive compounds. The stained section may also affect the emulsion despite the carbon layer interposed between stained section and emulsion. Thus, for any quantitative study, the sensitivity of the stained specimen should be checked against that of an unstained section. Furthermore, tissues can easily be overstained, the result being dense preparations in which the developed grains are hard to distinguish from the tissue components. The use of lead stains (Karnovsky, 1961; Reynolds, 1963) for autoradiographic specimens also frequently results in dense deposits on specific components such as chromatin, cellular filaments, mitochondrial membranes, and extracellular collagen (Fig. 10). Shortening the staining period tends to minimize the effect. Not all tissues are equally affected. Deposit-free specimens with excellent contrast can be obtained with some tissues after staining with Reynolds' lead citrate (Reynolds, 1963) for 10–30 minutes, or after staining 3–5 minutes in aqueous uranyl acetate followed by 5–30 minutes in Reynolds' lead citrate (Figs. 10, 12, and 13). Other tissues, such as muscle and nerve fibers or tissue rich in collagen, when similarly treated tend to have large areas covered by dense deposits and are, therefore, useless for scanning at low magnification. These preparations still have great value, however, for high magnification pictures, especially after fine-grained developing. A lead stain consisting of 0.25% lead citrate (K and K Laboratory, Plainview, New York) in 0.1 N NaOH may be less subject to such deposits. Uranyl acetate or uranyl nitrate for 3–6 hours gives good results in all these tissues (Fig. 14) although the contrast is lower than that obtained with lead.

It is advisable that the staining procedure for any tissue be tested under conditions equivalent to those of actual autoradiography. Several ribbons of tissue should be cut, each stained differently, and then all

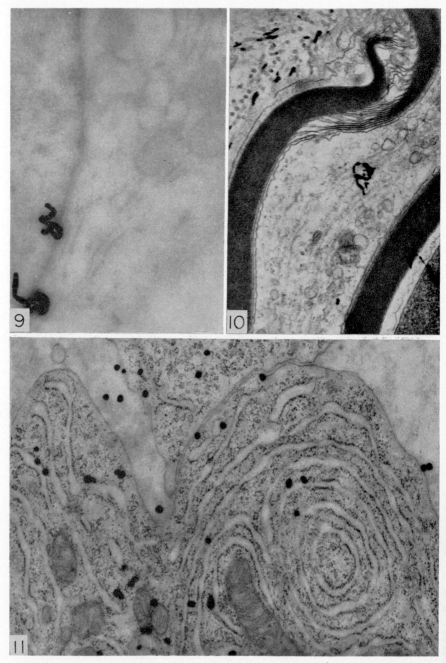

FIGS. 9–11. Autoradiograms using Ilford L4 emulsion with different embedding, staining and developing procedures. Sections (interference color—light gold) were

carbon-coated and emulsion-coated. Without any exposure period, the specimen should be developed and examined with the electron microscope to determine the optimum stain. Any emulsion serves equally well for this purpose.

C. Intermediate Layer

A thin (50–100 Å) carbon layer should be evaporated over the section after staining (Fig. 8C). [Union Carbide SPK spectroscopic carbon is recommended (National Carbon Co., Pittsburgh, Pa.)]. The carbon layer serves three functions: to protect the stained section from being destained by the developing fluids, to protect the emulsion from the stained section (see Section II,B, "Sensitivity"), and to provide a surface with uniform physical and chemical properties over which an even emulsion layer can be formed. If the carbon layer is too thin this fact can be detected by the mottled or honeycombed interference color pattern of the emulsion. Such a thin carbon layer is less effective in preserving sensitivity and contrast. On the other hand, it should be remembered that the thickness of the carbon is a factor in decreasing resolution due to geometric considerations and thus should not be made much thicker than necessary for producing an emulsion layer of uniform interference color.

D. Emulsion Coating

If the surface to be coated is of uniform physical properties, as is the case when the techniques described above are used, very uniform emulsion layers can be obtained when emulsions are applied either in liquid or pre-gelled form.

---

coated with a monolayer of Ilford L4 emulsion (emulsion interference color—purple). ca. × 30,000. (From Salpeter and Bachmann, 1965.)

FIG. 9. Tissue: mesenchymatous cell of adult newt, *Triturus*, labeled with thymidine-H[3]. Stain: section is unstained and has no carbon layer. Embedding medium: methacrylate. Development: Microdol-X for 3 minutes at 24°C.

FIG. 10. Tissue: axon of peripheral nerve in newt, *Triturus*, labeled with histidine H[3]. Stain: aqueous uranyl acetate for 5 minutes followed by lead citrate (Reynolds, 1963) for 15 minutes. Stained section was coated with a carbon layer (50 Å). Embedding medium: Epon 812. Development: Microdol-X for 3 minutes at 24°C. (Note dense stain deposits over collagen at *upper left*.)

FIG. 11. Tissue: mesenchymatous cell of the newt, *Triturus*, labeled with leucine-H[3]. Stain: lead cacodylate (Karnovsky, 1961). Stained section is coated with a carbon layer ~50 Å. Embedding medium: methacrylate. Development: *p*-phenylenediamine.

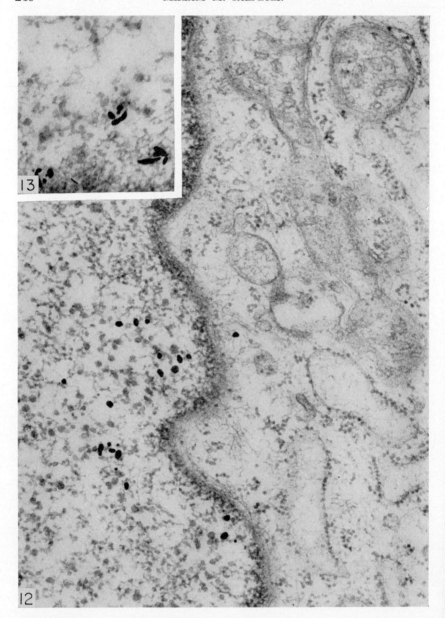

FIGS. 12–13. Autoradiograms using Kodak NTE emulsion. Sections (interference color—silver) were coated with a monolayer of Kodak NTE (emulsion interference color—silver). ca. × 60,000. (From Salpeter and Bachmann, 1965.)

FIG. 12. Tissue: mesenchymatous cell of the newt, *Triturus*, labeled with thymidine-H³. Stain: 2% uranyl acetate for 5 minutes followed by lead citrate for 10

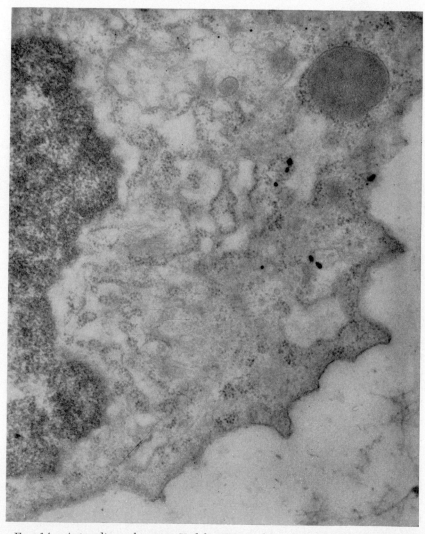

FIG. 14. Autoradiography using Kodak NTE emulsion. Section (interference color —silver) was coated with layer of Kodak NTE (interference color—silver-to-gold). Tissue: cartilage cell from regenerating limb of newt, *Triturus*, labeled with proline-H^3. Stain: 2% uranyl acetate for 3 hours. Stained section covered with 50 Å of carbon. Embedding medium: methacrylate. Development: gold latensification–Elon-ascorbic acid. ca. × 40,000.

minutes. Stained section covered with a carbon layer ∼50 Å. Embedding medium: methacrylate. Development: gold latensification–Elon-ascorbic acid.

FIG. 13.   Tissue: same as in Fig. 12 except developed by Dektol for 1 minute at 24°C.

1. Ilford L4

The Ilford L4 emulsion is diluted with distilled water, melted in a warm water bath at 40°–50°C, and gently stirred. The final dilution is determined empirically. To do this, carbon-coated test slides are coated with emulsion after successive dilutions until the desired interference color (examined in white light) is obtained (Fig. 2). To coat, the slides can be dipped into the emulsion, and drained and dried in a vertical position. Other coating procedures are equally acceptable as long as the area of the slide on which the specimens are mounted receives an emulsion layer of the desired interference color. If the dipping method is used, the final dilution is approximately 5 ml distilled water per gram of emulsion.

2. Kodak NTE

As presently supplied, the Kodak NTE emulsion has too much gelatin to make monolayers of closely packed silver halide crystals. To reduce the gelatin content, 1 gm of emulsion is dissolved in 10 ml distilled water in a water bath at 45°–60°C. The warm emulsion is then centrifuged until the supernatant is clear (at approximately 14,000 g for 10 minutes). To facilitate the separation of the silver halide from the gelatin, the rotor of the centrifuge is first heated. This is done easily by allowing the centrifuge to rotate slowly for about 20 minutes while a hot air gun is directed at its rotor. If a small cart holds the centrifuge it can be heated outside the darkroom and then wheeled into the darkroom just before use. After centrifugation, the bottom of each centrifuge tube is chilled briefly and the supernatant discarded. The remaining concentrated emulsion is reheated and resuspended in 1–2 ml water per gram of the original emulsion. The exact amount of water is determined empirically, depending on the emulsion thickness desired (i.e., by the interference colors of emulsion layers on test slides after successive dilutions). Coating can be accomplished either by dipping the slide with the specimen into the emulsion or by applying the liquid emulsion with a medicine dropper over the specimen held horizontally (Fig. 8D). In either case, the emulsion is then drained and air-dried vertically.

E. Storage

To avoid latent image fading which occurs with the Kodak NTE emulsion when specimens are stored in air during exposure, storage in helium is recommended. Any other inert gas should do as well.

The Ilford L4 emulsion does not exhibit the same fading and can be stored in the refrigerator during exposure. With both Kodak NTE and

Ilford L4, the specimens must be thoroughly dried before storing. The addition of Drierite to the storage containers is essential.

F. Development

Cleanliness is an important consideration during all the steps in development. All solutions should be made with distilled water. A distilled water rinse is used between consecutive steps. Figure 8F shows the recommended procedure. Slides are developed individually in 30- or 50-ml beakers at room temperature, 23°–24°C.

1. DEVELOPERS FOR ILFORD L4

a. Microdol-X (Pelc *et al.*, 1961; Caro and Van Tubergen, 1962) is a reliable standard developer. Three-minute development results in developed grains 2000–4000 Å in diameter (Figs. 9 and 10).

b. *p*-Phenylenediamine, the physical developer first recommended for electron microscope autoradiography by Caro and Van Tubergen (1962) (Caro, 1964), gives developed grains of 500–700 Å. These grains are compact rather than filamentous and obstruct the underlying fine structure of the specimen less than do the Microdol-X grains (Fig. 11). They also improve the over-all resolution (see discussion on resolution, Section, II,C). The developer is made by dissolving, in a water bath at 60°C, 1.1 gm *p*-phenylenediamine in 100 ml of a 12.6% solution of sodium sulfate. It is then cooled and filtered. The developer is unstable and should be made fresh on the day of use. For adequate sensitivity with this developer a carbon intermediate layer is essential (Bachmann and Salpeter, 1964). A 1-minute development is adequate.

2. DEVELOPERS FOR KODAK NTE

a. *Dektol.* The recommended standard developer for the Kodak NTE emulsion is Dektol (diluted 1 to 2 with distilled water). Developing for 1 minute results in 800–1200 Å developed grains which are either filamentous or appear as a cluster of small, unconnected spheres or rods (Fig. 13). Development for 2 minutes gives optimum sensitivity.

b. *Gold Latensification and Elon-Ascorbic Acid.* The procedure called "gold latensification," which deposits gold atoms on the latent image, enhances sensitivity by approximately a factor of 2 to 3 and also makes fine-grained development possible:

(*i*) *Gold latensification.* The sensitizing gold thiocyanate solution is made as follows (James, 1948): 0.5 ml of a 2% stock solution of gold chloride ($AuCl_3HCl \cdot 3H_2O$) (stock is stable for several months), 9.5 ml distilled water, 0.125 gm potassium thiocyanate, 0.15 gm potassium bromide, and bring to 250 ml with distilled H_2O. (The gold thiocyanate solution is unstable and should be used for only 1 day.)

(ii) *Elon-ascorbic acid.* A fine-grained developer, ineffective on its own but useful in conjunction with gold latensification, is Elon-ascorbic acid (Hamilton and Brady, 1959). It consists of 0.045 gm Elon (Metol), 0.3 gm ascorbic acid (Fisher Scientific Co.), 0.5 gm borax, 0.1 gm potassium bromide, and 100 ml distilled water. (The developer is unstable and decays rapidly during the first few hours after it is made, but much more slowly thereafter.)

(iii) *Development procedure.* Sensitivity is increased by gold latensification as a function of solution strength and of soaking time. An exposed slide is wetted in water, soaked in the gold thiocyanate solution (diluted 1 to 20 with distilled water) for 30 seconds, washed in a beaker of distilled water, and then transferred to the Elon-ascorbic acid developer for 8 minutes. The size of the developed grains is influenced by both the developing time and the age of the developer and can range from 100 Å to over 1000 Å. When the developer is 5–48 hours old, an 8-minute development results in developed grains roughly 500 Å in diameter (Fig. 12). This grain size is desirable because of the difficulty in differentiating much smaller developed grains from such tissue components as cross sections of collagen, and also because no significant gain in over-all resolution can be expected from developed grains which are much smaller than the silver halide crystals.

G. Final Steps in Photographic Processing

After development by any of the solutions described above, slides are rinsed in distilled water, dipped in a 3% acetic acid stop bath for 15 seconds, again rinsed in distilled water, and then fixed for 1 minute. To facilitate the final stripping of the specimen, a nonhardening fixer is recommended consisting of 20% sodium thiosulfate (hypo) and 2.5% potassium metabisulfite. After fixing, the slides are rinsed in several changes of distilled water.

H. Stripping and Mounting Specimens

After photographic processing the specimen sandwich (consisting of collodion film, section, carbon layer, and developed emulsion) is stripped onto a water surface. Metal grids are placed over the individual ribbons, which can be identified on the floating film by their interference colors, and the sandwich is picked up. One method of picking up the specimens is seen in Fig. 8H. Stripping the specimen from the slide can be a problem especially if a long staining period has been used (e.g., uranyl acetate for 3 hours). On the other hand, unstained specimens may float off prematurely if agitated roughly during developing and subsequent processing. If stripping is a problem, the following should be observed:

1. The collodion film may be made slightly thicker than is ordinarily used for section substrates. Adequate layers are obtained if slides are dipped in 0.5% collodion solution in amyl acetate. When stripped onto a water surface, such collodion films show a silver interference color.

2. After photographic processing, the slide should be left in the last distilled water rinse for 15 minutes to 2 hours and should not be allowed to dry before stripping.

3. A nonhardening fixer should be used.

4. Stripping should be done with the aid of a dissecting microscope or large magnifying glass and a teasing needle used to start and guide the process.

5. Not all carbon layers strip with equal ease. When Union Carbide spectroscopic carbon SPK is used, stripping is easier than with some other carbon.

IV. Concluding Remarks

The existence of emulsions having very small silver halide crystals (e.g., Kodak NTE, approximately 500 Å) and of a high sensitivity developing technique (gold latensification and Elon-ascorbic acid) for obtaining small developed grains ($\sim$500 Å) has brought the photographic resolution (i.e., resolution due to the size of silver halide crystals and developed grains) of electron microscope autoradiography to a new level, approximately by a factor of 2 better than previously attainable (Bachmann and Salpeter, 1965). These improvements bring with them problems of their own, such as greater instability, storage in helium, centrifugation of Kodak NTE emulsion, difficulties in seeing developed grains during quick low magnification scanning of tissue, and a frequently higher background than found in the Ilford L4 emulsion especially if gold latensification is used (Salpeter and Bachmann, 1964, 1965).

As was mentioned before (Section II,C), photographic resolution is, however, only a small part of the total autoradiographic resolution which is, with existing techniques, now primarily limited by geometric factors (i.e., specimen thickness). A significant gain in resolution thus requires the use of very thin tissue sections. This, of course, results in a decreased amount of radioactive material in the specimen—yet another price to pay in return for increased resolution.

Therefore, to make most effective use of the autoradiographic techniques available to date, it is reasonable first to obtain information with the light microscope regarding the gross distribution of radioactive material in the specimen. For this, 1μ sections are cut from the blocks to be used for electron microscope autoradiography and prepared as already suggested by others (Hay and Revel, 1963; Caro and Van Tubergen, 1962; or Caro, 1964). Such preparations provide information regarding the tissue area to be used for any electron microscope work and also regarding expected exposure time.

A second equally useful step is the examination of large random areas with the electron microscope at low magnification to obtain a quantitative determination of the distribution of radioactive material at the fine structural level. For this purpose a large developed grain is useful, and a high degree of resolution less important. Tissue sections (relatively thick, i.e., 1000 Å—gold interference color) coated with Ilford L4 emulsion and developed with Microdol-X are recommended for this step. It is, however, very essential for a meaningful quantitation to have a uniform emulsion layer of known thickness. (The thickness becomes more important if the isotopes used have higher energy decays than does tritium.) The use of emulsion layers of controlled interference color on flat substrates provides this information (Section II,A,3).

For many electron microscope autoradiographic problems the level of resolution obtained by the above procedure (i.e., 1500–2000 Å) is sufficient. If resolution better than 1500 Å is desired, the next step requires the use of much thinner sections (interference color gray or silver) in conjunction with a fine-grained emulsion (e.g., Kodak NTE) and developer. The information gained from the large area scanning (step 2) can profitably be used to define the tissue area to be examined, since any tissue scanning has now to be done at much higher magnification. A statistical analysis of the kind suggested by Bachmann and Salpeter (1965) (see Section II,C) can then be used to determine the zone (E_t), most likely to contain the radioactive source.

Most efficient utilization of autoradiography at this time combines all the available techniques—going from the light microscope level, with its relatively thick sections and thus high radioactivity but low resolution, through two stages of electron microscope autoradiography, gaining resolution via thinner sections at the expense of radioactivity, and via finer grained emulsions at the expense of stability and convenience. This last factor hopefully will be eliminated as commercially produced fine grained emulsions become more trouble-free.

ACKNOWLEDGMENT

This investigation was supported by United States Public Health Service Research Grant GM 10422 from the Division of General Medical Sciences, and by Career Development Award NB-K3-3738 from the Division of Neurological Diseases and Blindness.

REFERENCES

Bachmann, L., and Salpeter, M. M. (1964). *Naturwiss.* **51**, 237.
Bachmann, L., and Salpeter, M. M. (1965). *In* "Quantitative Electron Microscopy" (G. F. Bahr and E. H. Zeitler, eds.), pp. 303–315. Williams & Wilkins, Baltimore, Maryland *or Lab Invest.* **14**, 1041.
Caro, L. G. (1962). *J. Cell Biol.* **15**, 189.
Caro, L. G. (1964). *In* "Methods in Cell Physiology" (D. M. Prescott, ed.), Vol. I, pp. 327-363. Academic Press, New York.
Caro, L. G., and Van Tubergen, R. P. (1962). *J. Cell Biol.* **15**, 173.
Dohlman, G. F., Maunsbach, A. B., Hammerstrom, L., and Applegren, L. E. (1964). *J. Ultrastruct. Res.* **10**, 293.
Granboulan, P. (1963). *J. Roy. Microscop. Soc.* **81**, 165.
Hamilton, J. F., and Brady, L. E. (1959). *J. Appl. Phys.* **30**, 1893.
Hay, E. D., and Revel, J. P. (1963). *Develop. Biol.* **7**, 152.
James, T. H. (1948). *J. Colloid Sci.* **3**, 447.
Karnovsky, M. J. (1961). *J. Biophys. Biochem. Cytol.* **11**, 729.
Koehler, J. K., Mühlethaler, K., and Frey-Wyssling, A. (1963). *J. Cell Biol.* **16**, 73.
Moses, M. J. (1964). *J. Histochem. Cytochem.* **12**, 115.
Pelc, S. R. (1963). *J. Roy. Microscop. Soc.* **81**, 131.
Pelc, S. R., Coombes, J. D., and Budd, G. C. (1961). *Exptl. Cell Res.* **24**, 192.
Revel, J. P., and Hay, E. D. (1961). *Exptl. Cell Res.* **25**, 474.
Reynolds, E. S. (1963). *J. Cell Biol.* **17**, 208.
Salpeter, M. M., and Bachmann, L. (1964). *J. Cell Biol.* **22**, 469.
Salpeter, M. M., and Bachmann, L. (1965). *Symp. Intern. Soc. Cell Biol.* **4**, 23.
Young, B. A., and Kopriwa, B. M. (1964). *J. Histochem. Cytochem.* **12**, 438.

Chapter 10

High Resolution Autoradiography[1]

A. R. STEVENS

*Department of Anatomy, University of Colorado Medical Center,
Denver, Colorado*

I. Introduction

Biochemical cytology has made considerable progress in the last fifteen
years by the introduction of isotopically labeled compounds for the study
of the synthesis and/or localization of specific cellular constituents. For
example, one major area that has been exploited from the direct application of autoradiography to biological investigations has been that of
nucleic acid metabolism. Several of the more important findings that may
be cited are the demonstration that deoxyribonucleic acid (DNA) synthesis occurs in interphase of the cell cycle (Howard and Pelc, 1953),
the localization of ribonucleic acid (RNA) synthesis in the nucleus and
transport of RNA to the cytoplasm (Goldstein and Plaut, 1955), the dis-

[1] This work was supported by NSF Grant GB-1635 to D. M. Prescott.

covery that the DNA of chromosomes segregates in a semiconservative fashion (Taylor *et al.*, 1957), and the mutual exclusiveness of RNA and DNA synthesis (Prescott and Kimball, 1961).

Although significant observations concerning cellular activities were made soon after autoradiography was introduced as a cytochemical tool, autoradiography in its infancy did not escape the inherent problems that usually accompany a new technique. The necessity of using radioactive compounds that emitted high energy β-particles resulted in poor resolution of the incorporated label, since the distance that the emitted electrons could travel exceeded the thickness of the overlying emulsion (e.g., C^{14}—average range in water, 40μ). $P^{32}O_4$, a widely used isotope in the early tracer experiments, proved effective in some experiments but has the disadvantages of inadequate autoradiographic resolution and frequently insufficient biochemical specificity. Finally, the operations required for applying a sensitive layer of silver halide crystals over the preparation (stripping film method), to register the electrons emanating from the incorporated isotope, were tedious and time-consuming.

In recent years, however, the field of autoradiography has expanded rapidly due to the availability of tritiated compounds of very high radiochemical purity and the production of fine-grained nuclear emulsions. Investigators now have a more critical means for introducing radioactivity into specific cellular components in view of the many tritiated precursors that can be applied to biological investigations. Moreover, autoradiographic patterns can be analyzed more accurately because the short range of tritium radiation permits a resolution of 1μ with conventional methods at the light microscope level. Although the stripping film method still has some advantages in certain experimental situations, e.g., when a precisely uniform thickness of emulsion is required, the technique of applying liquid emulsion has proved to be a very convenient way of preparing large numbers of high quality autoradiographs quite suitable for quantitative autoradiography.

Even though the methodology of light microscope autoradiography has been considerably improved, it is realized that theoretically an autoradiographic resolution of less than 1μ could be achieved if extremely thin specimen preparations and emulsion layers are used. However, the limitations imposed by the optical properties of the light microscope itself have prevented the experimental demonstration of this theoretical resolution. It was fortuitous, therefore, that workers began to modify the standard autoradiographic techniques so that tracer experiments could be performed in conjunction with the electron microscope. Perfecting the technique for applying autoradiography to high resolution investigations could serve two purposes: (1) provide the higher resolution required for better interpretation of autoradiographs that had been observed with the light microscope, and (2) be an effective cytochemical method for studying the functional roles of many of the subcellular organelles that heretofore could not be investigated with usual electron microscope techniques.

Although Liquier-Milward (1956) and O'Brien and George (1959) are credited with the first observations of autoradiographs in the electron microscope, the resolution obtained in these experiments was not a significant improvement over that of light microscope autoradiography. The reason for this can be attributed to the use of high energy emitters such as polonium-210 (1959) and cobalt-60 (1956). On the contrary, van Tubergen (1961) devised a more sophisticated technique for applying a thin layer of emulsion to preparations and demonstrated that it was possible to localize incorporated thymidine-H³ in *Escherichia coli* to areas less than 0.5μ². Subsequent to van Tubergen's experiments, Caro (1962) calculated the expected distribution of exposed crystals around a point source of tritium and predicted from these data that an autoradiographic resolution of the order of 0.1μ could be attained in high resolution autoradiography. He verified this theoretical resolution from analyzing the grain pattern obtained in electron microscope autoradiographs of T2 bacteriophages and *Bacillus subtilis* labeled with thymidine-H³.

Numerous laboratories (Silk *et al.,* 1961; Caro and van Tubergen, 1962; Meek and Moses, 1963; Hay and Revel, 1963; Moses, 1964) have become engaged in perfecting effective methods for the combined use of tritiated compounds and thin layers of fine-grain emulsions with conventional electron microscope techniques. However, the technicalities of the procedures and the time involved for acquiring good electron microscope autoradiographs have discouraged many workers from using this important cytochemical tool in biological investigations. In this chapter, therefore, it is hoped that the novice in electron microscope autoradiography will be informed about (1) the important contributions that light microscope observations can make to high resolution autoradiography, (2) the incorporation of several steps in electron microscope autoradiographic experiments that will make operations more easily and quickly performed, and (3) the use of a simplified method for obtaining suitable electron microscope autoradiographs that will provide adequate resolution for critical evaluation.

II. Light Microscope Techniques

A. Usefulness of Thick Sections

When high resolution autoradiography is being considered as an experimental technique, it must be realized that, in addition to obtaining good ultrathin sections for observation with the electron microscope, suitable preparations must also be acquired for examination with the light microscope. It may be somewhat misleading to use the term "thick section" for the light microscope preparation if the reader is thinking in terms of the thicknesses ordinarily cut in routine histological sectioning (3–10μ). Throughout this chapter, however, the expression "thick section" will imply that the experimental material is to be sectioned at thicknesses of 0.12–0.5μ.

It is an established fact among microscopists that many problems pertaining to the interpretation of ultrastructure can best be approached by comparing structures in adjacent thick and thin sections with the light and electron microscopes, respectively (Runge et al., 1958). It should not be surprising, therefore, to learn that there are essential reasons why thick sections are useful at the light microscope level when tracer studies combined with electron microscopy are being performed. The main purposes for the thick sections are (1) to provide an estimate of the approximate autoradiographic exposure time for ultrathin sections, (2) to furnish light microscope autoradiographs for comparison with the electron microscope preparations, and (3) to determine if the structure or region being studied in the thin sections will be present.

Of the three reasons mentioned above, the first will be the most informative if high resolution autoradiography is to be employed. As Caro (1964) points out, "if a light microscope preparation from a 0.4μ section gives a suitable autoradiographic response in 1 week, a useful preparation for the electron microscope will need a 2–4 months exposure. The sensitivities of the two methods differ therefore by a factor of 10, approximately."

No definitive statement can be made concerning the length of time required for obtaining "a suitable autoradiographic response" because this will vary with each experimental situation. Some of the numerous factors contributing to the required length of time are the amount and specific activity of the isotope in the incubation medium, the amount of isotopically labeled material incorporated, the metabolic properties of the particular cell or tissue and the rate at which the chemical substance being labeled is synthesized, and the distribution of the labeled compound in the experimental material.

It is important to mention that the availability of H³-labeled compounds with unusually high specific activities has considerably enhanced the usefulness in recent years of light and electron microscope autoradiography. For example, until 1964 thymidine-H³ was obtainable at a specific activity of 6.7 C/mM (New England Nuclear) whereas it is now available at an activity of greater than 15 C/mM (Schwarz BioResearch,

Inc.). A word of warning should be given regarding the specific activity of the labeled compound that will be used in a particular experiment. Although it is most desirable to administer an isotope with the highest specific activity commercially available, this is not always experimentally possible. Some cells, for example HeLa, are relatively radiosensitive, and thus the amount of a specific isotope may have to be carefully regulated to avoid intolerable radiation damage. In contrast, protozoa have been found to be highly radioresistant, and radiation damage resulting from the incorporation of the isotope (even at very high levels) appears not to be significant. It is not within the scope of this chapter to discuss radiation damage in cells; however, for additional information on the subject, the reader is referred to J. D. Thrasher in this volume (Chapter 12).

The specific activity of the isotope is one of the factors determining the amount of incorporated radioactivity that is attained, but of equal importance for adequate incorporation is the experimental material, the synthesis rate of the substance being labeled, and the distribution of the labeled compound in the tissue or cell. For example, consider an experiment in which the precursor uridine-H^3 is used to label RNA in two cell types: (1) *Tetrahymena pyriformis*, and (2) *Amoeba proteus*. A 10-minute pulse of uridine-H^3 will produce an intense autoradiograph in *T. pyriformis* after an emulsion exposure time of only 1 week. By comparison, in *A. proteus* an autoradiograph of comparable intensity will be obtained with the same exposure time (1 week) only if the cells are incubated for many hours with uridine-H^3. The results reflect the fact that RNA is being synthesized at a much greater rate in the ciliate than in the amoeba. In addition, the sites of RNA synthesis appear to be more concentrated in *Tetrahymena* (which is probably related to the larger ratio of DNA to nuclear volume in the ciliate as compared to the amoeba) (D. M. Prescott, personal communication, 1965).

Other examples of such differences in autoradiographic responses could be cited to illustrate the difficulties of predicting appropriate conditions for new experimental material. Thick sections are therefore essential for the purpose of estimating the emulsion exposure time of ultrathin sections, and in addition provide information concerning the many factors of paramount importance when this estimation is made.

Before continuing the discussion on the uses of thick sections, it is necessary to say a few words about what is considered an acceptable autoradiographic response from a thick section. Figure 1 shows a light microscope preparation (approximately 0.3μ) of *A. proteus* that has been labeled with H^3-amino acids. This intense autoradiograph was obtained after an exposure time of only 1 day, and therefore an excellent autoradiograph would be expected with the electron microscope after an ultrathin preparation had exposed for about 2 weeks. Figure 2 is an electron micrograph of the experimental material and confirms the expected 2-week exposure time. On the other hand, Fig. 3 demonstrates that, after 2 weeks under emulsion, a 0.3μ section of *Stentor coerulis* labeled with thymidine-H³ (in an attempt to demonstrate DNA synthesis in the kineto-

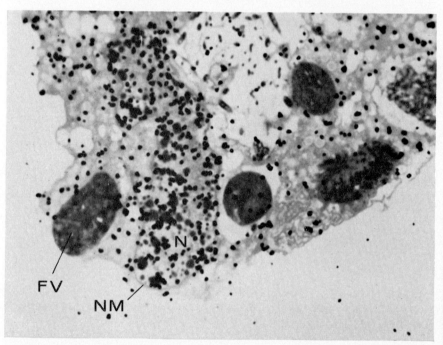

Fig. 1. Light microscope autoradiograph of 0.3μ section of *Amoeba proteus*. Label, H³-amino acids; stain, azure B; emulsion exposure time, 1 day. Although it may not be easily recognized in this photograph, there is some localization of the incorporated label over the nucleoli, located at the periphery of the nucleus. N = nucleus; NM = nuclear membrane; FV = food vacuole. ca. × 800.

somes) does not exhibit sufficient label to warrant using ultrathin sections for electron microscope autoradiographic purposes. Means must be found in the latter case to produce a greater incorporation of the isotope into the structure being studied.

From the above-mentioned results it can be stated that, if an autoradiograph comparable to that shown in Fig. 1 is obtained within a 2-week exposure period, the experimental material is usable for high resolution autoradiography; however, if a response as weak as that shown in Fig. 3 is obtained after 2 weeks, it would be inadvisable to progress to electron microscope autoradiography with that particular labeled material. Autoradiographic responses in the intermediate range between these two illustrations (e.g., the labeled macronucleus seen in Fig. 3) might be acceptable in some circumstances, but a careful judgment must be made by the operator.

We have routinely found that an ultrathin section can expose under a monolayer of emulsion for approximately 5 months before latent image

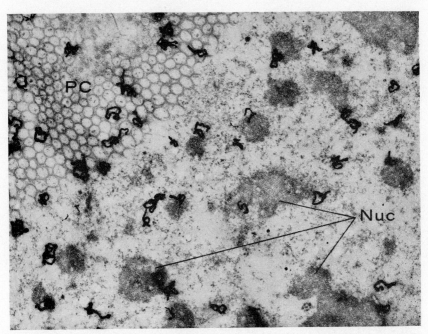

FIG. 2.   Electron microscope autoradiograph of A. *proteus* nucleus. Label, H³- amino acids; stain, uranyl acetate and lead citrate; embedding material, Epon 812; emulsion exposure time, 2 weeks. The incorporated label is localized somewhat over the nucleoli; labeled material is present within the pore complex. PC = pore complex; Nuc = nucleoli. ca × 10,000.

formation resulting from heat, light, mechanical, or chemical effects, etc. (background interference), will significantly interfere with the results. Taking into consideration that it will require about 10 times as long for an ultrathin preparation to expose as compared to a 0.4μ section, a period not exceeding 2 weeks should be the maximum time allowed for requiring a suitable light microscope autoradiographic response. This is a practical limit established by the rate of background accumulation of reduced silver grains in the emulsion.

The second reason that thick sections may be useful is that light micro-

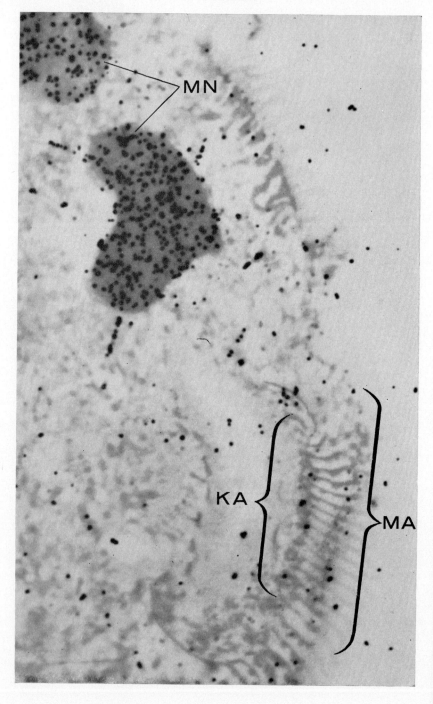

scope autoradiographs are often excellent comparative preparations for high resolution autoradiography. For example, in determining the site of DNA synthesis with thymidine-H³ in nuclei of *A. proteus*, it was apparent from the light microscope autoradiographs (thick sections approximately 0.3μ, exposure time 1 week) that the incorporated label was localized in the central portion of the nucleus (Fig. 4). After a 2-month exposure, the ultrathin sections produced electron microscope autoradiographs in which the majority of silver grains were associated with the central chromatin network, and confirmed the initial impression that little or no label was incorporated into the nucleoli (which are distributed around the periphery of the nucleus) (Fig. 5). The interpretation from the light and electron microscope results was in excellent agreement, i.e., the nucleoli contain little or no DNA, whereas the primary site of DNA synthesis is in the centrally located chromatin.

In many instances, the structure under study may be below the resolving power of the light microscope, and the advantage of light microscope autoradiographic preparations for comparative purposes may seem obviated. This is not entirely true, however, since it may be useful to compare the autoradiographic pattern obtained with the light microscope with the pattern of structures visible in the electron microscope. Thick section autoradiographs of the experimental material will indicate if the isotope has been incorporated into the cellular region where the particular synthesis is suspected to be taking place. A good example of this is the recent work performed on mitochondria. Stone and Miller (1965) administered thymidine-H³ to *T. pyriformis* for a period of 4 hours and obtained autoradiographs demonstrating not only heavy incorporation in the macronucleus, but also significant label in the cytoplasm. In the initial experiments, intact animals were used for the light microscope preparations but, in subsequent work, autoradiographs of thick sections of *T. pyriformis* (labeled with thymidine-H³ and embedded for electron microscopy) exhibited a labeling pattern identical to that of the whole mount preparations.

The point that can be made from the preceding discussion is that Stone and Miller progressed to high resolution autoradiography as a result of their light autoradiographic results. Electron microscope autoradiography

---

FIG. 3. Light microscope autoradiograph of 0.3μ section of *Stentor coerulis*. Label, thymidine-H³; stain, azure B; emulsion exposure time, ca. 2 weeks. The macronuclear nodes demonstrate good incorporation of thymidine-H³, but the area in which the kinetosomes are located does not show significant incorporation of the labeled precursor. MN = macronuclear nodes; MA = membranellar area; KA = kinetosome area. Experiment of A. R. Stevens and G. E. Stone. ca. × 700.

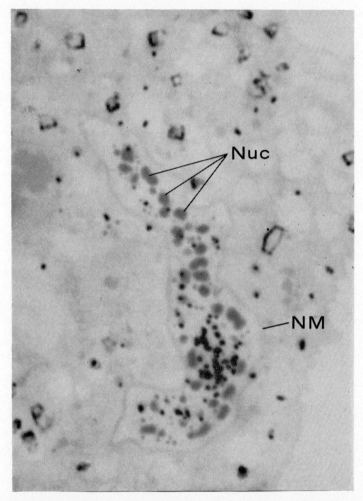

FIG. 4. Light microscope autoradiograph of 0.3μ section of *A. proteus*. Label, thymidine-H³; stain, azure B; emulsion exposure time, 1 week. Maximum incorporation is in the central portions of the nucleus, with little or no label associated with the nucleoli. Nuc = nucleoli; NM = nuclear membrane. ca. × 900.

provided the resolution necessary to localize the cytoplasmic label in the mitochondria.

Sections of thickness 1500–5000 Å are quite suitable for the light microscope comparative preparations. The thinner sections (1500 Å) will afford a somewhat greater resolution but less contrast at the light microscope level. On the other hand, the intensity of the autoradiograph and contrast of the image will be greater with the thicker sections (5000 Å), but the resolution attained will be decreased.

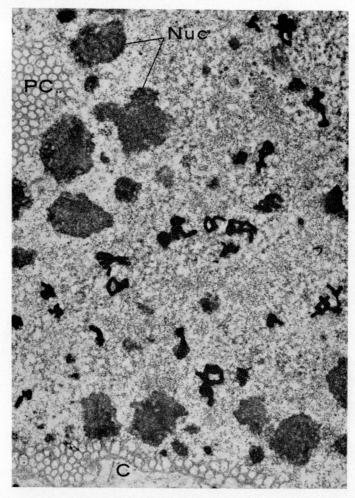

FIG. 5. Electron microscope autoradiograph of *A. proteus* nucleus. Label, thymidine-H³; stain, uranyl acetate and lead citrate; embedding material, Epon 812; emulsion exposure time, ca. 2 months. The incorporated label is localized in the central regions of the nucleus (chromatin area) with no apparent label associated with the nucleoli. PC = pore complex; Nuc = nucleoli; C = cytoplasm. ca. × 12,000.

The third advantage of thick sections, i.e., to determine if the specimen is being sectioned in the pertinent region, cannot be emphasized too strongly. In performing autoradiography with electron microscopy a great deal of time can be saved if the ultrathin sectioning process is monitored with the thick sections when specific structures or areas are not ordinarily in every thin section. For example, we have attempted to resolve incorporated thymidine-H³ within the macronuclear replication bands of

Euplotes eurystomus (Figs. 6 and 7) (Stevens, 1963; O. L. Miller, personal communication, 1962). This ciliate is approximately 120μ in length and 70μ in diameter. Although the macronucleus measures some 140μ in length and 6–8μ in diameter, each replication band constitutes only 1.5–2%. Every thin section taken from an embedded sample of euplotes,

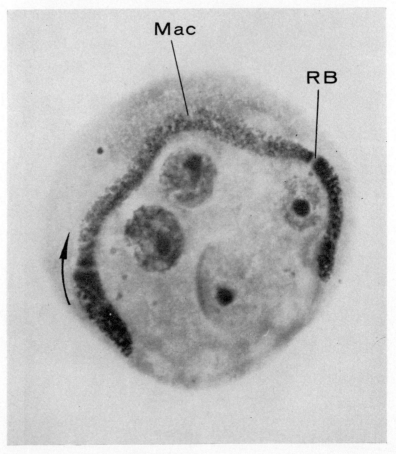

Fig. 6.   Acetocarmine squash of *Euplotes eurystomus*. Mac = macronucleus; RB = replication bands; arrow indicates direction of band movement. ca. × 500.

therefore, will certainly not contain a section of a replication band. In this experiment, to avoid a 2-month exposure time before determining if replication bands were present in the electron microscope autoradiographic preparations, the following simple check was made (O. L. Miller, personal communication, 1962). Thick sections (taken before and after

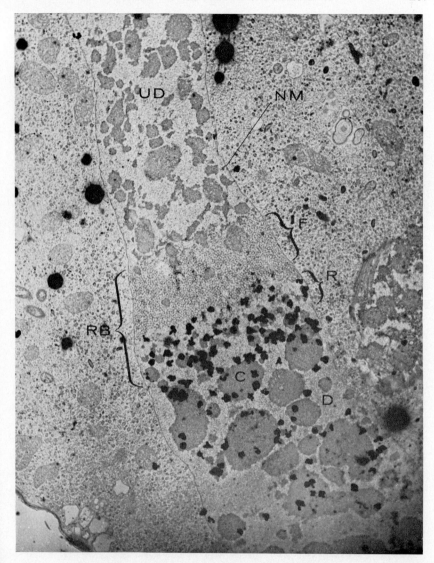

Fig. 7.   Electron microscope autoradiograph of *E. eurystomus*. Label, thymidine-H³; stain, uranyl acetate; embedding material, Epon 812; emulsion exposure time, ca. 2 months. The incorporated label is localized over the chromatin bodies in the duplicated region of the macronucleus; a single row of silver grains is seen at the interface between the forward and rear zones of the replication band and defines the region of DNA synthesis. RB = replication band; F = forward zone; R = rear zone; D = duplicated region; UD = unduplicated region; NM = nuclear membrane; C = chromatin body. The silver grains have been partially melted from overexposure to the electron beam. Experiment of O. L. Miller and A. R. Stevens. ca. × 16,000.

approximately 20–30 thin sections) were examined with the light microscope. If a replication band was identified in the thick sections, the
intervening thin sections were placed on a grid and used for autoradiography. If a replication band was not found, the thin sections were discarded and the procedure repeated until the appropriate number of
grids containing sections with replication bands was procured (Section
II,B,2).

When performing this preliminary check on thick sections, it is preferable to cut the material relatively thin, i.e., ca. 1200 Å. Less of the
material is wasted and the maximum amount of the particular region is
preserved for the subsequent thin sectioning. Moreover, the 1200-Å sections can be acquired under the same conditions necessary for ultrathin
sectioning, i.e., the same knife and block face can be used without damage to either.

In the preceding paragraphs, we have made a cursory examination of
the three main reasons why light microscope preparations are an important contribution to electron microscope autoradiography. Techniques
employed to obtain and handle thick sections are described in the next
section.

B. Obtaining Thick Sections

After the tracer experiment has been performed under conditions ensuring that the maximum amount of label has been incorporated without
deleterious side effects, the specimen is prepared according to conventional electron microscopic techniques. We routinely fix in buffered
osmium tetroxide (Palade, 1952), but any other preservative used for
electron microscopy may be employed with success. Following dehydration the sample can be embedded in methacrylate, the epoxy resins,
Epon (Luft, 1961) or Araldite (Glauert and Glauert, 1958), or the poly-
ester, Vestopal W (Ryter and Kellenberger, 1958). Caro (1964) states
that methacrylate is the preferred embedding material in his laboratory,
because "Methacrylate sections as thin as 0.2–0.3μ can be seen clearly in
phase contrast when the embedding medium is removed." Epon 812 has
been used primarily in our experiments, however, because of the advantages that this electron microscope embedding medium offers over methacrylate (i.e., absence of polymerization damage and stability of this resin
during irradiation with the electron beam) (Luft, 1961). Moreover,
Epon-embedded sections as thin as 1200 Å can be adequately observed with the light microscope when proper staining techniques are
employed (Section II,C).

METHODOLOGY

a. Obtaining Thick Sections for Determining the Approximate Exposure Time of Ultrathin Sections and Providing Light Microscope Autoradiographic Preparations for Comparative Purposes. When the experimental material is being sectioned for the above reasons, the block may be trimmed with a face much larger than that normally used for routine ultrathin sectioning. The larger face will facilitate handling the sections and allow a greater proportion of the specimen to be present in each section. A second feature that will allow easier handling of the sections is a block face trimmed to produce ribbons of sections.

A glass knife of much less perfection than required for ultrathin sectioning is made and adapted with a metal or tape boat. The investigator should not be concerned so much with a few scratches as with obtaining a knife with a straight, uninterrupted cutting edge free of contamination. The knife and block are mounted in the ultramicrotome, the water level is adjusted to give proper illumination for observation of the interference colors, and the section indicator set for the desired thickness (4000 Å for reason 1, 1500–5000 Å for reason 2, Section II,A). If the section indicator cannot be set to cut the thicker preparations, the knife must be advanced manually to attain the needed thickness.

After a ribbon of 4–5 sections of uniform thickness has been cut, a pre-subbed microscope slide [see Caro (1964) for subbing procedure] is placed on a flat surface and a small drop of distilled water added near one end. A very fine-haired nylon brush, thoroughly cleaned, is wetted and gently brought under the sections. The brush is then quickly transferred to the drop of water on the slide where a slight swirling action of the brush will tend to release the sections. If more sections are to be placed on the same drop of water, a certain amount of caution must be exercised. The sections present may adhere to the brush, thus making it impossible to remove additional sections from the brush tip. Usually sections greater than 2000 Å can be observed visually on the water, especially if the slide is resting on a black surface. With the thinner sections, it may be necessary to examine the slide with a dissecting microscope to ensure that the sections have been released from the brush.

Occasionally, sections will not float onto the water surface from the brush. There are two reasons why this may occur: (1) the brush has not been adequately cleaned prior to use, and/or (2) the sections have been placed too high on the brush instead of on the lower portion where the bristles are thinnest.

Once the sections are on the slide, the slide is labeled with a black pencil or crayon and dried on a hot plate (60°–80°C). The location of the sections is scored with a diamond pencil on the underside of the slide.

Even when sections are being used for light microscope autoradiographic purposes as opposed to electron microscope autoradiography, it is still advisable to check an occasional thick section to determine if the particular, desired region is being sectioned. For this preliminary check, a thick section (2000–5000 Å) may be mounted with a drop of water and a coverslip and observed with the phase-contrast microscope. This method of identifying defined regions offers the advantage that, after removal of the coverslip, the preparation may be used for autoradiography. Thinner sections (1500–2000 Å), embedded in Epon or Araldite will require staining to produce sufficient contrast for observation with the light microscope. Azure B is preferred in our laboratory for staining thin sections of osmium-fixed and epoxy-embedded material; however, other cytological stains (Section II,C) may be employed, but the choice will depend on (1) the particular tissue or cell, (2) the method of preservation, (3) the specific embedding medium, (4) the thickness of the preparation, and (5) the chemical substance or structure being examined.

It is important to mention at this point that certain stains [e.g., hemotoxylin, Feulgen, acetocarmine, periodic acid-Schiff (PAS)] can be used prior to or after autoradiography. Some of these stains will be effective in a particular experimental situation in order to give contrast to the light microscope preparations of the experimental material. The aniline dyes (e.g., azure B, toluidine blue, thionine, etc.), however, tend to activate the silver bromide crystals in the emulsion and therefore should be used prior to autoradiography only if the thick sections are expendable. In the event that every thick section needs to be preserved for autoradiography (e.g., if the amount of experimental material is small or serial sections are required), phase-contrast microscopy or one of the above mentioned dyes that does not expose the emulsion should be used when thick sections are being examined for identification of a particular structure.

Following preparation of the slides, autoradiography is carried out using any of the standard procedures. We routinely follow Prescott's method (1964) for autoradiography with liquid emulsion and consistently dilute the emulsion 1 : 1 with distilled water to acquire a thinner layer (this results in better optical properties).

The light microscope preparations (4000 Å) that will furnish information for determining the approximate exposure time for the electron microscope preparations should be developed at defined intervals (e.g., every 2 days) up to the maximum 2-week period. The exposure time of the preparations serving as the light microscope controls will vary

according to the thickness of the sections and the intensity of the autoradiograph. The appropriate length of time can be estimated from the
autoradiographic response obtained from the 0.4μ sections after approximately 2–4 days.

After developing, the slides are stained (unless this has been done
prior to autoradiography), mounted with a coverslip (water or a permanent mounting medium), and examined with the light microscope.

*b. Obtaining Thick Sections to Determine if the Specific Region Being
Studied with High Resolution Autoradiography Will Be Present in the
Ultrathin Preparations.* It may seem somewhat unnecessary to discuss
sectioning for the above reason separately from that of obtaining thick
sections for autoradiographic purposes, but one important difference
exists between the two methods. In the previous discussion, the microtome
was not set up under conditions favorable for ultrathin sectioning; however, this is imperative in order to monitor for the presence, in thin sections, of the specific region or structure under study.

The block is trimmed with a face compatible for ultrathin sectioning,
and a glass knife of superior quality is made and adapted with a boat.
For reasons that will become apparent, a metal boat is preferable to a
tape boat for the ensuing operations. If a diamond knife is ordinarily
used for routine ultrathin sectioning, this may be substituted for the glass
knife since no damage will be incurred to the edge when the 1200-Å
sections are cut.

After the block and knife are mounted in the microtome, the section
indicator is set for 1200 Å (gold interference color) and a single thick
section is cut. Although it is somewhat more difficult, the 1200-Å section
can be picked up with a brush and placed on a microscope slide, using
the method described (Section II,B,1). But an alternative method (which
probably will be mastered much sooner and will significantly decrease
the number of sections lost) can be utilized for placing the section on
the slide (Fig. 8 for illustration of method). This method can also be
used for picking up the sections for autoradiographic purposes (Section
II,B,1).

A wire loop (approximately 4 mm in diameter with an arm length of
40 mm) is made of nichrome wire (ca. 0.4 mm in diameter). The end of
the wire arm is wrapped around a piece of glass tubing, which serves as
a handle, or attached by annealing the end of the arm to the tubing in
an oxygen torch. The arm of the wire loop is bent at an angle allowing
the loop to be lowered parallel to the water surface with adequate clearance of the boat and the dissecting scope. This angle will depend upon
the relation between the particular microtome and dissecting scope
being used. After the single (1200 Å) section is cut, an eyelash (attached

to a wooden applicator with tape) is used to move the section away from the knife edge to the center of the boat. The wire loop, dipped in Desicote, dried, and rinsed thoroughly, is touched gently to the water surface in such a way that upon raising the loop the section will be cen-

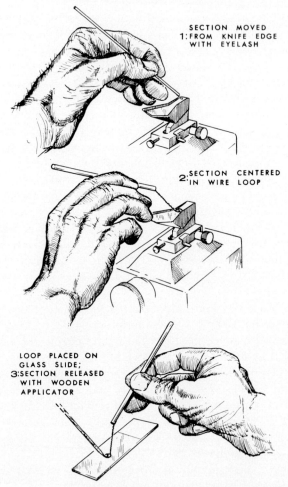

FIG. 8.   Loop method for picking up thick sections (0.12–0.5μ). See text for explanations.

tered in the thin film of water that forms the bubble in the loop. The loop is then placed flat on a dry slide (subbed or unsubbed) and a pointed wooden applicator is employed to release the drop of water containing the section. This is accomplished by touching the tip of the applicator to the edge of the water in any region around the circumference of the wire loop, while slowly raising the loop from the slide. The section can be observed on the drop of water either directly or with the aid of a dissecting scope. It may be necessary to repeat periodically the operation of coating the loop with Desicote to ensure consistent release of the section.

When a section has been mounted on a slide, the operator can return to the microtome and adjust the section indicator to the thickness desired for the ultrathin sectioning (e.g., 600–1000 Å). A suitable number of thin sections are cut, moved away from the knife edge with the eyelash to the rear of the boat, and the section indicator readjusted to 1200 Å. A second section (1200 Å) is taken and moved to a location on the water surface where it can be picked up with the wire loop without interference from the thin sections. The second 1200-Å section is placed on the same slide as the first (note should be taken of which section was cut prior to the thin sections, and which section was cut after the intervening thin sections), the slide dried on a hot plate (60°–80°C), and the preparation stained (Section II,C) to produce adequate contrast for examination with the light microscope.

If the two 1200-Å sections both reveal the structure or area being studied, the intervening thin sections are placed on a grid (or grids) and used for autoradiographic purposes. If both thick sections do not contain the desired area, the intervening thin sections are discarded and the procedure is repeated. In the event that the second thick section contains the appropriate area but the first does not, the operator should proceed to take additional thin sections and then a subsequent 1200-Å section to see if the area is still being sectioned. If this should be the case, the second set of thin sections can then be placed on a grid. The procedure of examining thick sections prior to and after a suitable number of thin sections should be repeated until the required number of grids is obtained. In this way, it is reasonably certain that every thin section placed under emulsion will contain the structure or area to be studied with electron microscope autoradiography.

We mentioned previously that a metal boat is preferable to a tape boat when a glass knife is used for the above mentioned operations. The metal boat or a boat of the size usually employed with diamond knives will allow the maximum space for picking up the 1200-Å sections with the 4-mm loop and also will ensure an adequate surface area for the thin sections to float freely.

Before leaving this section, one last point concerns the number of thin sections cut serially between the two 1200-Å sections. It is not possible to state the exact number since this will vary with the experimental situation; however, a few factors to be considered are (1) the plane of sectioning, (2) the size of the pertinent region, (3) the amount of experimental material, and (4) whether serial sections are required or whether sections with many different examples of the structure or area are needed.

C. Staining Thick Sections

Some difficulty is encountered when histochemical or cytochemical staining is attempted on electron microscopically prepared material to be observed with the light microscope. The factors that appear to be responsible for this difficulty are (1) the thinness of the preparation being examined with the light microscope, (2) the fixative used to preserve the material, and (3) the lower penetrability of the embedding media employed for electron microscopy.

As early as 1953 investigators began to examine and to modify if necessary the conventional cytochemical staining techniques applied to light microscope observations of electron microscope prepared material. Since there seems to be no comprehensive review in the literature concerning the various stains that can be used with success in such situations, a brief summary is given here of isolated reports on the particular dyes effective for different electron microscope embedding media (Table I). Most of these reports deal with particular tissues or cells that have been osmium-fixed (unless otherwise noted). Moreover, the perfected staining technique may be somewhat specific depending on the chemical substance or structure being studied. Thus each individual must decide on the stain suitable for the particular investigation, and to recheck and possibly to modify the technique if a different fixative and/or tissue (other than that listed) is being used. [For detailed information about the staining procedure and for characteristic appearance of the chemical substance or structure, the reader is referred to the literature.]

If sections thinner than those listed for a particular stain are required for light microscopy, the stain should not be passed over merely because it has been used only for thicker preparations (e.g., hematoxylin and eosin on sections 1–6μ). It may be possible to obtain adequate results with a thinner preparation by increasing the staining time and/or concentration of the specific dye.

As was mentioned, azure B is the preferred stain in our laboratory for light microscope preparations of material embedded for electron microscopy. Azure B produces excellent contrast in sections as thin as 0.1μ and can be used with ease and rapidity on all types of tissue, regardless of the fixative or embedding media employed (Table I). The following azure B staining procedure is a modification (A. E. Vatter, personal communication, 1963) of the thionine-azure-fuchsin technique of Tzitsikas *et al.* (1961) and is similar to the Dodge (1964) procedure for azure B staining of Dinoflagellates.

Azure B technique. (1) Dissolve 0.2 gm of azure B bromide (Matheson Coleman and Bell, Cincinnati, Ohio; Rutherford, N.J.) in 100 ml warm distilled water. Neutral phosphate or veronal buffer may be substituted for the distilled water. (2) After vigorous shaking, filter the solution and adjust the pH of the filtrate to 8.5 for Epon- or Araldite-embedded material or to 5.5 when Vestopal W- or methacrylate-embedded material is stained. (3) Cover the sections with a small volume of the staining solution and place the slide on a hot plate (60°–80°C) for a short time (2–20 seconds). Do not allow the stain to dry on the slide. (4) Rinse the slide free of the stain with distilled water, dry, and examine. (5) Repeat the staining procedure if sufficient contrast has not been produced for adequate examination with the light microscope. For extremely thin preparations, the procedure will probably need to be repeated 2–3 times. (6) Mount a coverslip with water or another suitable mounting medium over the sections before final observation.

Comment. Nucleoli—dark blue; chromatin—blue; cytoplasm—faint blue; vacuoles and carbohydrate-containing structures—clear. The staining solution may be stored at room temperature for an indefinite period, but the pH should be checked periodically. If a precipitate begins to form, the solution should be refiltered.

Another stain effective for Epon-, Araldite-, and Vestopal W-embedded material (Table I) is toluidine blue. The characteristic staining of cellular components is very similar to azure B, and little difference is found between the respective methods.

Toluidine blue technique. (1) Prepare a 0.25% solution of toluidine blue O (Fisher Scientific; Allied Chemical) in distilled water and allow the solution to stir overnight (do not filter). (2) Adjust the pH to approximately 10.0 with 0.1 N NaOH. (3) Add a few drops of the staining solution to the sections and place the slide on a hot plate (60°–80°C) for a short time (2–20 seconds). Do not let the stain dry on the slide. (4) Rinse the slide free of the stain with 95% alcohol and examine to determine if sufficient contrast has been obtained. The procedure may have to be repeated several times if extremely thin preparations are stained. (5) Mount the stained sections with a coverslip and with water or a permanent mounting medium before final observation.

Comments. Metachromasia is not demonstrable in electron microscope prepared material at this pH. Nuclei stain blue, with the nucleoli appearing somewhat darker than the chromatin; cytoplasm acquires a light blue color. Epon and Araldite have no tendency to pick up the stain, whereas Vestopal W will stain slightly in comparison to the tissue. A

TABLE I

SUMMARY OF STAINS EFFECTIVE ON VARIOUS EMBEDDING MATERIALS

Type of embedding material (reference)	Stains effective	Section thickness (μ)	Type of tissues or cells used
Methacrylate[a] (Houck and Dempsey, 1954)	Hematoxylin, PAS, Wilder's ammoniacal silver, Verhoeff's elastic	0.25–10	Various tissues from rabbit, pig, mouse, monkey, cat, and guinea pig
Methacrylate[a] (Moses, 1956)	Feulgen, PAS	1–2	Rat pancreas; grasshopper and crayfish testis
Methacrylate[a] (Weiss, 1957)	Harris' hematoxylin, PAS	2	Splenic sinuses in man and albino rat
Methacrylate (pre-fixed in formalin; post-fixed in OsO_4) (Churg et al., 1958)	Silver methenamine	0.5	Human kidney (pathological conditions)
Methacrylate[a] (Runge et al., 1958)	Gallocyanin-chromium-phloxine	0.5–1.5	Kidney, liver, brain, and pancreas of man and other animal species
Methacrylate[a] (Barton, 1959)	Polychrome methylene blue	0.1–2	—

Methacrylate[a] (Bencosme et al., 1959)	Toluidine blue	0.5–1	Rat kidney
Methacrylate (Farquhar et al., 1959)	Gallocyanin-chromium-phloxine, hematoxylin and eosin, fuchsin-alum-hematoxylin, Masson's trichrome, Mallory-Heidenhain azan, PAS	1–6	Renal tissue from patients with diabetic glomerulo-sclerosis
Methacrylate[a] (Jennings et al., 1959)	Fuchsin-alum-hematoxylin (new stain) Modified procedures: Mallory-Heidenhain azan, Masson's trichrome, colloidal iron, methyl violet Unmodified procedures: PAS, Feulgen, methenamine silver, von Kossa, gallocyanin-chromalum, Alcian blue, Gomori's aldehyde fuchsin, Weigert's resorcin fuchsin, Unna's orcein, Verhoeff's iodine-hematoxylin, iron-hematoxylin	1–6	Mainly renal tissues but some of the stains applicable to eyes, arteries, and developing teeth of rats

TABLE I (*Continued*)

Type of embedding material (reference)	Stains effective	Section thickness (μ)	Type of tissues or cells used
Methacrylate[a] (Moore *et al.*, 1960)	Crystal violet and/or basic fuchsin, PAS with or without the counterstains crystal violet, malachite green, or fast green	0.1–0.2	Rabbit tissues
Methacrylate (Suzuki and Sekiyama, 1961)	Silver methenamine	—	Rat kidney
Methacrylate (Peters, 1962)	Ehrlich's hematoxylin-triosin	1.5–2	—
Methacrylate (McGee-Russell and Smale, 1963)	Nile blue A	0.12–1	Krebs II ascites cells, cerebral ganglion of snail (*Helix pomatia*)
Methacrylate (formalin, Carnoy's or Bouin's fixatives, but not Zenker's or osmium) (Jones, 1957)	Periodic acid-methenamine	1–3	Renal tissue from patients with glomerulonephritis
Methacrylate[a] (de Harven, 1956)	PAS, Feulgen, hemalum-eosin, iron-hematoxylin, toluidine blue	1–4	Rat intestine

Method	Stain	Thickness	Tissue
Methacrylate (Richardson et al., 1960)	Mallory's azur II-methylene blue	0.25–1	Mouse brain
Methacrylate (Dodge, 1964)	Azure B, alkaline fast green	0.25–1	Dinoflagellates: *Prorocentrum micans, Peridinium trochoideum*
Methacrylate (osmium but not Orth's or formalin) (Movat, 1961)	Modified Gomori's PAS-methenamine	—	Kidney (pathological conditions); liver, small intestine of dog
Methacrylate (Welsh, 1962)	Slightly modified PAS	1	Human plasma cell
Methacrylate (Thoenes, 1960)	Giesma	—	—
Methacrylate (osmium prior to or after formalin) (Flax and Caulfield, 1963)	Giesma	0.5–2	Kidney, skin, lymph node, and thyroid tissues from various animals
Methacrylate[a] (Robinow, 1953)	Giesma	0.3–0.4	Spores of *Bacillus cereus* and *B. megaterium*
Methacrylate (Marinozzi, 1961)	Ammoniacal silver solution	(ultra thin)	Various animal tissues
Methacrylate[a] (Munger, 1961)	Ehrlich's hematoxylin and phloxine, PAS-hematoxylin, aldehyde fuchsin, trichrome	0.5–2	Pancreas and skin of man, liver and pancreas of rabbit

TABLE I (*Continued*)

Type of embedding material (reference)	Stains effective	Section thickness (μ)	Type of tissues or cells used
Methacrylate[a] (Pedersen, 1961)	PAS	—	*Planaria vitta*
Vestopal W (Tzitsikas et al., 1961)	Thionine-azure-fuchsin, thionine-azure counterstain for PAS	0.2–1	Liver, kidney, lung, spleen, skin, and nerve from several animal species
Vestopal W (Tzitsikas et al., 1962a)	Wilder's ammoniacal silver stain	0.1–2	Liver, kidney, spleen from pig, mouse, rat, dog
Vestopal W (Tzitsikas et al., 1962b)	Mallory's oil red O-thionine-azure	0.1–1	Adrenal cortex and liver of rat, liver and kidney of mouse and dog
Vestopal W (Schwalbach et al., 1963)	Mayer's acid hemalum, Hansen's iron trioxy-hematein, Heidenhain's iron-hematoxylin, Schneider's aceto-carmine, picrofuchsin, modified Schulemann and Wurmbach's Giesma	1–2	Tentacles of snail (*Helix pomatia*)

Vestopal W (McGee-Russell and Smale, 1963)	Nile blue A	0.12–1	Krebs II ascites cells, cerebral ganglion of snail (*Helix pomatia*)
Vestopal W (Gautier, 1960)	Chromotrope 2-R, toluidine blue, fast green, basic fuchsin, Sudan III, Lugol's solution, hemalum-eosin	—	Frog epidermis, cacao bean
Vestopal W (Thoenes, 1960)	Giesma	—	—
Vestopal W (Moe, 1962)	Cobalt sulfide, PAS, Feulgen, orcein, acid blue, Alcian blue, pyronin	1–1.5	Rat intestine
Araldite (osmium or Dalton's fixatives) (Richardson et al., 1960)	Mallory's azure II-methylene blue	0.25–1	Mouse brain
Araldite (Dalton's fixative) (Richardson et al., 1960)	Amethyst violet, crystal violet, new fuchsin, safranin O, thionine, PAS (counterstaining not possible)	0.25–1	Mouse brain

A. R. STEVENS

TABLE I (Continued)

Type of embedding material (reference)	Stains effective	Section thickness (μ)	Type of tissues or cells used
Araldite (Dodge, 1964)	Azure B, Feulgen	0.25–1	Dinoflagellates: *Prorocentrum micans, Peridinium trochoideum*
Araldite (McGee-Russell and Smale, 1963)	Nile blue A	0.12–1	Krebs II ascites cells, cerebral ganglion of snail (*Helix pomatia*)
Araldite (Lee and Hopper, 1965)	Modified Goodpasture's basic fuchsin combined with modified Stirling's crystal violet	0.3–0.6	Juxtaglomerular cells of rat kidney
Araldite (Zenker's and formalin are not usable) (Goldblatt and Trump, 1965)	del Rio Hortega's silver ammino-carbonate	0.5–1.5	Rat tissues
Epon 812 (Trump *et al.*, 1961)	Toluidine blue	0.5–2	Liver and kidney of rat
Epon 812 (Lafontaine and Chouinard, 1963)	Feulgen-methylene blue	0.25–1	*Vicia faba* root tips

Epon 812 (McGee-Russell and Smale, 1963)	Sudan black, Nile blue A	0.12–1	Krebs II ascites cells, cerebral ganglion of snail (*Helix pomatia*)
Epon 812 (Winkelstein *et al.*, 1963)	Basic fuchsin	1	Rat kidney
Epon 812 (Chandra and Skelton, 1964)	Toluidine blue with or without glove oil, basic fuchsin	1–2	Juxtoglomerular cells of rat kidney
Epon 812 (Dodge, 1964)	Feulgen	0.25–1	Dinoflagellates: *Prorocentrum micans, Peridinium trochoideum*
Epon 812 (pre-fixed in formalin, post-fixed in osmium) (Schultz, unpublished results, 1965)	Gomori's lead nitrate (for acid phosphatase), modified Vorbrodt's lead nitrate (for DNase)	0.25–0.5	Rat placenta
Epon 812 (Steven's, unpublished results, 1965; procedure given)	Toluidine blue	0.12–0.5	*Amoeba proteus, Stentor coerulis, Euplotes eurystomus, Tetrahymena pyriformis*

TABLE I (*Continued*)

Type of embedding material (reference)	Stains effective	Section thickness (μ)	Type of tissues or cells used
Epon 812 (A. E. Vatter, personal communication, 1963; procedure given)	Azure B	0.1–2	All types of tissue or cells, regardless of fixative or embedding media used
Epon 812 (Munger, 1961)	Ehrlich's hematoxylin and phloxine, PAS-hematoxylin, aldehyde fuchsin, trichrome	0.5–2	Pancreas and skin of man, liver and pancreas of rabbit
Epon 812 (Lee and Hopper, 1965)	Modified Goodpasture's basic fuchsin combined with modified Stirling's crystal violet	0.3–0.6	Juxtoglomerular granules of rat kidney
Epon 812 (Zenker's or formalin not usable) (Goldblatt and Trump, 1965)	del Rio Hortega's silver amminocarbonate	0.5–1.5	Rat tissues

a Methacrylate was removed prior to staining.

stock solution can be kept indefinitely if the pH is adjusted periodically.

The aniline dyes, azure B and toluidine blue, are excellent for staining thick sections being observed with the light microscope for the purpose of determining if the desired structure or region is being sectioned (Section II,B,2). Either of these stains will provide the necessary information with the minimum amount of time and work, thus allowing the operator to return almost immediately to the sectioning at hand. The main disadvantage of these stains is that they cannot be used prior to autoradiography since the aniline dyes tend to expose the emulsion. Either stain can of course be employed for light microscope autoradiography, if the staining procedures are performed after the autoradiograph has been developed; however, the stain selected for the final autoradiographic preparation will depend to a large extent on individual situations. Many stains other than azure B or toluidine blue can provide greater cytological specificity (Table I).

The discussion of the essential reasons for using thick sections of material prepared for high resolution autoradiography at the light microscope level may seem somewhat unnecessary at first glance. However, after the more tedious tasks of preparing ultrathin sections for electron microscope autoradiography have been performed, the value of the information acquired from the thick sections will become quite apparent. Moreover, the time spent in carrying out the operations on thick sections is slight in comparison to the time wasted if these operations are not performed prior to high resolution autoradiography.

III. Electron Microscope Techniques

Caro (1964) has emphasized three significant factors in obtaining adequate resolution for interpretation of electron microscope autoradiographs: (1) the use of an isotope that emits particles of low energy, thereby assuring a short range in the biological material, (2) the use of a highly sensitive emulsion with a small grain size (silver halide crystal), and (3) the application of a compact monolayer of silver halide crystals in close contact with the preparation. [For an excellent discussion of these and related factors pertaining to the problem of resolution, the reader is referred to Caro (1964).]

Since two of these important problems of autoradiography have been commercially alleviated for us, i.e., isotopically labeled compounds containing tritium are now available and nuclear emulsions with grain sizes 0.1μ or less in diameter can be purchased, we will be concerned in this section primarily with the application of emulsion. The importance of the emulsion thickness and to a lesser extent the thickness of the biological specimen has been discussed by Caro (1964), Moses (1964), and others. The application of the emulsion is the most critical step in electron microscope autoradiography; the sensitivity and resolution of the final autoradiographic preparation are directly dependent on the presence of a tightly packed layer of silver halide crystals.

Various methods have been devised for obtaining compact, uniform monolayers of emulsion over ultrathin preparations (Caro and van Tubergen, 1962; Hay and Revel, 1963; Moses, 1964). Although each of the published techniques seems to provide satisfactory results with adequate autoradiographic resolution, some of the procedures are quite laborious and time-consuming. For this reason we have devised a method whereby a compact monolayer of Ilford L-4 emulsion can be applied to grids in a rather short time, and the grids in turn can expose under storage conditions that are relatively simple. The method to be described is similar to that first proposed by Meek and Moses (1963); it has not been perfected for the new Kodak NTE emulsion, which has a smaller grain size (500 Å in diameter) [if NTE is the choice of emulsion, the reader is referred to M. M. Salpeter in this volume (Chapter 9) for the procedure used in applying the Kodak emulsion]. Before describing the method of application of the Ilford emulsion, a few words need to be said about the ultrathin preparations that will be used.

The thickness of the sections for electron microscope autoradiography is ordinarily the same as required for high resolution electron micrographs. The autoradiographic resolution will be enhanced with thinner preparations (500–800 Å), but the contrast of the final image and the sensitivity of the autoradiograph will be decreased. If one is willing to accept a somewhat decreased image and autoradiographic resolution in order to obtain better contrast and concomitantly a shorter exposure time, then sections of 800–1000 Å are quite permissible for electron microscope autoradiography.

Since the ultrathin sections are not observed with the electron microscope prior to application of the emulsion, it is imperative to obtain sections entirely free of chatter, contamination, and preferably knife marks. Furthermore, the sections should be carefully mounted to ensure flatness and the desired position on the grid. Electrolytic copper grids of the Athene type, regardless of the mesh size or diameter, are quite suitable for mounting the ultrathin sections; however, it is ordinarily mandatory to adapt the grids with a supporting film of plastic and carbon. Although the resolution of the final image is slightly lessened by the added thickness of the supporting film, the ancillary operations required for electron microscope autoradiography will be facilitated by the support provided by the film. The supporting film should be sufficiently thin to avoid significant interference with the resolution of the specimen, but thick enough to resist breakage during the autoradiographic procedures.

Early in our electron microscope autoradiographic observations, it was recognized that a 0.25% Parlodion-carbon coated film (ordinarily used in our electron microscope observations) was easily ruptured during the removal of the emulsion gelatin (Section III,E). For this reason preparations to be coated with emulsion are mounted on specimen grids (150-mesh) that have a somewhat thicker supporting film (stock solution of Parlodion, 0.5%). This increased thickness was found to reduce effectively (by approximately 95%) the number of grids discarded because of membrane damage. The procedure for applying the 0.5% film is the same as the microscope slide method given by Pease (1960).

After the grids have been coated with Parlodion (or another suitable plastic), a very thin layer of carbon is evaporated over the grid to further stabilize the plastic film. Before use, each grid should be checked to make sure the film is free of wrinkles and has not been ruptured in any area as a result of the carbon evaporation.

The number of sections placed on a grid is arbitrary. It will become apparent later in the discussion that the most suitable number of grids mounted with sections from a single experimental block should be 25 or some multiple of 25.

A. Application of Emulsion

The loop technique to be described requires several items that may not be found in conventional laboratories. The following list gives all the apparatus, glassware, and chemicals needed in the steps of the auto-radiographic procedure. In some instances the source of several items not available for purchase in the usual catalogs is listed.

Apparatus. (1) One wire loop (Techalloy A, ca. 0.6 mm in diameter, Techalloy Co., Inc., Rahns, Pennsylvania): 1.4-cm loop diameter; arm length ca. 2.4 cm; attached to a wooden probe ca. 10 cm in length and 7 mm in diameter (glass tubing of the same dimensions can be substituted for the wooden probe). This item can be made quite easily by the operator. (2) Plastic slide boxes—20-slide capacity with tightly fitting covers: inside length 9 cm; outside length 9.6 cm; inside width ca. 7.5 cm; outside width 8.2 cm; inside depth (without lid) 1.2 cm; outside depth (with lid) 3.2 cm. (3) Several wooden or plastic blocks (hereafter referred to as the filler blocks): 8.85 cm in length; 7 cm in width; 1.9 cm in depth; and drilled with 30 holes (arranged in rows of 5), 7 mm in diameter, 8 mm deep, and 7 mm apart. These can be made in any carpenter's shop. (4) Suitable number (ca. 20) of wooden or plastic blocks (hereafter referred to as storage blocks): 8.85 cm in length; 7 cm in width; 1.2 cm in depth, and drilled with 30 holes (arranged in rows of 5), 7 mm in diameter, that extend through depth of the block (i.e., are also 1.2 cm deep) and are 7 mm apart. These can also be made in a carpenter's shop. (5) Tape (masking and electrical); a pair of jeweler's forceps; a pair of curved forceps with blunt serrated tips wrapped with electrical tape; razor blades; kimwipes; and filter paper. (6) Timer without luminous dial; Kodak safelight (Wratten Series OA). (7) Two circular Technicon constant-temperature water baths (27 cm in diameter; 6 cm in depth). (8) Two ordinary clamps adapted with rubber jaws.

Glassware. (9) Glass rod (6 mm in diameter) cut into pieces, 1.9 cm in length. One end of each 1.9-cm piece should be rounded in an oxygen torch. These small pieces can be made by the operator and can be used over and over again if adequately cleaned after use. (10) A 5-or 10-ml graduated cylinder; one 500-ml beaker; two 10-ml beakers; adequate supply of watch glasses (20 mm in diameter; 8 mm deep; beveled surface (Arthur H. Thomas Co.); Petri dishes. (11) One porcelain spatula. (12) One glass stirring rod. (13) Two funnels (Büchner type with fritted disc, porosity M, 30–40-ml capacity).

Chemicals. (14) Ilford L-4 emulsion. (15) Drierite. (16) Distilled water (preferably in plastic squeeze bottle). (17) Alcohol (100%). (18) Kodak Microdol-X developer. (19) Sodium thiosulfate. (20) Acetic acid (0.5 N).

Figure 9 is pertinent to the following discussion. Prior to entering the darkroom, the small (1.9-cm long) pieces of glass rod, thoroughly cleaned and dried, are lifted from their storage container with the taped, curved forceps and placed one by one in the individual holes of the filler block (the rounded end of the rod on top). The number of pieces of glass rod inserted in the filler block corresponds to the number of grids that will be coated at one sitting. To decrease the risk of contaminating the grids, the glass rods should be kept covered at all times except when in use. Small pieces of gum scraped from the adhesive side of masking tape with a razor blade are affixed to the rods sitting in the filler block. A grid is taken from its container, checked to make sure that the side containing the sections is slightly convex, and laid gently on top of a glass rod. One corner of the grid should be tapped lightly with forceps in the vicinity of the piece of adhesive to attach the grid firmly. This operation is performed with each grid until all have been placed on individual glass rods.

In order to avoid confusion in the darkroom, some convenient method of identifying grids with sections from individual samples should be devised: e.g., either coat 25 (or some multiple of 25) grids from the same block since this number of grids can be kept together conveniently in one storage box (note: although each storage block can hold up to 30 glass rods, the last 5 holes will not be usable because they are filled with

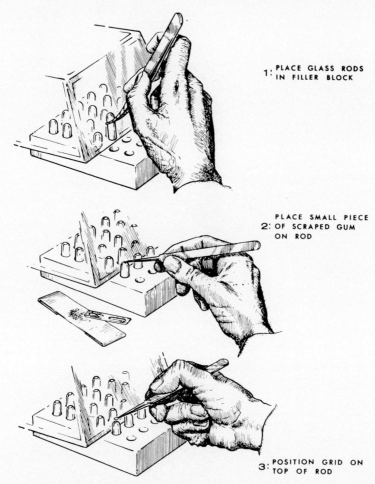

1: PLACE GLASS RODS IN FILLER BLOCK

2: PLACE SMALL PIECE OF SCRAPED GUM ON ROD

3: POSITION GRID ON TOP OF ROD

FIG. 9. Basic equipment and steps required for attaching grids to glass rods. See text for explanations.

Drierite), or leave a certain amount of space between different samples when placing the grids in the filler block and use identical spacing when the grids are placed in the storage block.

After each filler block contains the maximum number of grids, it should be kept covered to avoid contaminating the specimens (a large staining

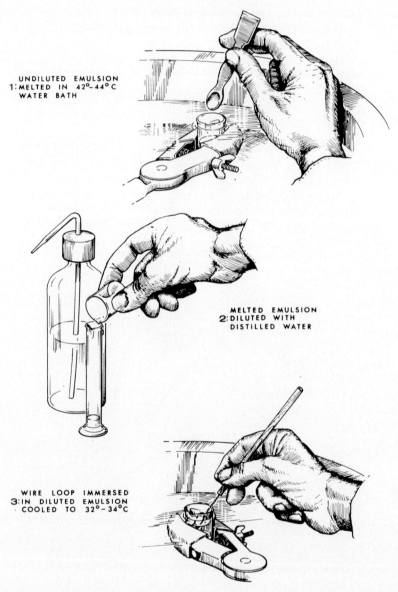

UNDILUTED EMULSION
1: MELTED IN 42°-44°C
WATER BATH

MELTED EMULSION
2: DILUTED WITH
DISTILLED WATER

WIRE LOOP IMMERSED
3: IN DILUTED EMULSION
COOLED TO 32°-34°C

FIG. 10. Loop technique for applying monolayer of Ilford L-4 emulsion to grids. See text for explanations.

dish inverted over the block suffices very well for this purpose). The filler block containing the grids is now ready to be taken into the darkroom.

When the grids are covered with emulsion, as opposed to when the

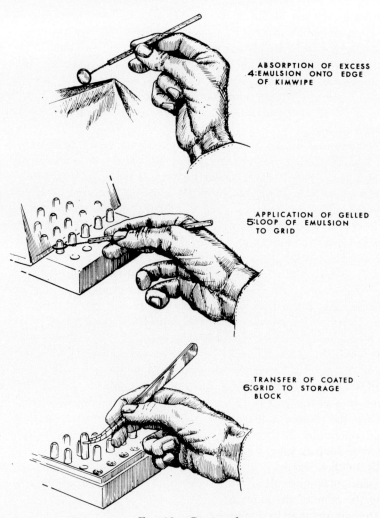

ABSORPTION OF EXCESS
4: EMULSION ONTO EDGE
OF KIMWIPE

APPLICATION OF GELLED
5: LOOP OF EMULSION
TO GRID

TRANSFER OF COATED
6: GRID TO STORAGE
BLOCK

FIG. 10. Continued.

grids are developed, fixed, etc., the following equipment should be located in the darkroom (taken from the detailed list of required equipment): (1) The wire loop. (2) The 5- or 10-ml graduate, marked in
advance with a black crayon pencil to indicate 2 ml and 4.5–5 ml. (3)
The two 10-ml beakers. (4) The glass stirring rod. (5) The 500-ml beaker
filled with icewater. (6) The two circular water baths (one should be
pre-set at 42°–44°C and the second at 32°–34°C). Each bath is filled to a
depth of 30 mm with water and allowed to reach the required temperature. The small indicator light on each bath should be tapped to avoid

accidental exposure of the emulsion. (7) The two rubber-jawed clamps. (8) The porcelain spatula. (9) A pair of jeweler's forceps and the curved forceps wrapped with electrical tape. (10) Electrical tape. (11) The necessary number of storage blocks, each placed in a plastic slide box. The bottom of each storage block is covered with masking tape and the last five holes in a storage block are filled with Drierite. (12) A pretested bottle of Ilford L-4 emulsion (Section III,C). (13) Distilled water. (14) Box of kimwipes.

All operations in the darkroom are performed under light from a 15-watt bulb filtered through a Kodak OA filter and at a working distance of 2.5–3 feet (Fig. 10 is pertinent to the following discussion).

After going into the darkroom, 2 gm Ilford L-4 emulsion (this amount is approximated and not weighed out in the darkroom) is taken from the bottle of emulsion and placed in one of the small beakers (10-ml capacity) with the porcelain spatula. The beaker containing the emulsion is locked in the rubber-jawed clamp and set in the water bath preheated to 42°C. After approximately 5 minutes, 2 ml melted emulsion is poured into the graduate, and distilled water is added to give a final volume of 4.5–5 ml. The diluted emulsion is thoroughly but gently mixed with a stirrring rod and subsequently poured into the second 10-ml beaker. To ensure adequate mixing, it is advisable to pour the emulsion from the beaker into the graduate and vice versa several times. The beaker (containing the diluted emulsion) is next held in the ice bath for several minutes to hasten the cooling process. The cooled emulsion is finally locked in a clamp and placed in the water bath set for 32°–34°C.

When the diluted emulsion has reached the desired temperature, the actual operation of applying the emulsion to the grids can commence. The wire loop is immersed in the emulsion, withdrawn, and held vertically to allow excess emulsion to be absorbed onto a piece of kimwipe. To assure maximum absorption, move the lower portion of the wire loop along the edge of the kimwipe.

After approximately 1 minute, interference colors will begin to form at the upper edge of the loop and move downward progressively as the loop of emulsion begins to gel. The upper half of the loop will exhibit a gold color and will be that portion of the emulsion corresponding to a compact, monolayer of silver halide crystals. After the loop of emulsion has gelled it is raised to a horizontal position, and the area of the emulsion exhibiting the gold interference color is brought down and over a grid (attached to the glass rod). If proper gelation has occurred, the loop of emulsion will automatically break when it has reached the midpoint of the rod.

The piece of rod holding the coated grid is then lifted from the filler block with the curved forceps, blown on gently to ensure that the emulsion adheres to all portions of the grid, and placed in one of the holes of the storage block. It is advisable to keep both the uncoated and coated grids situated in the filler and storage blocks, respectively, covered between operations.

As mentioned previously, the curved portion of the serrated forceps should be wrapped with electrical tape. This is done to prevent the forceps from slipping when they come in contact with a glass rod, and will readily allow a rod to be forced gently through the opening on the storage block until it adheres firmly to the taped bottom. The adhesive action of the tape (on the bottom of the storage block) keeps the rod pieces (and consequently the grids) from being dislodged in the event that the box is accidentally inverted or dropped.

Between successive operations the wire loop should be rinsed with distilled water, dried, and used to mix lightly the diluted emulsion before a fresh film is made. Subsequently the entire procedure is repeated until all the grids have been covered with emulsion and placed in the storage block. The plastic slide box holding the storage block is then carefully sealed with a double thickness of electrical tape, labeled for future reference (date, number of grids, etc.), and placed in a safe place to expose at room temperature.

B. Photographic Processing

The length of time necessary for adequate exposure of the ultrathin preparations is estimated from the light microscope autoradiographic response obtained from the 0.4μ sections (Section II,A). But in most cases this estimation will represent the minimum period required for a significant number of electrons to be registered in the emulsion; therefore, only one or two grids from a particular sample should be developed after the estimated exposure time to determine if the autoradiographic image is sufficiently intense for accurate interpretation. If the developed grids exhibit an autoradiographic response that is acceptable, the remaining grids can then be developed.

The final processing of the grids requires the following items (taken from the detailed list of required equipment): (1) The box of grids to be developed. (2) Six watch glasses, acid cleaned, thoroughly rinsed, and stored in a dust-free container until used. (3) Jeweler's forceps. (4) Timer. (5) Alcohol (100%). (6) Microdol-X developer (prepared according to directions on package) filtered through a Büchner funnel just before use. (7) Fresh 15% solution of sodium thiosulfate (pH 5.0–5.5) filtered through a Büchner funnel just before use. (8) Distilled water.

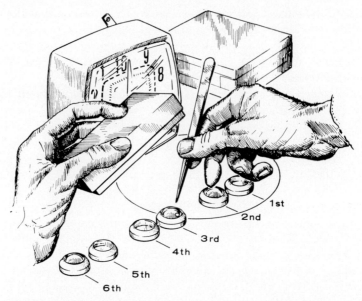

FIG. 11.   Photographic processing of grids. See text for explanations.

The series of watch glasses used for development and fixation are filled (from right to left, respectively) with the following solutions (kept at room temperature) (Fig. 11):

Watch glass	Solution
1	100% alcohol
2	Microdol-X; surface should be convex
3	Distilled water; surface should be convex
4	Sodium thiosulfate
5	Distilled water
6	Distilled water; surface should be convex

The entire setup is covered in a convenient manner at all times to avoid contamination of the solutions. The bottom of a plastic slide box inverted over the watch glasses can be used for this purpose and not only protects the solutions from dirt, etc., but also reduces the possibility that the grids may acquire an increased background during the photographic processing from overexposure to the safety light.

After entering the darkroom, the tape sealing the slide box is carefully removed and a grid, lifted from the top of its respective glass rod, is immersed (sections directed upward) in the 100% alcohol. If a second grid is being developed at the same time, it can be placed beside the first grid in the alcohol. During the alcohol treatment of the grids (3.5 minutes) the slide box should be retaped to avoid accidental exposure of the remaining grids.

The next step requires a certain amount of practice and careful attention because grids resting on the bottom of a watch glass are somewhat more difficult to pick up with forceps when the only illumination is that of a safelight. The Athene-type grids, however, make this step somewhat easier. Carefully lift the grids out of the alcohol, placing the forceps only in the region of the extreme outer edge, and set them on a piece of filter paper to dry. The dried grids are inverted and floated on the convex surface of the developer for 6 minutes (22°–24°C). After developing, the grids are quickly transferred to the watch glass containing the distilled water (left ca. 5 seconds) and subsequently immersed (sections directed upward) in the fixer solution. After 10 minutes the lights can be turned on and the grids lifted out of the fixer and immersed in the next watch glass (containing distilled water). The grids are rinsed for 5 minutes, withdrawn, flushed with a jet stream of distilled water from a plastic squeeze bottle (grids should be held vertically), dried, and inverted over the distilled water in the remaining watch glass for an additional 5 minutes. The rinsed grids are then removed, dried, and placed in a dust-free container until their final preparation (Section III,E).

C. Test for Background

It is necessary to test each new bottle of Ilford L-4 emulsion to determine if the background (latent image formation stemming from extraneous sources) is sufficiently low and thus will not interfere with interpretation of the autoradiographs. In comparison to the commercial emulsions normally used for light microscope autoradiography, and to Gevaert nuclear 307 used by some workers in electron microscope autoradiography, the background of a monolayer of Ilford L-4 is usually much lower.

Since the mean diameter of the silver halide crystals in Ilford L-4 is 0.12μ (thus below the resolving power of the light microscope), background checks must be made with the electron microscope. When a bottle is tested, the operations described for applying the emulsion (Section III,A) and processing the autoradiographs (Section III,B) are followed with three exceptions: (1) Parlodion-carbon coated grids, devoid of sections, are used, (2) the entire bottle of emulsion is melted at 42°–44°C and 2 ml poured into the graduate, and (3) the coated grid is developed almost immediately after the emulsion has been applied. We have routinely found that only 1–3 reduced grains will be encountered over a single grid square (150-mesh grid) in a monolayer of

Ilford L-4 emulsion, and that this number does not increase significantly when the preparations are exposed for periods up to 5 months in a light-tight box. Therefore eradication of the background on an autoradiographic preparation at the beginning of exposure with H_2O_2 (Caro, 1964) does not appear to be necessary if the usual precautions are taken when the emulsion is handled. That is, do not agitate the emulsion excessively or expose the emulsion to chemicals, dirty utensils, stray radioactivity, or prolonged periods in the darkroom in close proximity to the safelight.

A bottle of Ilford L-4 emulsion that exhibits low background in the initial test has been used up to periods of 6 months after the expiration date listed by the manufacturer; however, after several months of use a bottle should be retested to ensure that there has been no significant increase in background due to an accidental exposure.

D. Comments on Technique

The actual application of the emulsion described (Section III,A) has several features in common with previously published techniques (Meek and Moses, 1963; Caro and van Tubergen, 1962; Revel and Hay, 1961; Moses, 1964), and the monolayer of silver halide crystals (Fig. 12) that we obtain is quite comparable to those published by other workers. It is important, however, to point out several steps of this technique that should be performed with more than routine precaution and to discuss the reasons for using certain pieces of equipment and/or solutions.

The procedure for applying emulsion to a grid resting on a glass rod has been described by Moses (1964); however, after coating, the preparation was removed from the support and mounted on a glass slide to expose. It has been our experience that any operation involving direct manipulation of the grid in the darkroom should be avoided if at all possible. Hence, the grids are attached to small glass rods cut long enough for rapid application of the emulsion but short enough to hold the grids while exposing in the storage block. Since only the piece of rod (holding a grid) needs to be transferred to the storage container, the danger of accidentally dropping or contaminating a grid prior to exposure is reduced. Moreover, the coated preparations do not have to be re-mounted onto a glass slide, thus saving a great deal of time in the darkroom. By using this method of applying the emulsion and storing the grids, 40–50 grids can be coated in approximately 1 hour if all the steps of the technique are carefully followed.

It was stipulated several times in Section III,A (Application of Emulsion) that the loop of emulsion should be adequately gelled before it is applied to the grids. The importance of this step cannot be emphasized too strongly. If the emulsion is applied in the sol state, the silver halide crystals will become distributed over the sections in an uneven pattern and result in varying thicknesses in the monolayer of emulsion (Caro and van Tubergen, 1962). Furthermore, during the drying process there

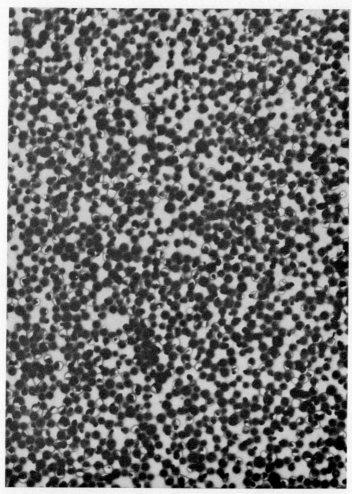

FIG. 12. Monolayer of Ilford L-4 emulsion obtained using technique described in this chapter. Some of the silver halide crystals have been melted from overexposure to electron beam.

may be a tendency for the grains to accumulate near the grid bars, leaving only thin patches of emulsion in the center of the grid squares (Moses, 1964).

Several ingenious procedures devised for ensuring uniformity and compactness in a monolayer of grains are (1) application of emulsion (in the sol state) to sections mounted on a Formvar-coated plastic slide (membrane method, see Budd, 1964), (2) use of a weak detergent in the emulsion and on the grids (Moses, 1964), (3) bromination of a film of silver evaporated over the grids (Silk *et al.*, 1961), and (4) application of a thin film of emulsion (in sol state) over a collodion-agar surface and subsequently placing this layer over the grids (agar method, see Caro and van Tubergen, 1962). However, in accordance with the observations of other investigators (van Tubergen, 1961; Caro and van Tubergen, 1962; M. M. Salpeter, this volume, Chapter 9), we have determined that no additional step is required in the loop technique (to ensure that the monolayer is uniform and compact) if the Ilford L-4 emulsion is adequately mixed and gelled before coating the preparations. Thus, since the validity of the interpretations made in high resolution autoradiography depends considerably on the presence of a uniform and tightly packed monolayer of emulsion, it is advisable for the operator periodically to examine a coated, undeveloped grid in the electron microscope to make absolutely certain that the loop of emulsion is being applied correctly.

Throughout the operations required for preparing electron microscope autoradiographs, the problem of contamination is of paramount importance. We have found that most of the dirt will be encountered during the photographic processing (if, of course, maximum precautions have been taken to protect the grids prior to this step). Two factors that reduce significantly the contamination during development are (1) using small volumes of solutions, and (2) filtering each solution just before use. Although there are several methods other than the technique given in this chapter for processing grids, which would make direct handling of the preparations unnecessary, they have not provided us with consistently satisfactory results. It is our opinion that the larger volumes of solutions required for developing the grids by other methods (i.e., developing grids attached to slides or to glass rods) enhance the problem of contamination. Initially a slight amount of difficulty may be experienced when the grids are transferred through the series of watch glasses; however, with practice 4–6 grids can be processed together quite easily at one time.

The first step in the photographic process (Section III,B), i.e., immersing the undeveloped grids in 100% alcohol, has not been described previously and therefore requires explanation. In our final autoradiographic preparations, it was occasionally noted that the number of grains present was less than expected. This apparent loss of grains occurred sometime during the photographic processing, since the decreased number was observed prior to removal of the gelatin (Section III,E). Subsequent to this observation it was realized that the initial contact of the grids with the aqueous developer tended to swell the emulsion, and this could result in a loss of grains (M. J. Moses, personal communication, 1965). By immersing the grids in 100% alcohol prior to development, the emulsion is hardened to an extent that appears to prevent this loss of grains in subsequent processing but does not prevent ultimate removal of the gelatin (Section III,E). In addition, pretreatment of the preparations with alcohol may affix the reduced grains more firmly to the sections and thus reduce the possibility that the grains are displaced or dislodged during gelatin removal. At this time such a contention is only hypothetical and not supported by critical experimental evidence.

A few words need to be said about the use of sodium thiosulfate as the clearing agent in the photographic processing. Although this fixer was not our original choice, it was introduced after we found that neither Kodak acid fixer (5–10-minute treatment) nor Kodak rapid fixer (5–10-minute treatment) was effectively removing all unexposed silver halide crystals. Since other workers had reported the use of sodium thiosulfate (Silk et al., 1961; Moses, 1964; Budd, 1964), this was tried on our preparations and discovered to be far superior to the other clearing agents. Moreover, fixation with sodium thiosulfate allows easier removal of the gelatin (Section III,E). This probably is related to the observation of Budd (1964), who discovered that the emulsion becomes more brittle and sometime fibrous in nature after clearing preparations in fixers containing hardener.

E. Final Preparation

The final preparation of the autoradiographs requires some effective means to improve contrast, since the gelatin of the emulsion will tend to obscure ultrastructural detail in the final image. A survey of the literature indicates several ways to approach this problem in electron microscope autoradiography: either the gelatin can be removed and, if necessary, the preparation stained in some conventional manner, or the gelation can be left intact and a strong staining procedure carried out on the preparation before or after application of the emulsion.

Regardless of the manner in which the contrast is improved, it will be obvious that the particular method employed depends in large part on the experimental material, the embedding medium, and the thickness of the sections.

Hampton and Quastler (1961) have obtained adequate contrast in unstained preparations by digesting the gelatin with acidic solutions of

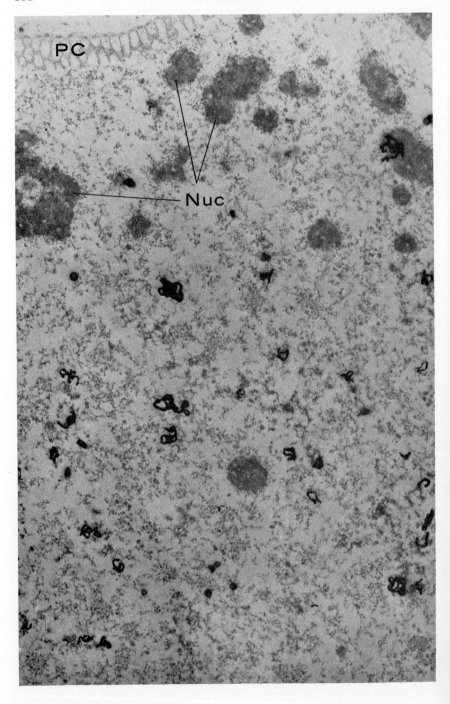

pepsin, but this method offers the distinct disadvantages of contaminating the specimen and causing partial loss of grains. In addition, proteolytic digestion of the gelatin makes accurate interpretation of the autoradiographs more difficult since fragments of undigested gelatin, of the same size as reduced silver grains, are occasionally present on the preparations. Silk *et al.* (1961), on the other hand, found that the gelatin could be effectively cleared by immersing the autoradiographs in a 37°C water bath for 16 hours. Although the preparations were unstained, adequate contrast was present for observation with the electron microscope and probably was partially a result of embedding the material in methacrylate. However, Silk *et al.* did not coat their ultrathin preparation with a commercial emulsion, and the removal of the gelatin of Ilford L-4 emulsion by prolonged exposure to water remains to be tested. More recently Hay and Revel (1963) have produced excellent autoradiographs, in which the contrast of the final image was greatly enhanced, by staining and removing the gelatin in one operation, using the Karnovsky (1961) lead staining procedure. According to Hay and Revel, their treatment "removes most of the visible gelatin and makes the specimen more transparent."

It is Caro's (1964) opinion, however, that all the procedures for enhancing contrast by gelatin removal have the distinct disadvantage of causing displacement or partial loss of silver grains. For this reason Caro has introduced contrast through the gelatin (in methacrylate-embedded bacterial preparations), using a strong staining solution of uranyl acetate prior to observation of the autoradiographs. An alternative to Caro's method is to stain the preparation before coating with emulsion (Moses, 1964; M. M. Salpeter, this volume, Chapter 9) (either during the dehydration of the specimen or just after sectioning), which apparently does not increase the background of the emulsion.

The main disadvantage of leaving the gelatin intact appears to be the danger of damaging the autoradiograph when observed in the electron microscope. The preparation will be relatively thick and thus easily ruptured by the electron beam; however, once the preparation is in the electron microscope, the gelatin can be successfully removed by sublimation if the specimen is exposed gradually to an increasingly intense electron beam (Moses, 1964). Although this method offers the advantage of being

---

Fig. 13. Electron microscope autoradiograph of *A. proteus* nucleus. Label, thymidine-H³; stain, uranyl acetate and lead citrate; embedding material, Epon 812; emulsion exposure time, ca. 3 months. The incorporated label is localized over the central regions of the nucleus (chromatin area) with no significant label over the nucleoli. Nuc = nucleoli; PC = pure complex. ca. × 12,500.

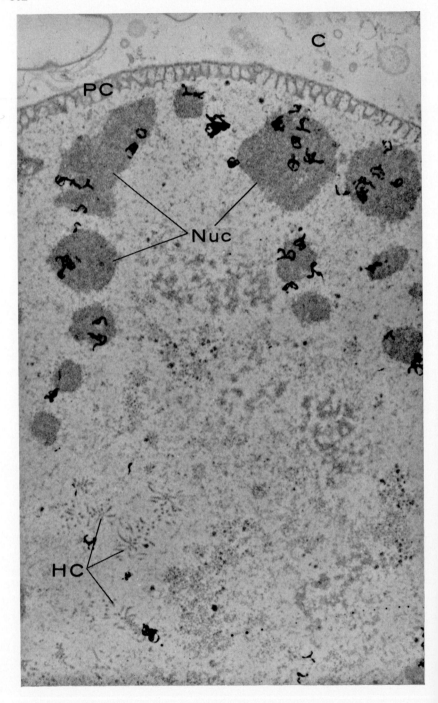

able to observe the area (and thus any artifacts produced) during gelatin removal, the damage that may occur to the specimen, the risk of melting and subliming the silver grains, and the necessity of staining the preparation prior to sublimation make the value of this method for improving contrast somewhat equivocal.

Early in our autoradiographic investigations, one or the other of the above mentioned procedures was attempted to improve contrast. It was found that removal of the gelatin by alkaline hydrolysis was extremely formidable because, for optimum results, a fresh solution of sodium hydroxide (made up in carbonate-free water) was required, and more than routine precaution was necessary to prevent deposits of sodium carbonate crystals on the preparations. Staining through the gelatin according to the technique of Caro (1964) likewise furnished unsatisfactory results on Epon-embedded, Ilford-coated sections. Moreover, since it was necessary to use extremely thin preparations (silver to gray sections) in our work, and the microscope could be operated only at low voltages, successful removal of the gelatin by sublimation was not effective. For these reasons we modified Granboulan's (1963) technique for removing the gelatin by acid hydrolysis and subsequently stained the preparations to enhance image contrast. The procedure has provided excellent results, been consistently successful, and thus far caused no detectable displacement or loss of the reduced grains.

Grids that have been developed, fixed, and rinsed are inverted and floated on a convex surface of distilled water (preheated to 37°C) and left for 30 minutes. This treatment causes the gelatin to swell and allows it to be readily removed when the autoradiographic preparation is quickly transferred to the surface of 0.5 N acetic acid solution and left for 15 minutes at 37°C.

The distilled water and acetic acid solutions should be placed in small and thoroughly clean containers, e.g., watch glasses, and kept in a covered vessel throughout the operation in the 37°C oven. If the grids are rinsed immediately with a jet stream of distilled water upon removal from the acetic acid and subsequently floated on fresh distilled water (room temperature) for an additional 10 minutes, the preparations will

---

FIG. 14. Electron microscope autoradiograph of A. proteus nucleus. Label, uridine-H³ and cytidine-H³; stain, uranyl acetate and lead citrate; embedding material, Epon 812; emulsion exposure time, ca. 5 months. Incorporated label is localized over the nucleoli; there appears to be some localization of the label over and around the nuclear helices. There is no significant label demonstrable in other areas of the nucleus. Nuc = nucleoli; PC = pore complex; C = cytoplasm; HC = helical clusters. ca. × 12,000.

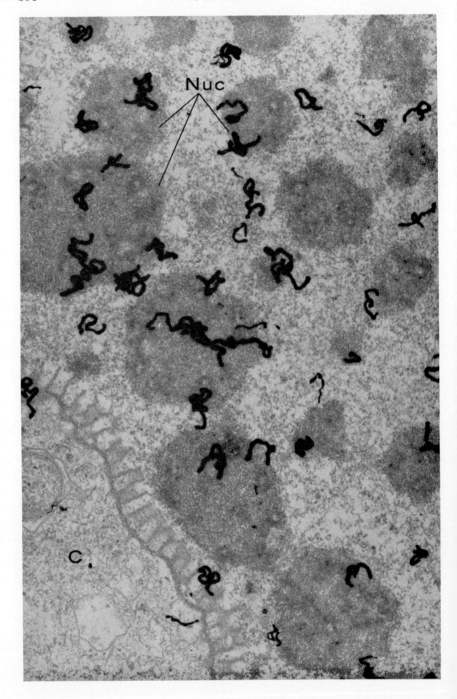

be extremely clean and easily stained with uranyl acetate and lead citrate.

Although some workers have achieved good image contrast in auto-radiographic preparations by employing only uranyl acetate after the gelatin was cleared (Moses, 1964), we have not found this to be the case. But even though uranyl acetate by itself produces only slight contrast in our sections, uranyl acetate staining appears to be necessary for optimum lead citrate staining. This could be a result of the uranyl acetate altering the binding characteristics of macromolecules, thus making them more accessible to the lead citrate.

The staining procedures are performed in a covered Petri dish with a smooth, shallow layer of dental wax on the bottom. The side of the grid holding the sections is first wetted with a small drop of distilled water and then inverted over a drop of filtered uranyl acetate (7.5% aqueous solution) for 20 minutes at 45°C. The wetting procedure of the grid reduces the contamination that results when surface-to-surface contact is made between the grid and the staining solution. Next the grid is held vertically and the uranyl acetate solution is flushed off with a jet stream of distilled water. After drying on a piece of clean filter paper, the grid is again wetted with distilled water and inverted on a drop of filtered lead citrate for varying lengths of time (1–15 minutes). The lead citrate staining solution is prepared according to the technique of Reynolds (1963) and used undiluted for optimum results. Occasionally, we have encountered dense lead deposits over membranous structures in the ultrathin preparations; however, this problem has been significantly reduced by (1) thoroughly rinsing the stained grid with distilled water, omitting the 0.02 N NaOH rinse suggested by Reynolds (1963), (2) filtering a small amount of the lead citrate solution through a Büchner-type funnel prior to each time it is used, and (3) using a lead citrate solution for only 3–4 weeks after it has been prepared. Recently, we have had success with the commercially available lead citrate stain (K and K Laboratories, Plainview, N.J.) used at a concentration of 0.20% for short staining periods (10–60 seconds) (Venable and Coggeshall, 1965). This stain offers the advantages of easier preparation, less contamination, and shorter staining times to achieve adequate contrast in the autoradiographic preparations. It seems to be more effective on

---

Fig. 15. Electron microscope autoradiograph of *A. proteus* nucleus. Label, uridine-H³ and cytidine-H³; stain, uranyl acetate and lead citrate; embedding material, Epon 812; emulsion exposure time, 2.5 months. The incorporated label is localized over and around the nucleoli. See text for explanations. Nuc = nucleoli; C = cytoplasm. ca. × 18,000.

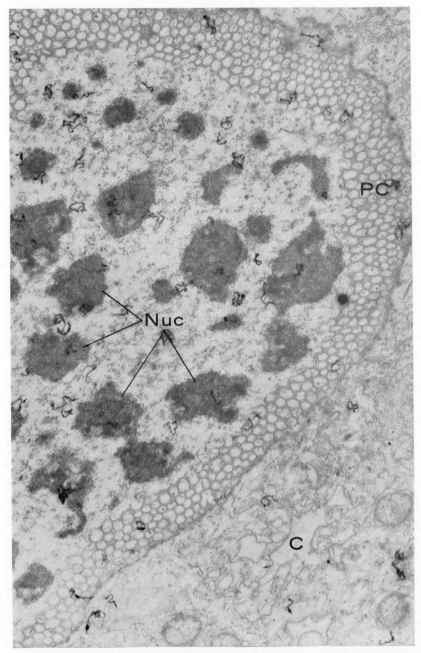

Fig. 16. Electron microscope autoradiograph of *A. proteus* nucleus and cytoplasm. Label, H³-amino acids; stain, uranyl acetate and lead citrate; embedding material,

autoradiographic preparations of Epon-embedded material than on Araldite-prepared material.

F. Examples of Technique

The techniques described in this chapter for obtaining high resolution autoradiographs have been applied mainly in our laboratory to investigations using the protozoan *Amoeba proteus*. The series of electron microscope autoradiographs shown in Figs. 13–16 illustrate the different types of experiment performed and the preliminary conclusions made from the interpretation of the autoradiographic image. The autoradiographic resolution obtained from using tritiated precursors and a monolayer of Ilford L-4 emulsion is of the order of 0.15μ.

Figure 13 shows an autoradiograph of an ultrathin section taken from a sample of amoebae labeled with thymidine-H^3. It is difficult to obtain an intense autoradiograph in amoebae when one is attempting to label the DNA, since this organism exhibits a DNA : nuclear volume ratio that is quite low in comparision to other organisms. For this reason a suitable ultrathin autoradiographic preparation was acquired only after an exposure time of 2–3 months. The labeled compound was incorporated over the more central regions of the nucleus, which correspond to the Feulgen-positive areas of the amoeba nucleus. Little or no label was associated with the nucleoli, which are more peripherally located.

In the next series of experiments, amoebae were labeled with cytidine-H^3 and uridine-H^3 for short periods to determine the primary sites of RNA synthesis and/or accumulation. The results of the first experiment, as shown in Fig. 14, were obtained after the ultrathin preparations had exposed for approximately 5 months and demonstrated good localization of the incorporated label over the nucleoli. We were also interested in this experiment in determining the macromolecular composition of the nuclear helices, structures first described by Pappas (1956). Although the results are only preliminary, there appears to be a slight localization of the tritiated precursors over the helices, suggesting that the amoeba helices may contain RNA.

Since the ultrathin preparations of amoebae labeled with the RNA precursors required such long exposure times (ca. 5 months) to produce suitable autoradiographic responses, the experimental conditions were

---

Araldite; emulsion exposure time, 1 month. The incorporated label is localized somewhat over the nucleoli; labeled material is present within the pore complex and in the cytoplasm. Nuc = nucleoli; PC = pore complex; C = cytoplasm. The silver grains are partially underdeveloped. ca. × 14,000.

modified in an attempt to obtain better incorporation and concomitantly shorter exposure times. In the first experiment, *Tetrahymena* (food for amoebae) were labeled for 1.5 hours with uridine-H^3 and cytidine-H^3 (low specific activities) prior to addition to the amoebae culture. The amoebae were allowed to feed on the radioactive food for approximately 3.5 hours before subsequent preparation. In the second experiment, *Tetrahymena* were labeled more intensely with uridine (high specific activity) that carried the tritium in the 5'-position. The use of uridine labeled in this position guarantees that, even up to a labeling time of 24 hours in *Tetrahymena*, the incorporated precursor is 99% digestible with RNase. The labeled food was then added along with the nucleoside cytidine-H^3 to the amoebae culture. After approximately 3 hours, the amoebae were prepared for electron microscopy.

The results of the second experiment, shown in Fig. 15, reflect the fact that, after an exposure time of only 10 weeks, an autoradiograph was acquired that was equivalent to if not greater than that obtained in the first experiment. Furthermore, the results of the two experiments using tritiated RNA precursors were in agreement: the incorporated label was localized over the nucleoli.

Figure 16 shows an electron microscope autoradiograph of an amoeba heavily labeled with H^3-amino acids. After the addition of labeled *Tetrahymena*, the amoebae were subjected to regular cytoplasmic amputations, alternated with periods of growth on nonradioactive nutrient. Since the amputations prevent cell division, the net effect of the procedure was to replace the original radioactive cytoplasm with nonradioactive cytoplasm without direct disturbance to the nucleus and its radioactive proteins. The high resolution autoradiograph (Fig. 16) is an example of an amoeba that had undergone three cytoplasmic amputations before preparation for electron microscopy, and demonstrates the radioactive protein in the nucleus with some apparent localization over the nucleoli.

The above descriptions illustrate the types of investigation that can be made by using tracer experiments in conjunction with electron microscopy; the details of these and related experiments will be published at a later date.

REFERENCES

Barton, A. A. (1959). *Stain Technol.* 34, 348.
Bencosme, S. A., Stone, R. S., Latta, H., and Madden, S. C. (1959). *J. Biophys. Biochem. Cytol.* 5, 508.
Budd, G. C. (1964). *Stain Technol.* 39, 295.
Caro, L. G. (1962). *J. Cell Biol.* 15, 189.
Caro, L. G. (1964). In "Methods in Cell Physiology" (D. M. Prescott, ed.), Vol. I, pp. 327-363. Academic Press, New York.

Caro, L. G., and van Tubergen, R. P. (1962). *J. Cell Biol.* **15**, 173.
Chandra, S., and Skelton, F. R. (1964). *Stain Technol.* **39**, 107.
Churg, J., Mautner, W., and Grishman, E. (1958). *J. Biophys. Biochem. Cytol.* **4**, 841.
de Harven, E. (1956). *Compt. Rend. Soc. Biol.* **150**, 63.
Dodge, J. D. (1964). *Stain Technol.* **39**, 381.
Farquhar, M. G., Hopper, J., and Moon, H. D. (1959). *Am. J. Pathol.* **35**, 721.
Flax, M. H., and Caulfield, J. B. (1963). *Arch. Pathol.* **74**, 387.
Gautier, A. (1960). *Experientia* **16**, 124.
Glauert, A. M., and Glauert, R. H. (1958). *J. Biophys. Biochem. Cytol.* **4**, 191.
Goldblatt, P. J., and Trump, B. F. (1965). *Stain Technol.* **40**, 105.
Goldstein, L., and Plaut, W. (1955). *Proc. Natl. Acad. Sci. U.S.* **41**, 874.
Granboulan, P. (1963). *J. Roy. Microscop. Soc.* **81**, 165.
Hampton, J. H., and Quastler, H. (1961). *J. Biophys. Biochem. Cytol.* **10**, 140.
Hay, E. D., and Revel, J. P. (1963). *Develop. Biol.* **7**, 152.
Houck, C. E., and Dempsey, E. W. (1954). *Stain Technol.* **29**, 207.
Howard, A., and Pelc, S. R. (1953). *Heredity* Suppl. **6**, 261.
Jennings, B. M., Farquhar, M. G., and Moon, H. D. (1959). *Am. J. Pathol.* **35**, 991.
Jones, D. B. (1957). *Am. J. Pathol.* **33**, 313.
Karnovsky, M. J. (1961). *J. Biophys. Biochem. Cytol.* **11**, 729.
Lafontaine, J. G., and Chouinard, L. A. (1963). *J. Cell Biol.* **17**, 167.
Lee, J. C., and Hopper, J. (1965). *Stain Technol.* **40**, 37.
Liquier-Milward, J. (1956). *Nature* **177**, 619.
Luft, J. H. (1961). *J. Biophys. Biochem. Cytol.* **9**, 409.
McGee-Russell, S. M., and Smale, N. B. (1963). *Quart. J. Microscop. Sci.* **104**, 109.
Marinozzi, V. (1961). *J. Biophys. Biochem. Cytol.* **9**, 121.
Meek, G. A., and Moses, M. J. (1963). *J. Roy. Microscop. Soc.* **81**, 187.
Moe, H. (1962). *Acta Anat.* **49**, 189.
Moore, R. D., Mumaw, V., and Schoenberg, M. D. (1960). *J. Ultrastruct. Res.* **4**, 113.
Moses, M. J. (1956). *J. Biophys. Biochem. Cytol.* **2**, Suppl., 397.
Moses, M. J. (1964). *J. Histochem. Cytochem.* **12**, 115.
Movat, H. Z. (1961). *Am. J. Clin. Pathol.* **35**, 528.
Munger, B. L. (1961). *J. Biophys. Biochem. Cytol.* **11**, 502.
O'Brien, R. T., and George, L. A., II (1959). *Nature* **183**, 1461.
Palade, G. E. (1952). *J. Exptl. Med.* **95**, 285.
Pappas, G. D. (1956). *J. Biophys. Biochem. Cytol.* **2**, 221.
Pease, D. C. (1960). "Histological Techniques for Electron Microscopy," 1st ed., pp. 150-155. Academic Press, New York (2nd ed., 1964).
Pedersen, K. J. (1961). *Z. Zellforsch. Mikroskop. Anat.* **53**, 569.
Peters, H. (1962). *Stain Technol.* **37**, 115.
Prescott, D. M. (1964). *In* "Methods in Cell Physiology" (D. M. Prescott, ed.), Vol. I, pp. 365-370. Academic Press, New York.
Prescott, D. M., and Kimball, R. F. (1961). *Proc. Natl. Acad. Sci. U.S.* **47**, 686.
Revel, J. P., and Hay, E. D. (1961). *Exptl. Cell Res.* **25**, 474.
Reynolds, E. S. (1963). *J. Cell Biol.* **17**, 208.
Richardson, K. C., Jarett, L., and Finke, E. H. (1960). *Stain Technol.* **35**, 313.
Robinow, C. F. (1953). *J. Bacteriol.* **66**, 300.
Runge, J., Vernier, R. L., and Hartmann, J. F. (1958). *J. Biophys. Biochem. Cytol.* **4**, 327.
Ryter, A., and Kellenberger, E. (1958). *J. Ultrastruct. Res.* **2**, 200.

Schwalbach, G., Lickfeld, K. G., and Hoffmeister, H. (1963). *Stain Technol.* **38**, 15.

Silk, M. H., Hawtrey, A. O., Spence, I. M., and Gear, J. H. S. (1961). *J. Biophys. Biochem. Cytol.* **10**, 577.

Stevens, A. R. (1963). *J. Cell Biol.* **19**, 67A.

Stone, G. E., and Miller, O. L. (1965). *J. Exptl. Zool.* **159**, 33.

Suzuki, T., and Sekiyama, S. (1961). *J. Electronmicroscopy (Tokyo)* **10**, 36.

Taylor, J. H., Woods, P. S., and Hughes, W. L. (1957). *Proc. Natl. Acad. Sci. U.S.* **43**, 122.

Thoenes, W. Z. (1960). *Z. Wiss. Mikroskopie* **64**, 406.

Trump, B. F., Smuckler, E. A., and Benditt, E. P. (1961). *J. Ultrastruct. Res.* **5**, 343.

Tzitsikas, H., Rdzok, E. J., and Vatter, A. E. (1961). *Stain Technol.* **36**, 355.

Tzitsikas, H., Rdzok, E. J., and Vatter, A. E. (1962a). *Stain Technol.* **37**, 293.

Tzitsikas, H., Rdzok, E. J., and Vatter, A. E. (1962b). *Stain Technol.* **37**, 299.

van Tubergen, R. P. (1961). *J. Biophys. Biochem. Cytol.* **9**, 219.

Venable, J. H., and Coggeshall, R. (1965). *J. Cell Biol.* **25**, 407.

Weiss, L. (1957). *J. Biophys. Biochem. Cytol.* **3**, 599.

Welsh, R. A. (1962). *Am. J. Pathol.* **40**, 285.

Winkelstein, J., Menefee, M. G., and Bell, A. (1963). *Stain Technol.* **38**, 202.

Chapter 11

Methods for Handling Small Numbers of Cells for Electron Microscopy

CHARLES J. FLICKINGER[1]

Department of Anatomy, University of Colorado Medical Center, Denver, Colorado

I. Introduction

Most investigations with the electron microscope have dealt with tissues or suspensions of large numbers of free cells or cell organelles. If free cells or parts thereof are available in sufficient quantity, they may be centrifuged into a pellet and subsequently processed like a piece of tissue. When, however, experiments have to deal with visually selected and/or manually manipulated cells, the electron microscopist is confronted with the task of keeping track of a microscopic amount of material through the prolonged processes of selection, fixation, dehydration, embedding, and sectioning.

This chapter will present several methods for handling a limited amount of material for electron microscopy. The techniques of preliminary embedding in agar, of microcentrifugation, and of embedding in a capillary tube are most useful for free cells such as protozoa; the various

[1] The author's work reported in this article was supported by NSF Grant GB-1635 and NIH Postdoctoral Fellowship 1-F2-GM-28, 214-01.

modifications of the flat, *in situ* embedding technique have been most widely used with cultured cells, grown and embedded on a flat surface.

II. Agar Embedding

With the technique of agar embedding, previously fixed material is embedded in a small cube of agar, which is easily visible and may then be handled like a small piece of tissue throughout subsequent dehydration and embedding. Although use has been made of agar embedding for aid in orientation of specimens for both light and electron microscopy (see for example Samuel, 1944; Richardson, 1958), the method was applied to the handling of small numbers of organisms by Kellenberger *et al.* (1958). Fixed bacteria were suspended in a small amount of agar which was allowed to solidify, cut into cubes, dehydrated, and embedded in Vestopal. Caro has reported the use of a similar technique to facilitate the autoradiography of bacteria by keeping organisms uniformly distributed (Caro *et al.*, 1958; Caro, 1961; Caro and Forro, 1961). Kimball and Perdue (1962) handled small numbers of paramecia for light microscope autoradiographic studies by placing them in a depression in a small agar block and covering them with an agar cap. Haller *et al.* (1961), working with the same organism, added cells to a drop of molten buffered agar on a glass slide. After solidification of the agar, the portion containing the cells was cut out, dehydrated, and embedded in Vestopal. Kimball and Perdue subsequently refined their technique for both light and electron microscopy (S. Perdue, personal communication, 1965; Stone and Cameron, 1964). Cells are pipetted onto the surface of agar in a glass capillary tube and covered with additional agar. The cylinder of agar containing the cells is pushed from the capillary and the portion containing the cells is dehydrated and embedded. The author has used a modification of this latter technique for handling 100–300 *Tetrahymena*, selected at similar stages of division, in studies of ultrastructure at various points in the cell life cycle. This procedure will now be described in detail. The technique is summarized in Fig. 1, and examples of the results are illustrated in Figs. 2 and 3.

The selected cells are fixed in suspension by pipetting them into approximately 1 ml of the usual osmium or aldehyde fixative contained in a 10-ml capillary-bottom centrifuge tube (centrifuge tube, reduced capillary tip, graduated, 10 ml, Hopkins Vaccine, Kimax, Kimble, No. 45225. Obtainable from Matheson Scientific, Inc., Kansas City, Missouri). If buffered osmium is used, it is necessary to filter or centrifuge the fixative immediately before use, and in addition better results may be obtained

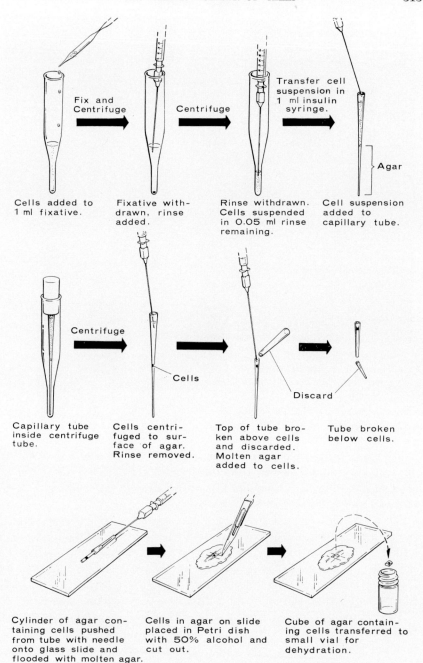

FIG. 1. Summary of the technique of agar embedding.

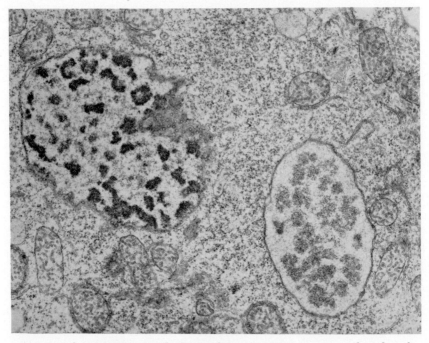

FIGS. 2 and 3. Several hundred *Tetrahymena pyriformis* were selected with a braking pipette at similar stages of cytokinesis, incubated, fixed in 1% OsO_4, and embedded as described in the text. Stained with uranyl acetate and lead citrate.

FIG. 2. Micronucleus, macronucleus, and several mitochondria at 80% completion of the cell cycle. The section has passed near the periphery of the macronucleus. × 12,000.

if fixation is carried out at room temperature to minimize the possibility of crystal formation. These precautions are useful because any debris or crystals present will subsequently be concentrated and embedded with the cells and will cause difficulty in sectioning.

After fixation for the desired length of time, the cells are concentrated by centrifugation, the fixative is withdrawn, and 1 ml of buffer rinse is added. A 1–2-ml syringe and 22-gauge 3-inch lumbar puncture needle have been found to be more satisfactory than a glass micropipette for changing fluids in order to eliminate the danger of breakage of a glass pipette and contamination of the cells with small bits of glass.

The cells are collected from the rinse fluid by centrifugation and all but 0.05 ml of rinse is withdrawn. The cells are suspended in the remaining rinse and transferred with a 1-ml insulin syringe and lumbar puncture needle to a previously prepared capillary tube.

Capillary tubes are prepared by breaking disposable glass Pasteur

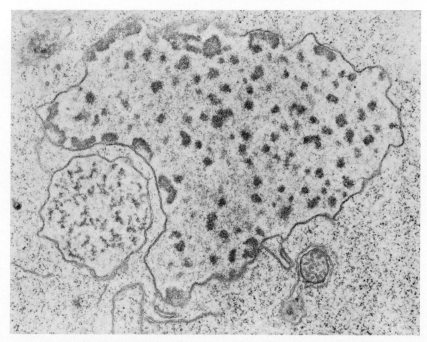

FIG. 3. Nuclei 5–10 minutes after cytokinesis. The micronucleus lies in a depression in the macronucleus. × 7000. (From Flickinger, 1965; reprinted by permission of the Rockefeller Univ. Press, from *J. Cell Biol.*)

pipettes 10–11 cm from the tip. The resulting capillary is cleaned and filled to 3–4 cm from the bottom with freshly prepared molten 2% bacteriological grade agar-agar. When the agar has solidified, the tips are heat-sealed in an alcohol lamp (Fig. 4).

The cell suspension is transferred via syringe and needle to the capillary tubes above the agar. The capillary tube is then placed inside a 10-ml capillary-bottom centrifuge tube and held in place with a disposable foam stopper or cotton wad. Centrifugation of the entire assembly will concentrate the cells on the surface of the agar. The excess rinse is withdrawn, the tube scored and broken above the cells, and molten 2% agar added directly to the surface of the cells with syringe and needle. The tube is scored below the cells and the agar cylinder containing the cells is pushed with a needle onto the surface of a glass slide, where it is surrounded by additional molten 2% agar. When the agar has solidified, the slide is immersed in 50% alcohol in a Petri dish to prevent drying.

Under a binocular dissecting microscope, a small cube of agar approximately 1 mm on a side containing the cells is cut from the slide and transferred to a small vial for dehydration.

The agar cube may then be handled like a block of tissue for dehydration and embedding. Kimball and Perdue have found methacrylate most satisfactory for embedding, as epoxy resins seemed not to penetrate the agar (S. Perdue, personal communication, 1965). The author, however, has had success with Epon embedding according to the method of Luft (1961), although attempts at embedding in Vestopal were unsuccessful due to poor penetration. As noted above, however, several workers (Kellenberger et al., 1958; Haller et al., 1961) have embedded agar in Vestopal successfully. It appears, therefore, that the individual investigator

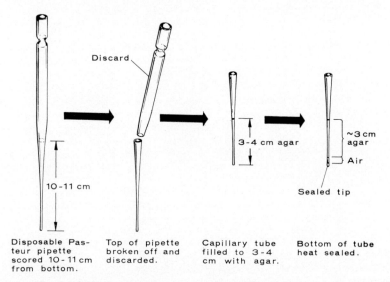

Fig. 4. The method of preparation of capillary tubes for use in the agar embedding technique.

will have to experiment to find that embedding medium which is most successful in his hands. If Epon is used, the material should remain in Epon at room temperature for at least 24 hours and at 35°C for 24 hours before subjecting it to higher temperatures to ensure adequate penetration of the agar with Epon. During the 24 hours at room temperature, the cube of agar, which initially floats on the Epon, will sink to the bottom of the capsule and may be positioned as desired.

The blocks are trimmed and sectioned in the usual manner. As noted above, unless care is taken to avoid the presence of foreign material—especially glass and crystals—about the cells, sectioning will be extremely difficult. In any event, sectioning of agar embedded material with glass knives may be anticipated to be more difficult than the sectioning of tissue specimens. Glass knives often seem to deteriorate rapidly and it is necessary to move frequently to an unused part of the knife. Preliminary observations on the use of diamond knives indicate that superior results may be obtained.

This technique presents a means for handling approximately fifty to several hundred free cells, e.g., protozoa synchronized for studies of the cell cycle by visual selection at a similar stage of division. The cells may be labeled with radioactive compounds and problems of localization of label which require the resolution of electron microscope autoradiography may be attacked with this method (see Chapter 10, this volume).

III. Flat Embedding

Cells grown on a flat surface or placed in a small depression slide may be fixed, dehydrated, and embedded *in situ*, either by flooding with embedding medium or by inversion of a capsule of medium over them. This technique has been applied primarily to tissue culture cells grown on slides or coverslips.

Borysko and Sapranauskas (1954) and Borysko (1956a) grew rat fibroblasts on glass coverslips mounted in a drilled depression in a stainless steel slide. The cells were fixed, dehydrated, and impregnated with methacrylate *in situ*. After polymerization, the coverslip was removed by chilling from the cells, now embedded in a flat piece of methacrylate. Borysko (1956b) showed that it was possible to obtain serial sections of an optically selected single cell or small group of cells with this technique. The flat embeddings were examined with a phase-contrast microscope and the desired cell(s) selected. With the aid of a micromanipulator and compound microscope, the flat embedding was trimmed for sectioning perpendicular to plane of the original coverslip. Phase-contrast micrographs were taken before and during sectioning to provide a record of those portions of the selected cell(s) included in the sections for electron microscopy.

Gay (1955), working with dipteran salivary gland chromosomes, described a method for embedding smears or squashed preparations for electron microscopy. The material on a coverslip was fixed, dehydrated, and embedded in methacrylate by inversion of a capsule of partially polymerized methacrylate over the material. After polymerization, the block, with the material at its surface, was examined with a compound microscope, and photographed. The desired area was selected, excess material was trimmed away, and the selected area was sectioned.

Nebel and Minick (1956) and Howatson and Almeida (1958) reported the application of a similar technique, i.e., utilizing inverted capsules of medium, to tissue culture cells grown on glass slides. It should be noted that the method of Borysko and Sapranauskas results in the sectioning of flat embeddings in a plane perpendicular to the surface on which they were grown, whereas the technique involving inversion of a capsule over smeared or cultured cells results in sectioning in the plane parallel to the surface on which the cells have been grown or smeared.

Many modifications of these basic techniques have appeared, concerned largely with one or more of the following problems: (1) decreasing the leakage of fluid embedding medium from a capsule inverted oved the cells, (2) increasing the accuracy of positioning desired cell(s) in the center of the inverted capsule, (3) facilitating the separation of cells from the surface they were grown on, after polymerization of the embedding medium.

Nebel and Minick (1956) described an apparatus to reduce leakage and aid in positioning, but only one capsule could be handled at a time. Nishiura and Rangan (1960) devised a stand in which several capsules could be clamped in place for inversion, and Micou *et al.* (1962) constructed a magnetic mounting holder for embedding in up to eight capsules simultaneously.

Latta (1959) substituted rings of solid methacrylate for capsules, and Rosen (1962) advocated the use of capsules with their ends clipped off to form gelatin cylinders. Both the methacrylate rings and the gelatin cylinders were placed around the desired cells and filled from their open ends, eliminating the need for inversion of a capsule of medium.

Bloom (1960) and Yusa and Apicella (1964) preferred to embed their material in a flat block similar to the method of Borysko, trim away excess material, and cement the remaining cube to a support block for sectioning. Bloom describes an apparatus to ensure that a previously located single cell is properly located in the final block.

More recently, Sparvoli *et al.* (1965) have described a method for selecting and orienting single cells in a small drop of embedding medium. Sheffield (1965) has reported an additional device for positioning and holding capsules over cells grown on a flat surface, and Sutton (1965) has given an account of a technique useful with cells grown in Leighton tubes.

Most of the investigators referred to above advocate chilling in ice water or with solid CO_2 as a means of separating the embedded cells from the glass surface on which they are grown. Bloom (1960) points out, however, that occasionally difficulty may be encountered in separation which may be aided by (1) culturing cells on mica, or (2) coating the glass with carbon before beginning the culture. Carbon coating appears to offer an additional advantage, as described by Robbins and Gonatas (1964). An optically selected cell may be circled with a slide marker and subsequently identified on the surface of the polymerized block by its location in the center of a circular break in the carbon. Alternatively, coating of the glass surface with silicone may facilitate separation of the cells (Micou *et al.*, 1962; Rosen, 1962). Heyner (1963) indicated that epoxy resins may adhere more firmly to glass than does methacrylate, and that separation is aided by prior coating of the glass with reconstituted rat tail collagen. These difficulties, of course, are not present with organisms that can be grown on agar, as reported by Koehler (1961).

In summary, it may be said that this method is most applicable to the embedding of optically selected single cells or groups of cells grown in tissue culture or smeared on a flat surface.

IV. Embedding in a Capillary Tube

Very small numbers of cells may be handled by the method described in the previous section if they grow on a flat surface, but it is less applicable to the manipulation of free cells. Deutsch and Dunn (1958), however, were able to examine single protozoa selected at a known stage of the mitotic cycle with the following technique.

The organism was fixed, dehydrated, and placed in methacrylate in a depression slide by means of a braking micropipette. It was then transferred to a small capillary tube with sealed tip, filled with methacrylate. Two micromanipulators, one for the braking micropipette and one for the capillary tube, were used to effect the transfer. The capillary was centrifuged to bring the cell to the tip, and the methacrylate was polymerized. The capillary tube was broken off and the resulting cylinder of methacrylate was cemented in a hole in a supporting rod for sectioning.

Baker and Pearson (1961) have described the application of a similar technique to the systematic sampling of cell suspensions, and point out its adaptation to handling small numbers of cells.

V. Use of a Microcentrifuge

The widely used technique of centrifugation of suspensions of many cells into a pellet has become applicable to smaller numbers of cells or isolated organelles with the availability of microcentrifuges with disposable polyethylene tubes (manufactured by Beckman Instruments, Inc.,

Spinco Division, Palo Alto, California; and Coleman Instruments, Inc., Maywood, Illinois).

As described by Malamed (1962, 1963), the material is added to the fixative in a microcentrifuge tube, and centrifuged into a small pellet. The fixative is withdrawn, the tube cut with a razor blade, and the pellet transferred to a vial for dehydration. Embedding is done either in another microcentrifuge tube or in a gelatin capsule. The method is simple and convenient. It appears, however, that smaller quantities of material can be handled with the techniques previously described.

VI. Conclusion

The techniques described have all been successfully used by various investigators for the handling of small numbers of cells for electron microscopy. In considering their use in a given instance, it should be noted that one is not confronted with true alternatives, for the methods are optimally effective in attacking different problems. If one wishes to examine optically selected single cells or small groups of cells that can be grown on a flat surface, the flat embedding technique is the method of choice. Single free cells, such as protozoa, may be embedded in a capillary tube. If more material is available, the use of a microcentrifuge is quick and efficient. Limitation of material to several hundred free cells, coupled with the desire to sample as many of these cells as possible, points to the use of the technique of preliminary embedding in agar.

REFERENCES

Baker, R. F., and Pearson, H. E. (1961). *J. Biophys. Biochem. Cytol.* 9, 217.
Bloom, W. (1960). *J. Biophys. Biochem. Cytol.* 7, 191.
Borysko, E. (1956a). *J. Biophys. Biochem. Cytol.* 2, Suppl., 3.
Borysko, E. (1956b). *J. Biophys. Biochem. Cytol.* 2, Suppl., 15.
Borysko, E., and Sapranauskas, P. (1954). *Bull. Johns Hopkins Hosp.* 95, 68.
Caro, L. G. (1961). *J. Biophys. Biochem. Cytol,* 9, 539.
Caro, L. G., and Forro, F. (1961). *J. Biophys. Biochem. Cytol.* 9, 555.
Caro, L. G., van Tubergen, R. P., and Forro, F. (1958). *J. Biophys. Biochem. Cytol.* 4, 491.
Deutsch, K., and Dunn, A. E. G. (1958). *J. Ultrastruct. Res.* 1, 307.
Flickinger, C. J. (1965). *J. Cell Biol.* 27, 519.
Gay, H. (1955). *Stain Technol.* 30, 239.
Haller, G., Ehret, C. F., and Naef, R. (1961). *Experientia* 17, 524.
Heyner, S. (1963). *Stain Technol.* 38, 335.
Howatson, A. F., and Almeida, J. D. (1958). *J. Biophys. Biochem. Cytol.* 4, 115.
Kellenberger, E., Ryter, A., and Sechaud, J. (1958). *J. Biophys. Biochem. Cytol.* 4, 671.

Kimball, R. F., and Perdue, S. W. (1962). *Exptl. Cell Res.* **27**, 405.

Koehler, J. K. (1961). *Stain Technol.* **36**, 94.

Latta, H. (1959). *J. Biophys. Biochem. Cytol.* **5**, 405.

Luft, J. H. (1961). *J. Biophys. Biochem. Cytol.* **9**, 409.

Malamed, S. (1962). *Proc. 5th Intern. Congr. Electron Microscopy, Philadelphia, 1962* Vol. 2, art. OO-4. Academic Press, New York.

Malamed, S. (1963). *J. Cell Biol.* **18**, 701.

Micou, J., Collins, C. C., and Crocker, T. T. (1962). *J. Cell Biol.* **12**, 195.

Nebel, B. R., and Minick, O. T. (1956). *J. Biophys. Biochem. Cytol.* **2**, Suppl., 61.

Nishiura, M., and Rangan, S. R. S., (1960). *J. Biophys. Biochem. Cytol.* **7**, 411.

Richardson, K. C. (1958). *Am. J. Anat.* **103**, 99.

Robbins, E., and Gonatas, N. K., (1964). *J. Cell Biol.* **20**, 356.

Rosen, S. I. (1962). *Stain Technol.* **37**, 195.

Samuel, D. M. (1944). *J. Anat.* **78**, 103.

Sheffield, H. G. (1965). *Stain Technol.* **40**, 143.

Sparvoli, E., Gay, H., and Kaufmann, B. P. (1965). *Stain Technol.* **40**, 83.

Stone, G. E., and Cameron, I. L. (1964). *In* "Methods in Cell Physiology" (D. M. Prescott, ed.), Vol. I, pp. 135-140. Academic Press, New York.

Sutton, J. S. (1965). *Stain Technol.* **40**, 151.

Yusa, A., and Apicella, J. V. (1964). *Stain Technol.* **39**, 60.

Chapter 12

Analysis of Renewing Epithelial Cell Populations

J. D. THRASHER[1]

*Department of Anatomy, University of Colorado Medical Center,
Denver, Colorado*

[1] Present address: Department of Anatomy, The Center for the Health Sciences, University of California, Los Angeles, California.

324   J. D. THRASHER

I. Introduction

The identification of mitotic figures, and the recognition that they represent a morphological event in the cell life cycle in a variety of organisms (Flemming, 1879), provided early appreciation of the role that daughter cells have in growth of tissues and organs. Soon it was recognized that mitotic activity in many epithelial cell populations was far greater than could be accounted for on the basis of growth alone (Bizzozero, 1894). Reappraisal of this mitotic activity led to the formulation of the concept of tissue renewal (Leblond and Stevens, 1948; Leblond and Walker, 1956).

After cessation of absolute growth in mammals, many tissues of the adult body maintain the capacity for proliferative activity, while other organs exhibit little or no cell division (Leblond and Walker, 1956). The mitotic activity of these tissues has been attributed to cell renewal, i.e., providing a continuous supply of differentiated cells to replace those lost as a result of normal function. It is the purpose of this chapter to review and describe the use of labeled thymidine and autoradiography in studying the kinetics of renewing cell populations, with particular reference to gut epithelium. Attention will be focused on identifying and characterizing the "G_1" population of cells, since the "G_2" population is adequately reviewed by Gelfant in this volume (Chapter 13). However, before proceeding into the techniques employed in this laboratory and others, it is felt that the terminology used in renewing cell populations should be defined for clarification of the text, and for those not entirely familiar with the field.

Renewing epithelial cell populations in the adult mammal are characterized by the constant addition of new cells which replace those presumably functionally incompetent cells lost from the population. On the whole, the rate of cell production exactly balances the rate of cell loss. In this manner, the structural and functional integrity of the epithelium is maintained in a *steady state*.

An epithelium in steady state is constantly undergoing replacement of its cells. The time taken to replace the number of cells equal to the total population is, then, the *turnover time* (formerly renewal time).

A renewing epithelium can contain any number of contiguous compartments. The cells in general migrate from the proliferative compart-

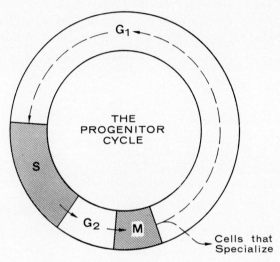

FIG. 1. Diagrammatic representation of the progenitor cell cycle for renewing epithelia. The diverging arrows indicate that some cells that are produced specialize and do not return to the cycle (see text).

ment to the functional region of the epithelium. During this migration a sequence of metabolic changes ensues, leading to eventual cellular specialization. Thus, the *transit time* refers to the period of time required for a given cell to migrate from one compartment of the population to the next. The total of all transit times is equal to the turnover time.

The generative (or stem) cells of an epithelium are referred to as *progenitor cells* and the *progenitor compartment*. The bulk of differentiated nonmitotic epithelial cells comprises the *functional compartment*. In general, the progenitor compartment contains the least specialized cells, whose major function is that of cell renewal. The functional compartment, therefore, contains several subdivisions, each of which represents a single morphological state in the life history of the cell population, e.g., stratified squamous epithelium.

The *progenitor cell cycle* is the average time elapsed between successive divisions of a progenitor cell. On the average, in order to maintain a steady state, one daughter cell of a progenitor division specializes and migrates out of the progenitor compartment, while the other remains behind to pass through another progenitor cell cycle (Fig. 1). For convenience, the progenitor cell cycle has previously been divided into four phases according to nuclear events: M, mitosis; G_1, pre-DNA (deoxyribonucleic acid) synthesis; S, DNA synthesis; and G_2, post-DNA synthesis (Howard and Pelc, 1953; Cameron and Greulich, 1963). The *DNA synthetic index* refers to the fraction of progenitor cells labeled within 90 minutes following tritium thymidine (hereafter termed T-H³) administration, and is generally expressed as a percentage.

II. Preparation and Handling of Tissues for Autoradiography

The procedures described here for preparation and handling tissues for autoradiography have been worked out by Mrs. Z. Trirogoff in the laboratory of R. C. Greulich. We have found that they gave us the best possible cytological preparations for autoradiographic observations.

A. Sacrifice and Fixation

The animals are sacrificed by a high cervical blow followed by decapitation. The thoracic and abdominal walls are opened and the tissues rapidly removed. Pieces of whole organs no larger than 0.5 cm square, or segments of the gut no longer than 0.5 cm in length, are fixed for 3 days in several volumes of Hollande's modification of Bouin's fluid: distilled water 100 ml, formaldehyde (40%) 10 ml, 2% trichloroacetic acid 1.5 ml, neutral copper acetate 4 gm, and picric acid 2.5 gm; filter before use.

B. Embedding and Sectioning

After fixation the tissues are washed in running tap water for 24 hours, trimmed to desired size, and treated as follows: 70% alcohol 1 hour or longer, 80% alcohol 1 hour or longer, three changes of 95% alcohol 1 hour each, two changes of absolute alcohol 1 hour each, two changes of xylene 1 hour each, xylene/paraffin (1 : 1) 1 hour, and two changes of paraffin 1½ hours each (paraffin melting point = 56°–58°C). The tissues are then embedded in 56°–58°C paraffin and oriented in the block for proper plane (cross, longitudinal, etc.) of sectioning required by the investigator. I have found that mouse tissues of similar structure and texture (e.g., gut) can be placed in the same block and sectioned with no difficulty. Therefore it is possible to obtain sections of the esophagus, small intestines, large intestines, etc., and autoradiograph them together.

The tissues are easily sectioned at 2–5µ on a rotary microtome to obtain a ribbon of several sections. If cold rooms are not available for serial sectioning, the paraffin block and microtome knife can be kept cool with ice cubes. The sections are then floated in a bath (40°–45°C) to remove wrinkles, and mounted on clean microscope slides (4–6 sections per slide). The slides are allowed to air-dry in a vertical position. The sections are deparaffinized in three changes of xylene, followed by three changes of absolute alcohol, and air-dried for autoradiography or can be stained prior to dipping in the liquid emulsion (see below).

C. Preparation of Autoradiograms

The autoradiograms are prepared by a slight modification of the method described by Messier and Leblond (1957).

All darkroom procedures are carried out in total darkness. We have found that the use of red safe light within a few feet of NTB-2 and -3 nuclear track emulsions (Eastman Kodak) significantly raises the background. The rest of the changes and precautions emphasized below also aid in reducing the background.

Nuclear track emulsion (NTB-2 or -3) is spooned into a 4-oz glass reservoir. The reservoir is fitted with a lid to prevent condensed water from diluting the emulsion. The reservoir is then immersed in a thermostatically controlled seriological bath at 40°C fitted with a light tight top. After melting (1–2 hours) the emulsion is transferred to a dipping vial and air bubbles removed by dipping 2 or 3 clean microscope slides into the emulsion. The histological slides with deparaffinized sections are dipped, drip drained, and allowed to dry in a vertical position at room temperature for 2 hours. To facilitate drying in a vertical position we have constructed wooden light tight boxes that are partitioned into squares by heavy gauge copper wire. The box can be of any desired size. Our boxes hold a little over one hundred slides. The only requirement is that the squares be slightly larger than the width of the slide. In addition, allow the slides to extent at least ½ inch above the wire so that they can be easily found in the darkroom.

After the slides have dried they are stored in light tight plastic slide boxes (3 × 4 inches) with a desiccant at 4°–6°C for the desired length of time. Test slides can be developed periodically to determine the preferred exposure time.

Developing is done in Dektol (Eastman Kodak) full strength for 2 minutes at 17°–18°C with agitation. The autoradiograms are placed in a stop bath (tap water at 21°–23°C), immediately followed by Eastman Kodak acid fixer (hardener) at 17°–20°C for 1 hour. The temperature can be dropped to desired level by using ice cubes or by storing the solutions in a refrigerator. The autoradiograms are then washed in running tap water for 2 hours.

D. Staining Procedures for Autoradiograms

We routinely use the stains listed below in the laboratory. Hematoxylin and eosin B or Y and periodic acid-Schiff have proven to be the most universal and desirable stains.

Hematoxylin and eosin B or Y: The autoradiograms are stained through the emulsion after development. The sections are stained in Harris' hematoxylin for 5 minutes, dip-washed in tap water (3–5 times), differentiated in 1% acid alcohol (1 N HCl:70% alcohol), and counterstained in 0.2% aqueous solution of eosin B or Y for 10–20 seconds. The slides are quickly transferred to 95% alcohol to prevent leaching of the eosin. This is followed by three changes each of absolute alcohol and xylene. The slides are coverslipped, using permount. I have recently found that absolute alcohols with Euparol as the mounting medium also give excellent preparations.

Periodic acid-Schiff: Deparaffinized sections before dipping into the liquid emulsion are overstained in 0.5% periodic acid and Schiff's reagent (5 and 15 minutes, respectively). After development the autoradiograms are counterstained through the emulsion with Harris' hematoxylin for 5 minutes, differentiated in 1% acid alcohol, dehydrated beginning with 50% alcohol, and coverslipped as above.

Feulgen: It is recommended that Carnoy's fluid be used as the fixative. The Feulgen reaction is carried out before application of the liquid emulsion. Overstaining is recommended. The Feulgen procedure can be found in any standard techniques textbook and need not be outlined here.

Toluidine blue: Staining is done through the emulsion after development. The procedure has been described by Stone and Cameron (1964).

III. Assumptions in the Use of Labeled Thymidine

Although a number of labeled nucleic acid precursors have been used for studying the kinetics of renewing cell populations, the most rewarding tool has been thymidine-H^3 (T-H^3) employed with high resolution autoradiography. From these investigations several basic assumptions regarding the use of labeled thymidine have been formulated. It is felt, therefore, that these assumptions should be briefly reviewed for the benefit of those not entirely familiar with them.

A. Thymidine, a Specific Precursor of DNA

Cells which are synthesizing DNA at the time of administration of labeled thymidine specifically incorporate the precursor (Reichard and Estborn, 1951; Hughes *et al.*, 1958; Amano *et al.*, 1959). Once incorporated into DNA the label is stable and is diluted only by subsequent mitoses (Leblond and Messier, 1958; Walker and Leblond, 1958; Painter *et al.*, 1958b).

B. Pulse Labeling in Mammals

Normal liver contains the enzyme system necessary for catabolism of thymidine and is apparently responsible for keeping the plasma free of the nucleoside (V. R. Potter, 1959). When labeled thymidine is injected into a mammal, 40–60% is rapidly incorporated into newly synthesized DNA. Unincorporated precursor is cleared from the blood stream in approximately 30 minutes and appears as labeled catabolic products (Hughes, 1957; Rubini *et al.*, 1958; V. R. Potter, 1959). The incorporated label can be detected in the nucleus within minutes after the injection, and maximum nuclear labeling occurs within 30–45 minutes. No new cells are labeled after 15 minutes (Thrasher, unpublished observations). Even the endogenous pool of thymidylic acid is small and turns over in 2–4 minutes (R. L. Potter and Nygaard, 1963). Thus, a single injection of thymidine can be considered a pulse label with a duration of approximately 30 minutes.

C. Biological Effects of Thymidine

Because of its role in the genetic material and its limited metabolic pool, the use of thymidine in studes of cell proliferation has been the subject of considerable speculation (Friedkin, 1959; O'Brien, 1962). The effects of unlabeled and labeled thymidine upon dividing cells are divisible into two general categories: (1) biochemical, and (2) radiation damage.

It is now quite apparent that thymidine induces thymidine kinase in mouse fibroblast cells (Littlefield, 1965). This induction appears to take place at least 1 hour before DNA synthesis normally begins in germinating wheat seeds (Hotta and Stern, 1965). In addition, thymidine stimulates thymidylic acid kinase activity (Hiatt and Bojarski, 1961), increases the duration of metaphase (Barr, 1963), and suppresses HeLa colony growth in a dose-dependent manner (Painter et al., 1964). Finally, Greulich et al. (1961) demonstrated that thymidine stimulates mitotic activity in the mouse duodenal epithelium. More recently, Greulich (1962) has suggested that thymidine should be administered to Swiss mice in concentrations less than 0.03 μg per gram body weight to avoid stimulatory effects of the nucleoside upon mitotic activity. However, it is felt by the author that these problems need to be studied in greater detail before definitive conclusions can be made regarding the effects of thymidine upon the proliferative activity of renewing cell populations.

Radiation damage to chromosomes resulting from incorporation of thymidine-H^3,C^{14} has been demonstrated by several investigators (McQuade et al., 1956; Wimber, 1959; Natarajan, 1961; Hsu and Zenzes, 1965). Long exposures of HeLa cells to low concentrations of T-H^3 suppress and can completely inhibit colony growth (Painter et al., 1958a). Moreover, Marin and Prescott (1964) have shown that exposure of Chinese hamster fibroblast cells for 1–2 hours to T-H^3 at concentrations of 1.2–2.7 μC per milliliter in the medium delays division, and significantly decreases the number of third and fourth division labeled metaphases. However, these effects were not seen with smaller concentrations of the labeled nucleoside. Concentrations of 4 μC per milliliter in the culture medium also increase the duration that cells of Tradescantia labeled with T-H^3 spend in the G_2-phase of the cell cycle (Wimber and Quastler, 1963). In mammals it appears that T-H^3 above 1 μC per gram body weight causes radiation damage to mouse spermatogonia, and that low concentrations (1 μC or less) are recommended for short-term experiments (Johnson and Cronkite, 1959). Thus, thymidine can be safely used as a tracer both in vivo and in vitro in mammalian cells when pulse labeling and low concentrations of the isotope are utilized (Painter et al., 1964; Marin and Prescott, 1964).

D. Advantages of Tritium Thymidine

Maximum resolution in autoradiography is obtained by using low energy isotopes, thin histological sections and nuclear track emulsions, and good surface contact between the photographic emulsion and the tissue. Thus, for maximum resolution liquid emulsions and histological sections of 0.5–5μ are recommended (Falk and King, 1963).

The advantages of tritium over P^{32} and C^{14} for autoradiography have been long recognized (Eidinoff et al., 1951; Fitzgerald et al., 1951). The β-particles emitted by tritium have a maximum energy of 0.018 Mev, while those of C^{14} and P^{32} are 0.155 and 1.7 Mev, respectively. The greater the energy of the β-particle the further it travels from its point of origin before losing sufficient energy to produce a latent image in a nuclear track emulsion (Boyd, 1955). Beta particles from C^{14} and P^{32}, therefore, have been measured in photographic emulsions to travel a maximum distance of 20 and 1400μ respectively (Gross *et al.*, 1951). Tritium, on the other hand, produces autoradiographic images 0.04–6.5μ from the point of origin, with 90% of the silver grains lying within 1μ of the source (Fitzgerald *et al.*, 1951; Perry, 1964). Furthermore, back scattering from tritium is nonexistent, especially in histological sections of 1μ or more. When C^{14} is used, care must be taken to avoid observational errors arising from its poor resolution. [For more detailed information on the characteristics of emissions from C^{14} and H^3, see Perry (1964).]

IV. Determination of the Progenitor Cell Cycle by Double Labeling with Thymidine-H^3 and Thymidine-C^{14}

This portion of the chapter is devoted to a brief description of the double labeling technique for determining the mean duration of the progenitor cell cycle and its phases. Here I will be mainly concerned with principle and application of the method. [For a more detailed discussion, see Koburg and Maurer (1962), Wimber and Quastler (1963), and Pilgrim and Maurer (1965).]

A. Assumptions and Criteria

Before the technique can be applied to an analysis of the kinetics of any given renewing cell population, the following assumptions and criteria should be met: (1) The cell population should be in a steady state, i.e., population size is constant, and cell death (or loss) is equal to cell birth. For all physiologically normal renewing epithelia of the adult mammal this state generally exists. (2) An asynchronous state must be present. All cells, then, are randomly distributed in the cell cycle. Environmental conditions, such as diurnal fluctuations, hormones, estrous cycle, temperature, etc., which cause partial periodic synchronizations must be measured and proper corrections made (Pilgrim and Maurer, 1965). (3) Birth and death (or loss) of cells in the population must occur after division and before DNA synthesis. Generally this is assumed to occur in early G_1 (see below). (4) Finally, the population must have minor variations in the progenitor cell cycle. Minor variations in the duration of the cell cycle and each of its phases exist only in rapidly renewing cell systems (Thrasher and Greulich, 1965a), while slowly renewing epithelia, such as stratified squamous epithelia and their derivatives, have large diurnal as well as cyclic fluctuations (Walker, 1960; Perrota, 1961; Pilgrim *et al.*, 1963).

B. Principle of Double Labeling

If an animal is injected with T-H³, and a short time later given T-C¹⁴ and sacrificed shortly after the second injection, two-emulsion autoradiography of renewing epithelia will reveal three groups of labeled progenitor cells (Fig. 2). Following administration of T-H³ a fraction of

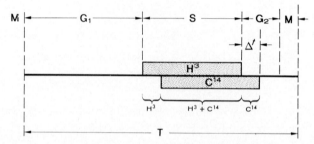

FIG. 2.   Diagram of the progenitor cell cycle demonstrating how progenitor cells become labeled following spaced injections of thymidine-H³,C¹⁴ (see text).

cells dependent upon the percentage of the progenitor cell cycle occupied by the S-phase duration (t_s) is labeled. During the interval between treatments (Δt) a certain fraction of the cells will have passed out of DNA synthesis into G_2, and an equal fraction will have entered the S-phase from G_1. Those progenitor cells which have left S will be labeled with T-H³ only. Another group of cells, which were in DNA synthesis at the time when both isotopes were administered, will be labeled with both C¹⁴ and H³. The number of doubly labeled cells is equal to $t_s - \Delta t$, provided that the interval between the two pulse labels is less than the S-phase duration. The last group of cells, which constitutes the fraction of cells that entered DNA synthesis during the interval between isotope administrations, will be labeled with C¹⁴ only. A note of caution must be inserted at this point. The injection of C¹⁴ and sacrifice should be carried out over a time period less than the duration of $G_2 + M$. If this interval is longer than the duration of $G_2 + M$, then divisions of H³-labeled progenitor cells will necessitate corrections in the data. Ideally, in mammalian cell populations this time interval should not exceed 100 minutes (Pilgrim and Maurer, 1965; Thrasher and Greulich, 1965a).

Preparations of two-emulsion autoradiograms may be accomplished by two different methods. Baserga (1962) described a technique which entails coating a histological slide with two layers of liquid emulsion separated by a thin layer of celloidin. The second method consists of double coating histological slides by the dipping technique. The photographic emulsion is allowed to dry between dippings in both procedures (Wimber and Quastler, 1963). In general, the β-particles of H^3 will produce latent images in the lower portions (about 1μ of the first layer of the emulsion. Beta particles of C^{14} will range up to 20μ from their source and will produce images in both layers of the emulsion. Thus, silver grains immediately above or adjacent to a nucleus indicate H^3 incorporation only, while all other nuclei will be labeled with C^{14} or $H^3 + C^{14}$.

The autoradiograms are then scored for total cells in the progenitor compartment labeled with T-H^3 only vs. all cells labeled with T-C^{14}. The DNA synthetic index can also be scored by determining the fraction of cells labeled with either C^{14} or H^3 for estimates of the progenitor cell cycle duration (see below). In practice it will be found that cells containing T-C^{14} only cannot be distinguished from doubly labeled cells. Once the total number of labeled cells have been enumerated, the mean duration of the S-phase is estimated from simple proportions:

$$\frac{H}{C} = \frac{\Delta t}{t_s}$$

where H = total number of tritium-labeled cells, C = total number of C^{14}-labeled cells, Δt = time interval between injection of isotopes, and t_s = duration of S-phase. Corrections for pulse labeling time of 30 minutes can be applied. However, it appears that reasonably accurate estimates of the S-phase duration can be obtained without the corrections (Wimber and Quastler, 1963; Pilgrim and Maurer, 1965).

In addition to estimating the mean duration of the S-phase, it has also been suggested that the entire nuclear cycle and its phases can be measured with this technique. This is accomplished by extending the interval of time between treatments (Wimber and Quastler, 1963).

The duration of mitosis (t_m) and its subdivisions can be obtained by enumerating the number of labeled mitoses in the progenitor population. The duration of the individual phases of mitosis is estimated by determining the number of labeled progenitor cells in each phase. The following ratio can then be applied:

$$\frac{C + H}{t_s + \Delta t} = \frac{M}{t_m}$$

where M = number of cells in mitosis, and t_m = duration of mitosis.

The duration of G_1 and the progenitor cell cycle can be estimated by extending the interval between treatments to greater time periods. In general it appears that as few as six labeling sequences are required to accomplish this type of analysis. As the time between injections increases the variability in the data will no doubt become larger, because the greatest amount of variability in the progenitor cell cycle arises from G_1. In addition, inherent errors will also arise through the yield of cells per mitosis. In renewing epithelia it has been generally assumed that the ratio of progenitor cell to specialized epithelial cell production with each progenitor cell division is 1 : 1. However, it will be demonstrated in the last section of this chapter that each progenitor cell divides to produce two new progenitor cells, while in order to maintain a steady state some other cell migrates out of the progenitor compartment to specialize. Thus, before an analysis of G_1 duration and the whole progenitor cell cycle can be undertaken, the sources of error just mentioned must be taken into consideration. [For further information regarding the formulas applicable to this type of analysis, see Wimber and Quastler (1963).]

C. Application and Limitations of Double Labeling

The double labeling technique has several advantages and is applicable to several types of experimental procedure. In general, it greatly simplifies the work necessary to determine the durations of the S-phase and mitosis. It also reduces several times the number of animals needed when compared to the conventional labeled mitosis technique (see below). In addition, the double labeling method can be adapted to studies on the effects of alterations in microenvironmental factors of the cell system (hormones, temperature, metabolic inhibitors, radiation, etc.). Finally, it makes it possible to measure the S-phase duration in a variety of cell types which heretofore have escaped analysis because of difficulties in the maturation series of the cell line, e.g., myeloid, thymic, and extrathymic tissues.

The double labeling technique has several limitations. The next section of this chapter deals with a technique for experiments to which the double labeling method cannot be appropriately applied. In general, these are estimates of turnover rates and transit times, analysis of size and distribution of the progenitor compartment, role of individual cells in renewing epithelia, unequal distribution of daughter cells resulting from progenitor divisions, estimates of the entire progenitor cell cycle and its phases, and many others regarding histogenesis of developing tissues.

V. Analysis of Renewing Epithelia Using Single Injection of Thymidine-H[3]

The most widely used autoradiographic method for studying the kinetics of renewing cell populations in mammals has been that of serial sacrifice of a group of animals following a single injection of T-H[3]. It is the purpose of this section to discuss the principle and application of the technique and to provide references which will give the reader a general background in the field.

A. Methodology

The experimental animals have ranged from mouse to man, with economy of the isotope being the main consideration more than anything else. In this laboratory a group of mice (approximately 36) is injected with a single dose of T-H[3] at a concentration of 0.5 μC per gram body weight in a volume of 10 μC/0.1 ml of physiological saline (specific activity, 6.7 mC per millimole). The injection may be administered subcutaneously, intravenously, or intraperitoneally. The animals are then sacrificed in pairs from $\frac{1}{4}$ hour to any time after the initial injection that is required by the experimental design. Ideally, I have found the following time intervals to give excellent results for analyzing the progenitor cell cycle and its phases: $\frac{1}{4}$, $\frac{1}{2}$, $\frac{3}{4}$, 1, 1 $\frac{1}{2}$, 2, 2 $\frac{3}{4}$, 3 $\frac{1}{2}$, 4, 5, 6, 8, 9, 10, 11, 12, and every 2 hours after 12 (see Fig. 4). After sacrifice the tissues are removed, fixed, sectioned, and autoradiographed as described above.

To avoid diurnal fluctuations characteristic of slowly renewing cell populations the following experimental design may be employed. All animals are sacrificed at the same time. The time of sacrifice is set when the greatest frequency of mitotic figures occurs during the day (4–10 A.M.). The time of injection of the isotope is serially moved back so that the progenitor cell cycle can be reconstructed. Thus, the scoring of labeled mitotic figures will be done at the time of day when the highest mitotic activity is present.

Microscopic observations of the autoradiograms will reveal a cohort of labeled cells proportional to the percentage of the cell cycle occupied by the S-phase duration. The number of cells in S-phase depends upon the mean duration of the progenitor cell cycle, and will vary from tissue to tissue.

Observations of the autoradiograms at the various time intervals after T-H³ administration reveals the orderly progression of labeled cells through the cell cycle (Fig. 3). Initially, all cells in the S-phase incorporate T-H³ (Fig. 3, 1) and will appear as labeled interphase nuclei. The labeled cells then leave S and enter G_2 and eventually appear as labeled mitotic figures (Fig. 3, 2). The labeled cells progress through

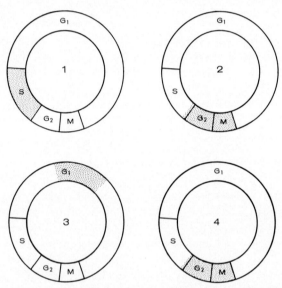

Fig. 3. Diagram showing how a cohort of progenitor cells initially in S incorporates T-H³ and progressively passes through the cell cycle with time (see text).

G_1 (Fig. 3, 3), and at some indefinite time later appear again as labeled mitotic figures (Fig. 3, 4) completing one progenitor cell cycle.

B. Background Determination

Before describing the procedures for analyzing labeled cells in the progenitor compartment, a few comments concerning the measurement of background are necessary. The largest single problem is deciding what intensity of labeling constitutes incorporation or background.

Individuals who have worked with autoradiography can qualitatively assess background labeling by routine observations of the autoradiograms. However, a few accepted methods for determining background have been used by various investigators.

The specificity of T-H³ incorporation into newly synthesized DNA and background can be checked by deoxyribonuclease (DNase) treatment of histological sections. In general, deparaffinized histological sections are incubated at 37°C for 24 hours in DNase (approximately 0.05 mg per milliliter) buffered with phosphate buffers at pH 6.2–8.0. Addition of magnesium ions (0.02 M) has been recommended. Control slides are incubated under identical conditions, except the enzyme is absent. The efficiency of digestion can be tested by staining with either 0.1% toluidine blue or Feulgen. The treated, control, and normal slides are covered with emulsion and developed at the same time. *Do not* stain slides for autoradiography with toluidine blue since aniline dyes will produce latent

TABLE I

COUNTS DEMONSTRATING THE NUMBER OF NUCLEI WITH 0–10 OR MORE GRAINS PER NUCLEUS IN THE MOUSE DUODENAL VILLUS EPITHELIUM VS. CRYPTAL EPITHELIUM AT ¾(A) AND 2¾(B) HOURS AFTER T-H³ ADMINISTRATION

Grains per nucleus	A		B	
	Villus cells	Crypt cells	Villus cells	Crypt cells
0	886	595	892	659
1	89	89	87	59
2	16	20	14	12
3	5	6	6	7
4	3	18	1	13
5	1	14	0	8
6	0	9	0	4
7	0	10	0	3
8	0	3	0	3
9	0	4	0	3
10 (or more)	0	236	0	193
Total cells	1000	1004	1000	964

images. Nonspecific incorporation of T-H³ is determined by quantitative grain counts. Background grains are enumerated on the DNase-digested slides. Generally this is done per unit area or per nucleus, and any background is then subtracted from nuclear labeling. A low background is usually found (0.1–0.05 grain per nucleus). However, this does not assure the investigator that nuclei with grain counts of 1–4 are positively labeled, particularly where mean grain counts of 20 or more are present.

Another method is counting the number of grains per unit area in a portion of the autoradiogram where no tissues are present. This figure is subtracted from any quantitative grain counts made on the material. However, this procedure does not take into account the possibility of the production of latent images by chemical compounds in the tissue section (Tonna, 1958).

The last approach, which I routinely use, enumerates the number of silver grains per nucleus in a portion of the tissue which does not have cells in DNA synthesis (villus epithelium) vs. that portion of the tissue that has large numbers of cells in DNA synthesis (cryptal epithelium). The analysis is done on 1000 nuclei in each region of the epithelium. Only nuclei with silver grains immediately above or adjacent to the nucleus are considered as labeled. The results of such a count are given in Table I. It is apparent from the data that the number of nuclei with 1–3 grains is identical in each portion of the tissue. Thus, only nuclei with 4 or more grains are considered to have incorporated T-H³. Errors resulting from an incomplete analysis can also be illustrated from the data. If the counts had been restricted to the villus epithelium only, and the results expressed as average grain counts per nucleus (approximately 0.1 grain per nucleus), then any nucleus with one or more silver grains in the cryptal epithelium would have to be considered as positively labeled. This could produce considerable error in the analysis of labeled mitotic figures.

C. Analysis of Labeled Progenitor Cells

Appropriate quantitation of the autoradiograms will give an indirect measurement of the progenitor cell cycle and its phases. The two methods used by various investigators have been the scoring of the percentage of either labeled mitoses (all stages of division) or labeled metaphases at each time interval after T-H³ administration. At least 100 labeled and unlabeled mitoses per animal are scored. Only mitotic figures with 4 or more silver grains are considered as labeled. In tissues with a slow renewal rate it is difficult to obtain 100 mitoses. Therefore, the minimum statistically acceptable number for determining a percent is 20, provided a sufficient number of time intervals and animals are used. The data are then plotted and measurements of the constructed mitotic curves give indirect estimates of the mean duration of the progenitor cell cycle and its phases (see below).

Scoring the percentage of labeled progenitor cells per 1000 cells in the progenitor compartment is the accepted approach for obtaining the DNA synthetic index. In many tissues it will be necessary to define what cells constitute the progenitor compartment before accurate assessment of the index is possible. The size and/or distribution of the progenitor compartment can be assessed by mapping the position of labeled cells in the epithelium (see below). The DNA synthetic index is used to calculate the mean duration of the progenitor cell cycle.

Other indices are generally employed to study cellular renewal. The mitotic index of the progenitor compartment is the percentage of cells in division based on an analysis of 1000 progenitor cells. This can also be used to estimate the progenitor cell cycle duration, provided the duration of mitosis is known. The mitotic and DNA synthetic indices of the total population, i.e., functional plus progenitor compartments, are used to determine tissue turnover (see below).

D. Measurement of the Progenitor Cell Cycle and S-Phase from Metaphase Curves

The selection of only one easily recognizable morphological event (metaphase) in the progenitor cell cycle circumvents the difficulties that arise by scoring the percentage of labeled mitoses. In many tissues it is

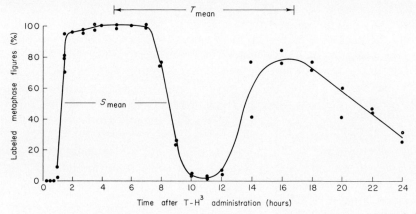

FIG. 4. Curve describing the rate of appearance and disappearance of labeled metaphases following a single injection of T-H³ in the Swiss mouse duodenum. Each symbol represents at least one animal, and in some cases two (see text).

difficult to ascertain the beginning of prophase and the ending of telophase. In a cell population which has a rapid turnover (small and large intestine), sufficient numbers of metaphases are present so that a statistically significant quantity can be enumerated.

A curve describing the rate of appearance and disappearance of labeled metaphase figures from ¼ to 24 hours after T-H³ injection in the mouse duodenum is presented in Fig. 4. The curve is typical of the kinetics of labeled metaphases for most renewing epithelia. Labeled metaphases make their appearance at 1 hour after injection of T-H³. Their frequency rises thereafter, attaining levels of approximately 100% by 2 hours. Labeling remains at this plateau until 7–8 hours, falling to about 5% by 10 hours. At 14 hours the incidence of labeled metaphases rises again, and attains a new plateau of 80% at 16–18 hours, falling thereafter to about 30% by 24 hours. In general, the time sequence for the first wave of labeled metaphases varies little from tissue to tissue. The second wave does vary, depending upon the mean duration of the progenitor cell cycle.

The interval of time between injection and the first appearance of labeled metaphases is the minimum duration of G_2 + prophase. The maximum duration of G_2 + prophase is defined by the point at which metaphases reach 100% labeling. These measurements are predicated on the basis that those cells which initially appear as labeled metaphases must have been at or near the end of DNA synthesis at the time of administration T-H³.

The mean duration of the S-phase is obtained from the first wave of labeled metaphases. The ascending limb describes the rate at which cells leave G_2 + prophase and enter metaphase. The shape of the transition from the peak (7–8 hours) to the trough (10 hours) has been attributed to the rate at which cells enter the S-phase (Quastler and Sherman, 1959). If the duration of the S-phase was constant, then the slope of the descending limb would be the inverse of the ascending portion of the curve. Since it is less steep, S-phase durations of individual progenitor cells have been considered to be of unequal lengths. However, it will be shown below that the asymmetry of the curve may be ascribed to the variability existing in the durations of G_2 and M. Therefore, the only meaningful measurement of S duration is taken from the difference between the times at which 50% labeling occurs, i.e., 1.3 and 8.8 hours, or a mean duration of 7.5 hours.

The mean duration of the progenitor cell cycle (T) is also obtained from the curve. Since any two corresponding points on the two waves of labeled metaphases indicate that the cohort of cells has passed through two identical consecutive morphological and physiological events in the cycle, then the time elapsed between these two events is a measurement of the duration of the progenitor cell cycle. Inspection of Fig. 4 reveals that the second wave of labeled metaphases is not a duplicate of the first. This lack of a well-defined second peak results from any one or all of the following reasons: (1) Variability does exist in the duration of the progenitor cell cycle and in each one of the phases. Thus, with completion of mitoses the cohort of labeled cells becomes asynchronous and the alternation of phases falls out of rhythm with time. For example, the variability existing in S, G_2, and M is indicated by the fact that labeled metaphases never come back to 0% labeling. The variation of G_1 and the entire cycle is manifested in the fact that metaphases never attain 100% labeling in the second wave. (2) Labeling time is not a pulse label.

Nuclear labeling may also be affected by the thymidylic acid pool. This is currently being investigated in this laboratory. (3) With each division the amount of T-H^3 per nucleus is halved. Hence, if sufficient isotope is not present the cells will not register autoradiographically. (4) False negatives (cells which are too far away from the emulsion) also will produce observational errors. (5) Unequal distribution of daughter cells with each progenitor division could possibly influence the data (see below). Since each progenitor cell produces two new progenitor cells, each with its own biological variation, the second wave of labeled metaphases becomes more exaggerated with time. (6) Diurnal fluctuations and a host of other mechanisms described by Leblond and Walker (1956) and Bullough (1962) do induce biological variability. It is possible, however, to obtain a good estimate of the mean duration of the progenitor cell cycle by measuring the distance (time elapsed) between the midpoints of the first and second waves of labeled metaphases. This is, in general, taken as being representative of the cycle duration since these points are least affected by most of the factors mentioned above.

E. Measurement of the Progenitor Cell Cycle and Its Phases from Mitotic Curves

Reasonably accurate measurements of the progenitor cell cycle and its phases can be derived from a curve describing the rate of appearance and disappearance of labeled mitoses (Quastler and Sherman, 1959). In Figs. 5–8 the open circles represent the percentage of labeled mitotic figures, while closed circles are the percentage of labeled metaphases in four renewing epithelial cell populations in the adult mouse intestines.

Restricting attention to the labeled mitotic curves for each cell system analyzed, it is apparent that little variation exists in the duration of time that it takes labeled mitotic figures to first appear (G_2), which ranges from ½ to ¾ hour in the various cell systems. The interval of time between the first appearance of labeled mitoses and when they attain 100% labeling is a good estimate of the mean duration of mitosis, and ranges from 1 hour (duodenum) to 2¾ hours (esophagus). Again, as in the metaphase curve analysis, the mean duration of the S-phase is obtained from the difference between 50% labeling on the ascending and descending limbs. This estimate ranges from 7.8 hours (duodenum) to 8.8 hours (esophagus). Although not shown, the progenitor cell cycle duration is derived from the distance between the midpoints of the two waves of labeled mitoses. The mean duration of G_1 is equal to T less $G_2 + M + S$ durations.

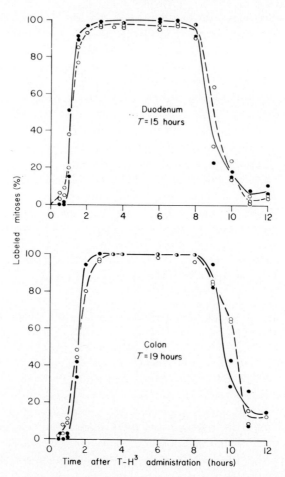

Figs. 5 (*above*) and 6 (*below*). See legend on opposite page.

F. Analysis of Mitotic Variation and Duration

A composite curve describing the rate of entry of labeled progenitor cells into the four phases of division is illustrated in Fig. 9 for the duodenum, colon, and esophagus. At 30 minutes postinjection a few labeled prophases first appear, and rise thereafter to approximately 100% at different times for each tissue. The orderly progression of each successive stage of division (metaphase, anaphase, telophase) occurs as anticipated, i.e., after a prescribed interval of time. Thus, the initial appearance of each stage occurs in a customary manner that would be expected following a pulse label of T-H³. However, considerable varia-

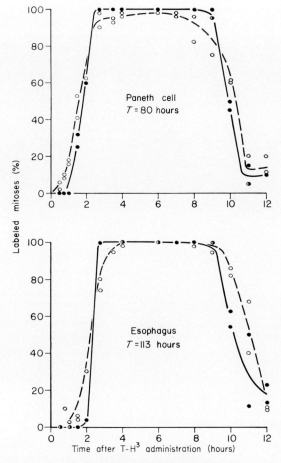

Figs. 7 (*above*) and 8 (*below*).

Figs. 5–8. Curves describing the rate of appearance and disappearance of labeled mitotic figures (*open circles*) and labeled metaphases (*closed circles*) in four renewing epithelial cell populations in the adult mouse. T is the mean duration of the progenitor cell cycle for each cell system. Each symbol represents at least one animal.

tion exists in the time that an individual progenitor cell requires to pass through each phase of division. Note the rate of accumulation of labeled prophases and metaphases in each tissue. It is apparent that, when 50% labeling of metaphases occurs, prophases have not attained 100% labeling. This could occur only if variation in the time that cells spend in G_2 and prophase is present. Telophase labeling has greater variability than any of the preceding phases, and is particularly evident in the esophagus.

It may also be seen from the data that the longer the progenitor cell cycle duration, the greater is the variability in the duration of each phase of mitosis.

In spite of the individual cell variation, statistical analysis of the data using quartiles can be applied to obtain meaningful information. The quartile least affected by variability is the interquartile (Q_2) range. Its

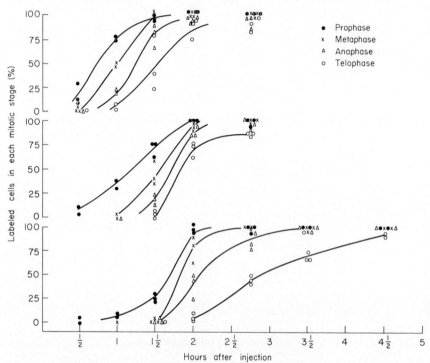

FIG. 9. Curves demonstrating the progression of labeled progenitor cells through prophase ($\bullet$), metaphase ($\times$), anaphase ($\triangle$), and telophase ($\bigcirc$) in the Swiss mouse duodenum (*upper figure*), colon (*middle figure*), and esophagus (*lower figure*) after T-H³ administration (see text for fuller explanation). (Courtesy of I. L. Cameron.)

use will allow assessment of the time elapsed between the midpoint of one to the midpoint of the next stage of division. For example, the time elapsed between Q_2 of prophase to that of metaphase is an estimate of the average interval between midprophase and midmetaphase. This type of analysis may allow investigations on the effects of various agents (chemicals, radiation, etc.) on each phase of division. Description of quartiles and their use for measures akin to the median can be found in any standard textbook of statistics.

G. Comparison of Metaphase and Mitotic Techniques

It was mentioned earlier in the chapter that variability existed in the duration of all phases of the progenitor cell cycle. As a result, the asymmetry of the first wave of labeled mitoses could be caused by the variability in G_2 and M. In the section above, the individual variation in the time it takes a progenitor cell to pass through each phase of mitosis was demonstrated. It is now apparent that the asymmetry of the mitotic-labeling curves can be ascribed to this variation, particularly that which exists in prophase and telophase. The asymmetry of the metaphase-labeling curve results from variations in G_2 and prophase. Thus, measurements of the mean duration of S are most accurate when based on metaphase curves rather than on mitotic curves. In conclusion, the S-phase duration can be considered a unit constant of approximately 7.5 hours based upon metaphase labeling. Inspection of Figs. 5–8 will verify this conclusion.

H. Estimates of the Progenitor Cell Cycle Duration from the Ratio of S-Phase to the DNA Synthetic Index

The premise for this method is based upon a constant S-phase duration of approximately 7.5 hours in most renewing epithelia (Koburg and Maurer, 1962; Cameron and Greulich, 1963), in a variety of mammalian cell types (Pilgrim and Maurer, 1965), in aging mice (Thrasher and Greulich, 1965a), and in a heterotypic cell population (Thrasher and Greulich, 1966).

Estimates of the mean progenitor cell cycle are made according to the method suggested by Quastler and Sherman (1959), which is analogous to the procedure for estimating tissue turnover from the mitotic duration and mitotic index (Leblond and Walker, 1956). The fact that the S-phase is some 7 times longer than mitosis (usually assumed to be 1 hour) suggests that T-H[3] labeling should permit an approximately 7-fold increase in sensitivity for estimating the progenitor cell cycle, as well as tissue turnover. Estimates of proliferative activity in the mouse duodenum by both DNA synthetic and mitotic indices have confirmed this suggestion (Thrasher and Greulich, 1965a,b). In addition, this technique will allow the analysis of proliferative activity in various tissues with a minimum (approximately 8) number of experimental animals.

The percentage of labeled interphase nuclei at 30–90 minutes after T-H[3] administration is a reliable indicator of the relative size of the progenitor population in the S-phase. Thus, if I is the percentage of cells in S-phase, S is the mean duration of S-phase, and T is the mean duration of the progenitor cell cycle, then

$$T = \frac{S/I}{100}$$

Before leaving this subject a few points of fundamental consideration should be made. One of the major problems in using this technique is that of defining what cells constitute the progenitor compartment. Before an accurate estimate of the DNA synthetic index can be obtained, precise information concerning the size and/or distribution of the progenitor compartment is needed (Thrasher and Greulich, 1965b). For example, the proliferative cells of the duodenal progenitor compartment are limited to the central region of the crypt, while those for the colon are found in the bottom twenty cells of the crypt. Finally, once the DNA synthetic index has been measured, it should be corrected for distribution of ages in the population before reasonably accurate values for the progenitor cell cycle can be made (Sisken and Kinosita, 1961; Thrasher and Greulich, 1965b).

VI. Measurement of Tissue Turnover and Transit Times

The earliest methods used for evaluating tissue turnover involved the mitotic index or the colchicine method (Leblond and Walker, 1956). More recently, T-H[3] has provided a highly sensitive tool for analyzing tissue turnover.

A. Mitotic Index

The mitotic index is determined for the entire cell population by scoring the number of dividing cells per 1000 nuclei or more in histological sections. The counts can be corrected by means of Abercrombie's formulas (1946), which give the true number of nuclei present in the histological section. However, the correction is necessary only if mitotic cell sizes differ from nondividing nuclear sizes (Storey and Leblond, 1950).

Estimates of tissue turnover are based on the assumption that mitotic duration is approximately 1 hour. Mitotic durations apparently can range from 15 to 150 minutes (Leblond and Walker, 1956) (Figs. 5–8). Assuming a 1-hour duration, tissue turnover time (T) is

$$T = \frac{\text{mitotic duration}}{\text{mitotic index}}$$

B. Colchicine Method

Colchicine has the property of blocking dividing mammalian cells in metaphase. The drug is generally administered to rats and mice at concentrations of 1 mg per kilogram body weight. The drug has variable effects upon different tissues [for its proper use, see Leblond and Walker (1956), and Leblond (1959)].

In general, colchicine is administered 4–6 hours prior to sacrifice. Turnover time is then calculated by dividing the duration of time used (4–6 hours) by the tissue mitotic index. The optimum time for administration before sacrifice can be assessed by using several time intervals between colchicine injection and sacrifice, and then enumerating the number of arrested metaphases at each time interval. The drug should not appreciably alter the number of prophases in the tissue section.

C. DNA Synthetic Index

Although this method has not been used to a great extent, it is apparent that it would give greater sensitivity to estimates of tissue turnover.

The DNA synthetic index is determined on the total cell population 30–90 minutes following T-H³ administration. Dividing this index into the mean duration of the S-phase (7.5 hours) will give a reliable estimate of tissue turnover. The only caution required in the use of this method involves diurnal fluctuations in the numbers of cells synthesizing DNA in slowly renewing cell population (Pilgrim et al., 1963). In these tissues the average DNA synthetic index over a 24-hour period should be used.

D. Estimates of Tissue Turnover from Cell Migration

Progenitor cells that have incorporated T-H³ divide and, on the average, one daughter cell of each division specializes and migrates from the progenitor to the functional compartment. The turnover time of a cell population can be obtained either by measuring the rate at which these cells are produced, or by measuring the time required for migration from one compartment to the next. The latter method is the easiest since it entails scoring the position of labeled cells in the epithelium at various times after T-H³ administration. The last time interval should extend to at least 200 hours after injection of T-H³ because the turnover time of tissues varies from 1.5 to several days (Leblond and Walker, 1956). The analysis for transit times and tissue turnover is dictated by the morphology of the cell system. Thus, experimental procedures for the small intestine (Lesher *et al.*, 1961; Fry *et al.*, 1961, 1962) are different than those for stratified squamous epithelia (Leblond *et al.*, 1964; Greulich, 1964).

The measurement of transit times has been adequately described by Fry *et al.* (1962) for the small intestine. The crypt transit time was defined as the time required for a cell to migrate from its site of origin (crypt) to the villus base. The minimum transit time was taken when labeled cells first appeared at the villus base, the average transit time when 50% of the bases were labeled, and total transit time when all villi had labeled cells at the base. The villus transit time was estimated by the same procedure, and was defined as migration from villus base to extrusion zone. The total transit time was equal to the crypt plus villus transit lines.

VII. Relationship between Cell Proliferation and Specialization in Adult Renewing Epithelia

The steady state concept which characterizes renewing epithelia has been formulated on a statistical average. Thus, in order to maintain a steady state, one daughter cell of a progenitor division differentiates and migrates into the functional compartment while the other daughter cell remains behind to pass through another progenitor cell cycle.

Such a concept, however, does not allow for considerable variation in the life history of individual cells. It seems quite feasible that the role of individual cells in tissue renewal can vary within wide limits, while the population on the whole behaves according to the statistical average. Even the morphology of renewing epithelia suggests that individual cellular behavior cannot be predicted when based upon the statistical concept of steady state (Leblond *et al.*, 1964; Greulich, 1964; Thrasher and Greulich, 1965b, 1966).

It is apparent that knowledge of the degree to which individual variations occur in the life history of epithelial cells would be of considerable value in the interpretation of pathological as well as experimentally induced aberrations of tissue renewal. In addition, such information can also provide insight into the relationship between cell proliferation, migration, and specialization, and their respective roles in maintaining the normal functional integrity of the epithelium (Greulich, 1964).

The purpose of this section is, then, to describe procedures which will allow assessment of the role of individual cells in tissue renewal. In general, the analyses have been made on the esophagus (Leblond *et al.*, 1964; Greulich, 1964) and the duodenum (Thrasher, 1966) of the rat and Swiss mouse. Presumably, they could be extended to other renewing epithelia as well as to other species. The procedures are divided into (1) an analysis of size and distribution of the doudenal progenitor population, and (2) the relationship between daughter cells of a progenitor division and their subsequent migration and specialization in the duodenum and esophagus.

A. Duodenal Progenitor Population; Size and Distribution

The duodenal progenitor population offers an excellent morphological and physiological system in which the causal relationships existing between the positioning of the progenitor cells in the epithelium and the duodenal microenvironment can be investigated. This work has been published in greater detail elsewhere (Thrasher and Greulich, 1965b, 1966), but is described here to demonstrate how knowledge of the size and distribution of progenitor population can lead to considerable information regarding cell specialization in the whole epithelium.

To minimize systematic errors in identification of the duodenal progenitor population, analyses were restricted to crypts of Lieberkühn sectioned throughout their entire length, exhibiting at least three fourths of the cryptal lumen, and possessing opposite sides which were approximately parallel. The point of junction between the crypt and villus walls (cryptovillal junction) was determined as precisely as possible, and the first cell below this junction was designated cell position 1. Beginning with cell number 1, labeled and unlabeled cells in the cryptal epithelium were scored in a consecutive fashion ending with the last cell (Paneth) at the cryptal base at 30, 45, 60, and 90 minutes following T-H³ administration. The number of cells constituting tthe length of the crypt wall, i.e., cryptal depth, averaged 18 cells. All crypts were converted to this mean cryptal depth by simple proportions so that evaluation of the results could be made. If the cryptal depth was less than 18 cells, for instance 17, each cell position was multiplied by the fraction 18/17. If the depth was larger than 18 cells, for instance 19, the fraction 18/19 was used in the conversions.

Figure 10 summarizes the results obtained from an analysis on 8 animals and 20 crypts per animal. Progenitor cells in DNA synthesis rose rapidly from a minimum at cell position 1 to a maximum at cells 7–8. The percentage of labeled cells remained at a plateau (maximum labeling) between cells 8–14, and fell at cells 15–18. In general, the cells which incorporated T-H³ extended from the first cell below the cryptovillal junction to the last cell (Paneth) at the cryptal base, so suggesting that all cryptal cells possess the capacity for division. However, labeling occurred most frequently in cells positioned in the central region of the crypt. From the data it became possible to subdivide the crypt into three zones according to this distribution: (1) zone of minimum cellular proliferation (cells 1–7), (2) zone of maximum cellular proliferation (cells 8–14), and (3) zone of Paneth cells (15–18).

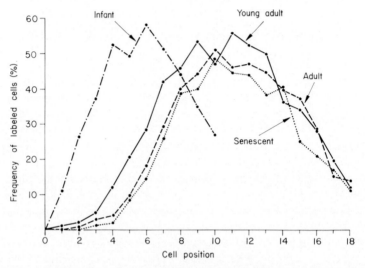

Fig. 10. Curve showing the distribution of labeled progenitor cells in the mouse duodenal crypt from the first cell below the cryptovillal junction (cell position 1) to the last cell at the cryptal base (cell 18) at 30–90 minutes following T-H³ administration (see text for fuller explanation) in three different age groups (the infant group has been neglected since its mean cryptal depth is only 10 cells). (Thrasher and Greulich, 1965b.)

Once the size and distribution of the progenitor population have been defined, it becomes relatively easy to speculate on the role that the duodenal microenvironment (all factors impinging upon the progenitor cells) has in influencing cell proliferation, migration, and specialization.

Generally the microenvironment contains varying concentrations of chemical factors which affect and regulate cellular processes. The transition of a progenitor cell into a functional villus cell involves profound changes in its biology, i.e., from the role of cell proliferation to that of absorption or secretion. The compartmentalization of the duodenal epithelium into its various zones is apparent, if it is assumed that the environment provided by the intestinal chyme would lead to the death of sensitive progenitor cells. Thus, as the crypt cell migrates from the optimal environment provided by the lamina propria in the zone of maximum cellular proliferation to the unfavorable environment surrounding the villus epithelium, any one of a number of factors brings about specialization into functional surface absorptive or goblet cells.

This hypothesis can now be further tested by determining changes in RNA and protein synthesis in relation to the distribution of proliferating cells, as well as by examining what effects various experimentally induced states have on this distribution, e.g., hormones (Moog, 1953; Leblond and Carrière, 1955), germ-free animals (Abrams *et al.*, 1963), partial resection of the intestine (Loran and Crocker, 1963), isolated intestinal loops (Hooper, 1956), artificial lesions (McMinn and Mitchell, 1954), etc.

B. Relationship between Daughter Cells, Migration, and Specialization in the Duodenum

Cellular specialization in the small intestine apparently results from some inherent mechanism related to cell migration from the crypt onto the villus epithelium. Implicit in understanding this phenomenon is knowledge of the relationship between daughter cells of a progenitor division and their subsequent migration. Since the active progenitor compartment is found in the central region of the crypt, it would appear to be a physical impossibility for daughter cells to behave according to the statistical concept of cell renewal. Moreover, if microscopic observations are made on a dividing epithelial cell, it may be seen to move away from the basement membrane although remaining attached to it by a "foot," divide, and both daughter cells move back to assume the position of the original mother cell. Thus, it appears that the statistical concept of tissue renewal is not the process by which a renewing epithelium maintains a steady state. Indeed, this type of reasoning would infer a differential mitosis (Rolshoven, 1951), for which there is no evidence.

Figure 11 (modified from Greulich, 1964) summarizes the various relationships between dividing progenitor cells (P) and specialized epithelial cells (S) that could account for the maintenance of the steady state. The statistical scheme describes what has been previously stated, i.e., a progenitor cell divides to give rise to one daughter cell that specializes, while the other daughter cell remains behind as a progenitor cell.

Alternative schemes, which by working solely or in concert to maintain the steady state, are also presented in Fig. 11. Scheme I is just a restatement of the statistical scheme. The second alternative (II) simply implies that a progenitor cell divides to produce two specialized cells. In order to maintain the steady state another progenitor cell divides at the same time to give rise to two new progenitor cells. The last alternative (III) suggests that a progenitor cell always divides to produce two progenitor cells. However, in order to be compatible with the steady state, another progenitor cell migrates out of the progenitor compartment and specializes.

To test which scheme may be operative in the mouse duodenum, an analysis was undertaken of the rate of migration of labeled daughter cells from the crypt to the villus. These results were then compared to a

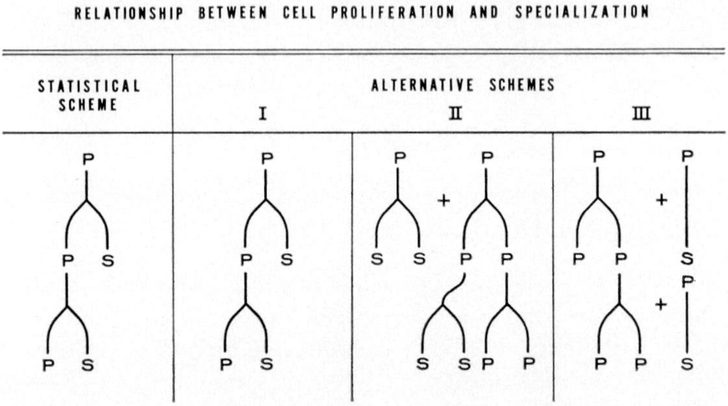

RELATIONSHIP BETWEEN CELL PROLIFERATION AND SPECIALIZATION

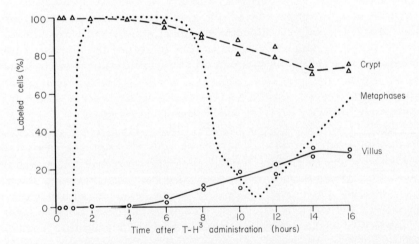

Fig. 11. Schematic representations of the relationship between cell proliferation and specialization in adult renewing cell populations which can account for the steady state (see text for a more detailed discussion).

Fig. 12. Curves demonstrating the location of labeled cells in the mouse duodenum as a function of time after T-H³ administration. Superimposed over these curves is a curve which describes the percentage of labeled metaphases in the same animals (see text).

metaphase labeling curve in order to correlate the relationship between cell specialization and proliferation. Autoradiograms were examined at 15 minutes to 16 hours after T-H^3 administration. A total of 1000 labeled cells for each animal at the various time intervals were scored over the villus and cryptal epithelium. The percentage of the labeled cells found on the villus vs. those present in the crypt was determined and the results are presented in Fig. 12. Superimposed over these data is a typical curve describing the rate of appearance and disappearance of labeled metaphases in the same animals. It may be seen that for a period up to 8 hours very few of the labeled daughter cells have migrated out of the crypt. Significant quantities of labeled cells appeared on the villus at 8–10 hours, and rose thereafter. Comparing this to the metaphase curve, it can be seen that 8–9 hours is the period when almost all initially labeled progenitor cells have passed through division. At this time interval, progenitor cells which were in G_1 at the time of T-H^3 administration are beginning to pass through division. Consequently, since several hours passed before labeled cells appeared on the villus, scheme III seems most appropriate for describing the relationship between cell proliferation and cell specialization.

C. Relationship between Daughter Cells, Migration, and Specialization in the Esophagus

Figure 13 represents the migration of labeled cells from their site of origin in the progenitor compartment (stratum basale) to the contiguous layers of specialized cells (stratum spinosum and stratum granulosum) in the mouse esophageal epithelium from $\frac{1}{4}$ to 100 hours after T-H^3 administration. It can be seen that 100% of the labeled cells remain as progenitor (basal) cells for a period up to 20 hours. Significant numbers of labeled cells begin to appear as spinous cells at 20 hours, while labeled granular cells do not appear for almost 50 hours.

Figure 14 describes the rate of appearance and disappearance of labeled metaphases in the same animals. Note that labeled cells have passed through division by 15 hours, the time at which all cells originally in S have divided and are in G_1 of the next progenitor cell cycle. Drawing a correlation between Figs. 13 and 14 leads to the inference that progenitor cells divide to produce two daughter cells which return to the basal layer of the epithelium. Since as much as 20 hours pass before labeled spinous cells are seen, scheme III is most appropriate for describing the relationship between cell proliferation and specialization in the esophagus also. Confirmation of this conclusion has been made by Leblond et al. (1964), who examined labeled daughter cells in the rat esophagus by (1) physical proximity, (2) comparable grain counts, and (3) isolation of observed pairs from other labeled nuclei with comparable grain counts.

In conclusion, it appears that progenitor cells of the mouse duodenum and esophagus divide to produce more progenitor cells. Migration and

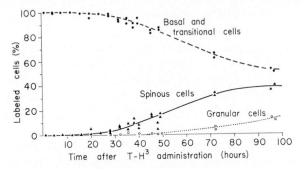

Fɪɢ. 13. Curves describing the location of labeled cells in the mouse esophagus at times after T-H³ administration (see text for fuller explanation). (Courtesy of R. C. Greulich.)

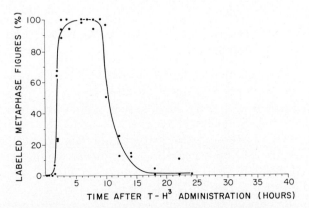

Fɪɢ. 14. Curve showing the percentage of labeled metaphases as a function of time after T-H³ administration in the mouse esophagus. Each symbol represents a single animal (see text for correlation with the data presented in Fig. 13). (Courtesy of R. C. Greulich.)

subsequent specialization of progenitor cells into cells of the contiguous layers of the epithelium occur in a random fashion. Moreover, this occurrence appears essentially to be independent of the mitosis from which migrating cells took their origin.

Finally, in reference to both epithelia, it appears that migration itself is sufficient to produce specialization of progenitor cells into the functional components of the epithelium. Thus, environmental changes that occur over the short distance between the progenitor compartment and the functional compartment are sufficient to elicit changes in the metabolic pathways of the cell which are requisite for cell specialization and loss of proliferative function. Apparently these changes must take place any time between early G_1 and the beginning of S.

ADDENDUM

Since the preparation of this review two articles have appeared regarding cell population kinetics in the rat jejunum (Cairnie et al. (1965a,b). These authors have carefully described the importance of understanding the relationship of the progenitor compartment to the remainder of the epithelium. Their data is in agreement with that presented in this review. In addition, they have analyzed the relationship between cell proliferation and cell specialization in the jejunum. Their interpretation of the data favors the second alternative scheme presented in Fig. 11. However, it appears that more information is needed before scheme II or III can be verified.

ACKNOWLEDGMENTS

This work was supported in part by Research Grant GB-1635, National Science Foundation, supervised by David M. Prescott. Grateful acknowledgment is made of the skilled technical assistance of Mrs. Z. Trirogoff. I particularly wish to thank Drs. R. C. Greulich and I. L. Cameron for their contributions to this manuscript.

REFERENCES

Abercrombie, M. (1946). Anat. Record 94, 288.
Abrams, G. D., Bauer, H., and Sprinz, H. (1963). Lab. Invest. 12, 355.
Amano, M., Messier, B., and Leblond, C. P. (1959). J. Histochem. Cytochem. 7, 153.
Barr, H. J. (1963). J. Cellular Comp. Physiol. 61, 119.
Baserga, R. (1962). J. Cell Biol. 12, 633.
Bizzozero, G. (1894). Brit. Med. J. I, 728.
Boyd, G. A. (1955). "Autoradiography in Biology and Medicine." Academic Press, New York.
Bullough, W. S. (1962). Biol. Rev. Cambridge Phil. Soc. 37, 307.
Cairnie, A. B., Lamerton, L. F., and Steel, G. G. (1965a). Exptl. Cell Res. 39, 528.
Cairnie, A. B., Lamerton, L. F., and Steel, G. G. (1965b). Exptl. Cell Res. 39, 539.
Cameron, I. L., and Greulich, R. C. (1963). J. Cell Biol. 18, 31.
Eidinoff, M. L., Fitzgerald, P. J., Simmel, E. B., and Knoll, J. E. (1951). Proc. Soc. Exptl. Biol. Med. 77, 225.
Falk, G. J., and King, R. C. (1963). Radiation Res. 20, 466.
Fitzgerald, P. J., Eidinoff, M. L., Knoll, J. E., and Simmel, E. B. (1951). Science 114, 494.
Flemming, W. (1879). Arch. Mikroskop. Anat. Entwicklungsmech. 16, 302.
Friedkin, M. (1959). In "The Kinetics of Cellular Proliferation" (F. Stohlman, ed.), p. 97. Grune & Stratton, New York.
Fry, R. J. M., Lesher, S., and Kohn, H. I. (1961). Am. J. Physiol. 201, 213.
Fry, R. J. M., Lesher, S., and Kohn, H. I. (1962). Lab. Invest. 11, 289.
Greulich, R. C. (1962). Anat. Record 142, 327.

356   J. D. THRASHER

Greulich, R. C. (1964). In "The Epidermis" (W. Montagna and W. C. Lobitz, eds.), p. 117. Academic Press, New York.

Greulich, R. C., Cameron, I. L., and Thrasher, J. D. (1961). Proc. Natl. Acad. Sci. U.S. 47, 743.

Gross, J., Bogoroch, R., Nadler, N. J., and Leblond, C. P. (1951). Am. J. Roetgenol. Radium Therapy 65, 420.

Hiatt, H. H., and Bojarski, T. B. (1961). Cold Spring Harbor Symp. Quant. Biol. 26, 367.

Hooper, C. E. S. (1956). J. Histochem. Cytochem. 4, 531.

Hotta, Y., and Stern, H. (1965). J. Cell Biol. 25, 99.

Howard, A., and Pelc, S. R. (1953). Heredity 6 (Suppl. 1), 261.

Hsu, T. C., and Zenzes, M. T. (1965). In "Cellular Radiation Biology," M. D. Anderson Hospital Symp., p. 404. Williams & Wilkins, Baltimore, Maryland.

Hughes, W. L. (1957). Brookhaven Natl. Lab. Quart. Progr. Rept. No. BNL-439 (S-35):35.

Hughes, W. L., Bond, V. P., Brecher, G., Cronkite, E. P., Painter, R. B., Quastler, H., and Sherman, F. G. (1958). Proc. Natl. Acad. Sci. U.S. 44, 476.

Johnson, H. A., and Cronkite, E. P. (1959). Radiation Res. 11, 825.

Koburg, E., and Maurer, W. (1962). Biochim. Biophys. Acta 61, 229.

Leblond, C. P. (1959). In "The Kinetics of Cellular Proliferation" (F. Stohlman, ed.), pp. 31–47. Grune & Stratton, New York.

Leblond, C. P., and Carrière, R. (1955). Endocrinology 56, 261.

Leblond, C. P., and Messier, B. (1958). Anat. Record 132, 247.

Leblond, C. P., and Stevens, C. E. (1948). Anat. Record 100, 357.

Leblond, C. P., and Walker, B. E. (1956). Physiol Rev. 36, 255.

Leblond, C. P., Greulich, R. C., and Pereira, J. D. M. (1964). In "Advances in Biology of the Skin" (W. Montagna and R. E. Billingham, eds.) Vol. I, pp. 39–67. Macmillan, New York.

Lesher, S., Fry, R. J. M., and Kohn, H. I. (1961). Lab. Invest. 10, 291.

Littlefield, J. W. (1965). Biochim. Biophys. Acta 95, 14.

Loran, M. R., and Crocker, T. T. (1963). J. Cell Biol. 19, 285.

Marin, G., and Prescott, D. M. (1964). J. Cell Biol. 21, 159.

McMinn, R. M. H., and Mitchell, J. E. (1954). J. Anat. (London) 88, 99.

McQuade, H. A., Friedkin, M. and Atchison, A. A. (1956). Exptl. Cell Res. 11, 249.

Messier, B., and Leblond, C. P. (1957). Proc. Soc. Exptl. Biol. Med. 96, 7.

Moog, F. (1953). J. Exptl. Zool. 124, 329.

Natarajan, A. T. (1961). Exptl. Cell Res. 22, 275.

O'Brien, J. S. (1962). Cancer Res. 22, 267.

Painter, R. B., Drew, R. M., and Hughes, W. L. (1958a). Science 127, 1244.

Painter, R. B., Forro, F., Jr., and Hughes, W. L. (1958b). Nature 181, 328.

Painter, R. B., Drew, R. M., and Rasmussen, R. E. (1964). Radiation Res. 21, 355.

Perrota, C. A. (1961). Anat. Record 139, 239.

Perry, R. P. (1964). In "Methods in Cell Physiology" (D. M. Prescott, ed.), Vol. I, pp. 305-326. Academic Press, New York.

Pilgrim, C., and Maurer, W. (1965). Exptl. Cell Res. 37, 183.

Pilgrim, C., Erb, W., and Maurer, W. (1963). Nature 199, 863.

Potter, R. L., and Nygaard, O. F. (1963). J. Biol. Chem. 238, 2150.

Potter, V. R. (1959). In "The Kinetics of Cellular Proliferation" (F. Stohlman, ed.), p. 104. Grune & Stratton, New York.

Quastler, H., and Sherman, F. G. (1959). Exptl. Cell Res. 17, 420.

Reichard, P., and Estborn, B. (1951). *J. Biol. Chem.* **188**, 839.
Rolshoven, E. (1951). *Verhandl. Anat. Ges.* (*Jena*) **49**, 189.
Rubini, J. R., Cronkite, E. P., Bond, V. P., Fliedner, T. M., and Hughes, W. L. (1958). *Clin. Res.* **6**, 267.
Sisken, J. E., and Kinosita, R. (1961). *J. Biophys. Biochem. Cytol.* **9**, 509.
Stone, G. E., and Cameron, I. L. (1964). *In* "Methods in Cell Physiology" (D. M. Prescott, ed.), Vol. I, pp. 127-140. Academic Press, New York.
Storey, W. F., and C. P. Leblond (1950). *Ann. N.Y. Acad. Sci.* **53**, 537.
Thrasher, J. D., and Greulich, R. C. (1965a). *J. Exptl. Zool.* **159**, 39.
Thrasher, J. D. (1966). Submitted to *J. Exptl. Zool.*
Thrasher, J. D., and Greulich, R. C. (1965b). *J. Exptl. Zool.* **159**, 385.
Thrasher, J. D., and Greulich, R. C. (1966). *J. Exptl. Zool.* **161**, 9.
Tonna, E. A. (1958). *Stain Technol.* **33**, 255.
Walker, B. E. (1960). *Am. J. Anat.* **107**, 95.
Walker, B. E., and Leblond, C. P. (1958). *Exptl. Cell Res.* **14**, 510.
Wimber, D. E. (1959). *Proc. Natl. Acad. Sci. U.S.* **45**, 839.
Wimber, D. E., and Quastler, H. (1963). *Exptl. Cell Res.* **30**, 8.

Chapter 13

Patterns of Cell Division: The Demonstration of Discrete Cell Populations[1]

SEYMOUR GELFANT

Department of Zoology, Syracuse University, Syracuse, New York

I. Introduction

In previous reports (Gelfant, 1962, 1963a,b) it was shown that a single layer of epidermal cells in the same tissue can be subdivided into distinct

---

[1] This investigation was supported (in part) by a research grant, GB-2803 from the National Science Foundation.

and separate cell types, which differ in their patterns of behavior during
the cell division cycle or in their physiological requirements for mitosis.
Two major cell types that exist within the basal layer of epidermis were
defined and named the "G_1" and the "G_2" epidermal cell populations. G_1
cells have a long G_1 period, a short G_2 period, and move through the cell
cycle ($G_1 \rightarrow S \rightarrow G_2 \rightarrow M$) in the usual manner. The designation "G_1
population" refers to the fact that mitosis is first initiated in these cells

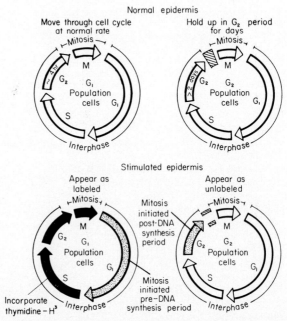

Fig. 1.   Diagrams of the cell division cycle characterizing the differences between
G_1 and G_2 cell populations in normal and in stimulated mouse epidermis (after Gelfant, 1962, 1963a,b).

from the G_1 period of the cycle. Experimentally stimulated G_1 cells enter
S, incorporate thymidine-H^3, move through G_2, and appear as *labeled*
mitoses in autoradiographs. Most epidermal cells are G_1 cells.
    In contrast, G_2 cells do not automatically pass through the cell cycle
in normal epidermis. Rather they hold up in the G_2 period for as long as
several days. The designation "G_2 population" refers to the fact that these
cells have an extraordinarily long G_2 period and that mitosis is initiated
from the G_2 period of the cycle. When these cells are stimulated experimentally, they do not incorporate thymidine-H^3 because they are in the
post-DNA synthesis period, and therefore appear as *unlabeled* mitoses in
autoradiographs. About 5–10% of the epidermal cells in the basal layer

are G_2 cells. See Fig. 1 for an illustrated description of G_1 and G_2 population cells in normal and in stimulated epidermis.

In vitro evidence was also presented (Gelfant, 1963a) for the existence of separate sugar-responding, sodium ion-responding, and potassium ion-responding epidermal cells. These "physiological subpopulations" belong to the major G_2 category of epidermal cells; they display their specific physiological requirements for mitosis during the G_2 period of the cell cycle.

This chapter will explain the principles and the procedures used to demonstrate these various cell populations in mouse epidermis. In addition, emphasis will be placed on the methods for demonstrating and distinguishing G_1 and G_2 cells within the same tissues in a variety of animal and plant experimental systems.

II. Demonstration of G_1 and G_2 Cell Populations in Mouse Epidermis

A. Materials and Methods

1. ANIMALS

Adult male (C57BL/6J) mice (ca. 25 gm, obtained from the Jackson Laboratory, Bar Harbor, Maine) are used in groups of five. The animal-room temperature is maintained at 22–23°C. Temperatures above 26°C stimulate mitotic activity in mouse ear epidermis.

2. EXPERIMENTAL SYSTEMS

Five experimental systems can be studied in each animal. (1) *Wounded ear epidermis:* One ear is wounded by making two radial cuts through the ear with a No. 11 surgical blade. (2) *Intact ear epidermis:* The other ear remains intact throughout the experimental period. (3) *Plucked body-skin epidermis:* The hairs are plucked from a dorsolateral area of the body. (4) *Wounded body-skin epidermis:* The hairs are gently clipped on the other side of the body and a 1-cm-long cut is made through the skin. (5) *Intact body-skin epidermis:* An adjacent area of clipped skin remains intact.

At the end of the experimental period, the wounded ear fragment is removed, a comparable sized fragment is cut from the intact ear, and rectangular areas of plucked, wounded, and intact body skin are cut out for fixation. The pieces of body skin are flattened on cards to prevent curling during fixation and to facilitate orientation for sectioning. Tissues are separated from cards prior to paraffin infiltration.

3. Autoradiographic and Counting Procedures

Tissues are fixed in Carnoy's acetic acid–ethyl alcohol, 1:3. Paraffin sections are cut at 4μ. Kodak NTB liquid emulsion is used for autoradiography. The slides are exposed at 4°C for 20 days, developed for 6 minutes in Kodak D-19 developer, rinsed in distilled water, fixed for 10 minutes in Kodak acid fixer, washed in running tap water for 30 minutes, stained with Ehrlich's acid hematoxylin and eosin, and cover glasses applied using Technicon mounting medium. Cell nuclei are first overstained

Experimental systems

■ Areas of epidermis counted in histological sections

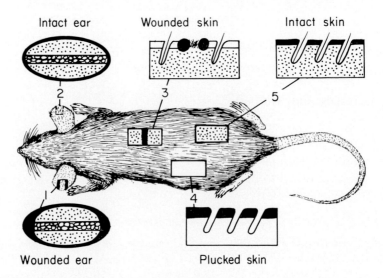

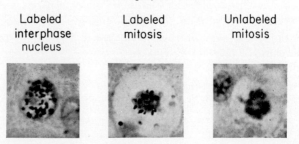

FIG. 2.   Procedures for studying G_1 and G_2 cell populations in mouse epidermis *in vivo.*

with hematoxylin, then destained to the desired intensity with acid alcohol. Acid alcohol also removes any nonspecific background staining of the emulsion. Poorly stained sections can be restained at any time by removing the cover glasses in xylene and running the slides down to 70% alcohol. (For a detailed description of autoradiography with liquid emulsion, see Prescott, 1964; Kopriwa and Leblond, 1962; the instruction manuals issued by Schwarz Bio Research, Inc., Orangeburg, New York, and by Controls for Radiation, Inc., Cambridge, Massachusetts.)

The ear fragments are sectioned transversely, the rectangular pieces of body skin are sectioned longitudinally, and autoradiographic counts are made in the basal cell layer of epidermis in areas extending around the ear fragments, 1 mm on either side of the wounded body skin, and throughout the section lengths of plucked and of intact body skin (see diagrams of histological sections and areas of epidermis counted, Fig. 2). Counts of labeled interphase nuclei and of labeled and unlabeled mitoses are made with a bright-field oil immersion objective. Counts can be recorded either in unit lengths of 1 cm of epidermis (by calibrating the number of oil immersion fields per centimeter) or as the number of labeled or unlabeled mitoses per 100 mitoses for each animal. Counts are made in every third section on the slide to avoid recording the same nucleus twice. Unlabeled mitoses can be verified by checking adjacent serial sections. Figure 2 illustrates the methods of setting up the experimental systems, the section areas of epidermis in which autoradiographic counts are made, and the various autoradiographic labeling situations that can be recorded in each experiment.

4. COMPOUNDS USED

Thymidine-H^3 (TdR-H^3) of specific activity 1.9 c per millimole (from Schwarz Bio Research, Inc., Orangeburg, New York) is injected intraperitoneally in concentrations of 10–100 μc/0.1 ml saline or distilled water. In most experiments, TdR-H^3 is injected in a concentration of 20 μc, which is the equivalent of 0.8 μc TdR-H^3 per gram body weight. Uniform concentrations of ready-to-inject TdR-H^3 may be obtained (from Schwarz Bio Research, as a special vialing service). Some experiments may require C^{14}-labeled thymidine. I have used thymidine-2-C^{14} of specific activity 25 mc per millimole (from New England Nuclear Corp., Boston, Massachusetts) in a concentration of 5 μc/0.1 ml saline per intraperitoneal injection.

Epidermal mitoses are collected *in vivo* by arresting dividing cells in metaphase with colchicine. Colcemid (from Ciba Laboratories, Ltd., Summit, New Jersey) is injected intraperitoneally in a concentration of 0.1 mg/0.1 ml saline (or 0.004 mg Colcemid per gram body weight) 5 hours before killing the animals by cervical dislocation. Vincaleukoblastine (VLB) (from Eli Lilly & Co., Indianapolis, Indiana) is also a powerful metaphase inhibitor (Gelfant, 1963c) and can be used in the same way as Colcemid.

B. Experimental Designs and Explanation of Results

1. Wounding or Hair Plucking *in Vivo*

Relatively few epidermal cells are engaged in the process of cell division at the time the epidermis is stimulated by wounding or by hair plucking. Only 1.7% of the cells are in S, and less than 0.1% are in M, in normal unstimulated epidermis (Gelfant, 1963a, 1962), so that about

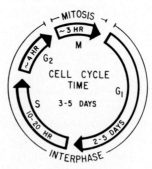

Fig. 3. Average cell cycle, mouse ear and body skin epidermis *in vivo* (Gelfant, unpublished results, 1965).

98% of the cells in the basal layer of epidermis are not moving through the cell cycle when the experiment begins. Wounding or hair plucking initiates cell division in part of this large fraction of "resting" cells. Epidermal cells stimulated by cutting the ear or the body skin, however, do not actively enter mitosis until about 48 hours after wounding. Up until that time the epidermis is involved in the process of wound closure. A similar delayed response in epidermal mitotic activity also occurs when body-skin epidermis is stimulated by hair plucking (for references, see Gelfant, 1963a). The purpose of the experiment is to determine the history and to demonstrate the pattern of cell division of those epidermal cells that are entering mitosis for the first time—about 48 hours after stimulation by wounding or by hair plucking.

Figure 4 illustrates the over-all experimental design and the kinds of result one obtains from this type of an experiment. The epidermis is stimulated at zero hours. Epidermal cells are then exposed to the continuous presence of TdR-H³ for 48 hours. The procedure of injecting TdR-H³

(20 μc) every 6 hours is considered continuous exposure, because 6 hours is less than the S period in mouse epidermis (see Fig. 3). Thus no cell could move into or out of S without being exposed to TdR-H³. Epidermal mitoses are collected with colchicine during the last 5 hours of the experimental period. Autoradiographic counts are made of the numbers of labeled and unlabeled mitoses *arrested by colchicine*, and the results in Fig. 4 are interpreted as follows:

a. The high degree of mitotic activity in stimulated epidermis (both labeled and unlabeled mitoses) shows that epidermal cells were activated to move through the cell division cycle. This is also demonstrated by the increased number of labeled interphase nuclei after stimulation by wounding or by hair plucking (Gelfant, 1963a).

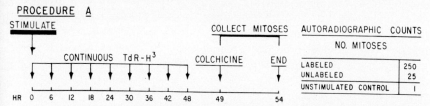

FIG. 4. Experimental design for demonstrating G_1 and G_2 cell populations in mouse epidermis stimulated by wounding or by hair plucking.

b. *Labeled mitoses* are cells that were "resting" in the G_1 period of the cycle when the experiment began. These epidermal cells were stimulated to enter S where they incorporated TdR-H³ and became labeled. They have passed through S and have moved through G_2 bearing the label; they now appear as *labeled mitoses*, having entered mitosis for the first time, 49–54 hours after stimulation. Epidermal cells stimulated from the G_1 period and that go through this pattern of cell division are called G_1 population cells.

c. The small number of mitoses collected by colchicine in the unstimulated control epidermis also represent G_1 population cells which were moving through the cell cycle in normal epidermis (see Fig. 1). It should be pointed out that this G_1 pattern of behavior in both stimulated and unstimulated systems had been considered the general rule of cell division for all higher animal and plant cells capable of renewal (for references, see Gelfant, 1962).

d. *Unlabeled mitoses* represent epidermal cells that were "resting" in the G_2 period when the experiment began. They could not have entered mitosis unlabeled from any other stage of the cell cycle; they were probably "resting" in G_2 because of the relatively large number of cells activated. Thus, the unlabeled mitoses collected in stimulated epidermis are cells which were in the G_2 period at zero time. *These cells have remained in the G_2 period for at least 48 hours*—as measured by the presence of TdR-H³ throughout the experiment. They were stimulated to enter mitosis from the G_2 period (the mitotic stimulus produced by wounding or by hair plucking can remain effective *in vivo* for 4 or 5 days) (Gelfant, 1962, 1963b), and now appear as *unlabeled mitoses*, 49–54 hours after wounding or hair plucking. Epidermal cells that behave in this manner are called G_2 population cells.

It should be emphasized that the numbers of labeled and unlabeled cells that enter mitosis after wounding or hair plucking *in vivo* do not reflect the relative proportions of G_1 and G_2 population cells actually present in normal unstimulated epidermis. It will be shown that these numbers vary with and are a function of the particular experimental design.

2. COMBINED *in Vivo–in Vitro* PROCEDURES

Cutting the ear into small pieces prior to incubation specifically induces epidermal cells to enter mitosis *in vitro* (Gelfant, 1959a). In contrast to wounding *in vivo*, epidermal cells enter mitosis *in vitro* without delay, and mitotic activity occurs uniformly throughout the ear fragment. This system, combined with pretreatment *in vivo*, is particularly useful for demonstrating the existence of G_2 population cells in mouse ear epidermis. Table I outlines an experiment and Fig. 5 illustrates the principles involved in this method.

Epidermal cells are first exposed to the continuous presence of TdR-H³ for a rather long period of time *in vivo*. The ears are removed. One ear is cut into fragments. The other ear serves as a control and remains intact. Fragments or whole ears are incubated in Warburg flasks at 38°C in a saline medium containing glucose. After 1 hour, to allow all mitoses originally present in the epidermis to pass beyond metaphase, colchicine is tipped into the main vessels of the Warburg flasks and incubation continued further for 4 hours. The whole ears are cut into fragments to facilitate fixation and to study areas of epidermis which are comparable to the experimentally cultured ear fragments. All tissues are sectioned longitudinally and autoradiographed to see if the mitoses collected by colchicine are labeled or unlabeled. (For further information on the *in vitro* technique, see Section V,A.)

The results in Table I show that cutting the ear into fragments stimulates epidermal cells to enter mitosis *in vitro*. Cells do not enter mitosis in comparable areas of epidermis when the ear is cultured intact. These results establish the cause and set the time of stimulation to the point immediately prior to incubation (see Fig. 5).

In view of the short time between stimulus and response, all epidermal cells that enter mitosis *in vitro* must have been in the G_2 period of the

TABLE I

In Vivo-in Vitro EXPERIMENT DEMONSTRATING G_2 POPULATION CELLS IN MOUSE EAR EPIDERMIS

Experimental design
(5 adult male mice used)

1. TdR-H^3 (20 μc) *in vivo* every 6 hours for 18 hours
2. Animals killed
3. One ear cut into fragments; other ear whole
4. Ear fragments and whole ears maintained *in vitro*
5. Mitoses collected with colchicine, 4 hours
6. Experiment ends; whole ear cut into fragments for fixation

Autoradiographic results

Mitoses (per cm unit length of epidermis)	Ear fragments (A)	Whole ears (B)
Labeled	1	0
Unlabeled	10	0

Ear fragments   Whole ear

(A)   (B)

cell cycle when the ear was stimulated by cutting. Since TdR-H^3 was continuously available *in vivo*, any cycling cell which passed through S would reach the G_2 period labeled. And, upon stimulation, this cell would enter mitosis *in vitro labeled*. By the same reasoning, any cell that enters mitosis *in vitro unlabeled* was not cycling *in vivo*; moreover, it must have been "resting" in G_2 throughout the entire experimental period during which TdR-H^3 was present *in vivo*. Labeled mitoses therefore represent G_1 population cells in the sense that they were moving through the cell cycle *in vivo*. Unlabeled mitoses, on the other hand, represent G_2 population cells which had held up in the G_2 period for at least 18 or 48 hours, as shown by the experiments in Table I and Fig. 5.

This procedure shows that G_2 population cells exist in mouse ear epidermis *in vivo*, and that they can be made to enter mitosis without delay under the appropriate experimental conditions. If there were no G_2 cells, all mitoses would come through *in vitro* labeled. As it is, most of the epidermal cells that enter mitosis *in vitro* are unlabeled. The few labeled mitoses represent the small number of cycling cells usually present in normal unstimulated epidermis (see Fig. 1). Thus, this combined *in vivo–in vitro* procedure selectively demonstrates G_2 population

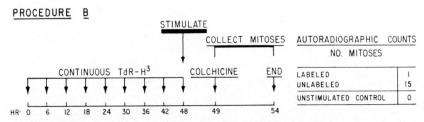

PROCEDURE B

1. Labeled mitoses = G_1 population cells; were passing through normal cell cycle in presence of TdR-H³; stimulated to enter M, appear as labeled mitoses.
2. Unlabeled mitoses = G_2 population cells; were in G_2 period for at least 48 hours as measured by presence of TdR-H³; stimulated to enter M, appear as unlabeled mitoses.

FIG. 5. *In vivo–in vitro* procedure for distinguishing G_1 and G_2 cell populations in mouse ear epidermis. TdR-H³ is administered *in vivo*; the ear is stimulated by cutting; epidermal mitoses are collected *in vitro*.

cells in mouse ear epidermis. It should be pointed out again that the ratio of labeled to unlabeled mitoses scored in these experiments does not reflect the actual proportion of G_1 and G_2 cells as they exist in normal ear epidermis.

The demonstration of G_2 population cells in mouse ear epidermis after wounding *in vivo* depends upon the inherent delay of about 48 hours between the time the epidermis is wounded and the time epidermal cells first enter mitosis. It was postulated that the original wounding stimulus was sustained *in vivo* throughout this interval, and the delay in time was attributed to the process of wound closure (see discussion of Fig. 3). Since there is no delay in mitotic activity when cut ear fragments are placed *in vitro* (as was shown above), a combined *in vivo–in vitro* experiment can be designed to study these questions (see Table II).

The ear is wounded by cutting a semi-isolated fragment which remains attached to the ear (see Fig. 2), and the animals are exposed to the continuous presence of TdR-H³ for 18 hours *in vivo*. At this stage of wound healing, there are as yet no signs of wound closure or mitotic stimulation (Bullough and Laurence, 1957) or of increased deoxyribonucleic acid (DNA) synthesis (Gelfant, 1963a) in epidermal cells near the wounded edges of the ear fragment. So as far as epidermal cell division is concerned, the wounded fragments resemble unstimulated epidermis when they are removed after 18 hours and placed *in vitro*.

TABLE II

MODIFIED *in Vivo–in Vitro* EXPERIMENT DEMONSTRATING THE MOBILIZATION OF G_2 POPULATION CELLS AFTER WOUNDING

Experimental design (5 adult male mice used)
1. Ears wounded by two radial cuts (see Fig. 2)
2. TdR-H³ (20 μc) *in vivo* every 6 hours for 18 hours
3. Animals killed
4. Wounded ear fragments removed, maintained *in vitro*
5. Mitoses collected with colchicine, 4 hours

Autoradiographic results	
Mitoses (per cm unit length of epidermis)	Wounded ear fragments (18 hours *in vivo* → 5 hours *in vitro*)
Labeled	2
Unlabeled	30

The results in Table II show that the original wounding stimulus is sustained *in vivo*, and that mitotic activity does not depend upon wound closure. A relatively large number of epidermal cells enter mitosis when the wounded ear fragments are transferred to *in vitro*. Most of these are G_2 population cells as defined by the presence of TdR-H³ throughout the experimental period *in vivo*. There is a substantial increase in the number of G_2 cells that are activated when the epidermis is wounded and allowed to remain *in vivo* for 18 hours (compared to the situation in Table I where mitoses are collected *in vitro* directly after the ear fragments are cut). Thus, G_2 population cells are progressively mobilized with time after wounding *in vivo*. This can be demonstrated by allowing these cells to enter mitosis prematurely when the wounded ear fragments are placed *in vitro*.

G_1 cells also are mobilized by wounding, but have a latent response of about 24 hours before moving into S *in vivo* (Gelfant, 1963a). Therefore, they do not show up *in vitro* in either of the experiments (Tables I and II). The few labeled mitoses that do appear come from the unstimulated pool of cycling G_1 cells always present in normal epidermis (Fig. 1).

Wound-activated G_1 population cells can be demonstrated *in vitro*, however, if the epidermis is stimulated and then placed *in vitro* in the presence of TdR-H^3, as shown in Table III. The *in vivo* latent response period is eliminated under these conditions. G_1 cells enter S *in vitro* shortly after stimulation. They incorporate TdR-H^3 and appear as labeled interphases in sectioned autoradiographs of the cultured ear fragments.

The results in Table III illustrate and summarize the main points covered so far. Wounding initiates DNA synthesis and mitosis in mouse epidermis. Both reactions are usually delayed *in vivo*. However, epi-

TABLE III

DEMONSTRATION OF BOTH G_1 AND G_2 CELL POPULATIONS
IN MOUSE EAR EPIDERMIS *in Vitro*

Experimental design (5 adult male mice used)		
1. Animals killed; one ear cut into fragments, other ear whole		
2. Ear fragments and whole ears maintained *in vitro* in the presence of TdR-H^3 (1 µc/ml)		
3. Mitoses collected with colchicine, 4 hours		

Autoradiographic results		
Counts (per cm unit length of epidermis)	Ear fragments	Whole ears
Labeled interphase nuclei	75.0	0.5
Labeled mitoses	0.5	0
Unlabeled mitoses	6.5	0

dermal cells enter DNA synthesis or mitosis without delay if wounded skin is placed *in vitro*. Epidermal cells resting in the G_1 period *in vivo* are activated by cutting the ear prior to incubation. These cells are free to enter S *in vitro* where they incorporate TdR-H^3, and can be detected as labeled interphase nuclei in autoradiographs. Other epidermal cells, resting in the G_2 period *in vivo*, also are activated by cutting the ear. These cells enter mitosis where they are arrested by colchicine *in vitro*, and appear as unlabeled mitoses in autoradiographs. (The experimental design in Table III could be improved on this point by previous administration of TdR-H^3 *in vivo*, as was done in Table I.) Table III also demonstrates the small fraction of normally cycling G_1 cells. These are represented by the few labeled interphases recorded in the unstimulated whole ear and by the few labeled mitoses that came through in the stimulated ear fragments *in vitro*. Both cases indicate cycling G_1 cells which were in the S period at the time the tissues were placed *in vitro*.

The cells that wound up as labeled mitoses must have been in late S to have incorporated TdR-H^3, moved through G$_2$, and entered mitosis labeled, within 5 hours *in vitro*.

Thus G$_1$ and G$_2$ epidermal cell populations can be activated and distinguished by a number of experimental procedures, both *in vivo* and *in vitro*.

III. Demonstration of G$_1$ and G$_2$ Cell Populations in Other Experimental Systems

The following account, drawn from the literature, shows that G$_1$ and G$_2$ population cells exist in a wide variety of tissues and can be demonstrated by the same procedures used for mouse epidermis (see Fig. 6). The examples have been organized and will be discussed according to the experimental designs and procedures outlined in Fig. 6.

A. Tobacco Pith Tissue *in Vitro* (Procedure A)

The report by Patau and Das (1961) on tobacco pith tissue presents the most convincing evidence for the existence of G$_1$ and G$_2$ cells in a differentiated plant tissue. Tobacco pith tissue is composed of polyploid cells (containing 2C, 4C, or 8C DNA content nuclei) which do not divide *in vivo*. Cell division is initiated by cutting a cylinder of pith lengthwise and placing it with the cut surface down on a nutrient medium containing kinetin and indoleacetic acid. Pieces of pith tissue are cultured under these conditions in the continuous presence of TdR-H^3 for 16 days. Samples are removed at various intervals and mitotic nuclei are examined first by Feulgen microspectrophotometry and then by autoradiography. In addition, "new cells," i.e., cells which have undergone one or more divisions in culture, can be distinguished morphologically from "old cells" which were entering mitosis for the first time *in vitro*. In this way, one can accurately relate the *in vivo* history to the *in vitro* performance of any pith cell in mitosis.

Their results showed two populations of "old cells" (the cells we are concerned with here) whose interphase DNA contents at the beginning of culture could be related to their corresponding mitotic DNA contents after culture, and to whether or not these cells had undergone DNA synthesis *in vitro* in preparation for mitosis (i.e., whether mitoses were labeled or unlabeled).

One group of "old cells," with interphase DNA contents of 2C, 4C,

PROCEDURE A

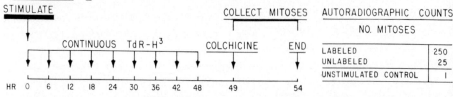

1. Labeled mitoses = G_1 population cells: were in G_1 period at 0 hours; stimulated to enter S, incorporate TdR-H^3, move through short G_2 period, and appear as labeled mitoses.
2. Unlabeled mitoses = G_2 population cells: were in G_2 period at 0 hours and for at least 48 hours as measured by presence of TdR-H^3; stimulated to enter M, appear as unlabeled mitoses.

PROCEDURE B

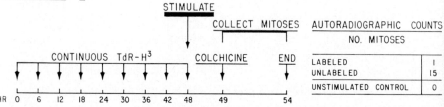

1. Labeled mitoses = G_1 population cells: were passing through normal cell cycle in presence of TdR-H^3; stimulated to enter M, appear as labeled mitoses.
2. Unlabeled mitoses = G_2 population cells: were in G_2 period for at least 48 hours as measured by presence of TdR-H^3; stimulated to enter M, appear as unlabeled mitoses.

PROCEDURE C

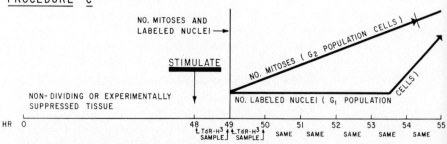

Evidence for two cell populations: increase in no. mitoses occurs promptly and before increase in no. labeled nuclei.

G_2 population cells: were in G_2 period for at least 48 hours in non-dividing or suppressed tissue; enter mitosis shortly after stimulation.

G_1 population cells: were in G_1 period in non-dividing or suppressed tissue; stimulated to move into S after a delay of several hours; will pass through short G_2 period and enter mitosis.

FIG. 6. Explanation and comparison of the main procedures used to demonstrate G_1 and G_2 cell populations in animal and in plant tissues. In principle, these procedures have been used (A) in mouse epidermis and kidney *in vivo*; tobacco pith tissue *in vitro*; germinating wheat, pea, lettuce, and corn seeds; human peripheral leucocytes *in vitro*; (B) in mouse epidermis *in vivo–in vitro* and mouse kidney *in vivo*; (C) in chicken esophagus and rat kidney *in vivo*.

and 8C, synthesized DNA in culture and appeared as labeled mitoses containing the corresponding double DNA polyploid values of 4C, 8C, and 16C. These were G_1 population cells which had been in the G_1 period of interphase *in vivo* when the experiment began. They were stimulated to enter S, incorporated TdR-H³ *in vitro*, and then came through as labeled mitoses, essentially in the manner outlined (Procedure A, Fig. 6).

Another group of "old cells" with interphase DNA contents of 4C and 8C entered mitosis *in vitro* without synthesizing DNA during the experimental period. These mitotic cells were unlabeled and contained the original tetraploid and octoploid interphase DNA values of 4C and 8C. They were G_2 population cells which had already synthesized DNA and were resting in the G_2 period of interphase *in vivo*. Upon stimulation, these cells went directly from G_2 into mitosis *in vitro*. They did not incorporate TdR-H³ in culture and therefore appeared as unlabeled mitoses (refer to Procedure A, Fig. 6).

It should be pointed out that unlabeled mitoses ("old cells") were observed after 8 and even 16 days of culture in the continuous presence of TdR-H³, thus demonstrating that tobacco pith G_2 population cells can hold up in the G_2 period for at least 16 days under experimental culture conditions—and probably much longer under natural conditions *in vivo*. This report also demonstrates that G_2 population cells can exist as polyploid cells.

B. Germinating Wheat, Pea, Lettuce, and Corn Seeds (Procedure A)

The following reports show that G_1 and G_2 population cells (1) exist in various embryonic tissues of dry seeds, (2) hold up in the G_1 or the G_2 periods of the cell cycle for months or even years in dormancy, (3) can be stimulated to enter mitosis by germination, and (4) can be distinguished from one another by germinating seeds in the presence of thymidine-H³ and applying autoradiography.

The study by Avanzi *et al.* (1963) uses Feulgen microspectrophotometry, thymidine-H³ autoradiography, and chromosome breakage by X-irradiation to demonstrate G_1 and G_2 population cells in the radicle meristem of dry *durum* wheat seeds. They first examine radicle tips of dry seeds and find 2C and 4C interphase nuclei present in a 1 : 2 ratio. These two classes of cells are then activated and traced by germinating other seeds for 24 hours in the presence of TdR-H³. Both labeled and unlabeled mitoses came through in a 4 : 6 ratio, showing that the original 2C and 4C nuclei represented G_1 and G_2 population cells (see Procedure

A for explanation). These results are confirmed in another way. Dry seeds
are irradiated, germinated in the presence of TdR-H³, and anaphase
chromosomes are examined for aberrations and for autoradiographic
labeling. Anaphases with chromosome breaks were all labeled, showing
that they had originally been hit by radiation in the unsplit condition,
i.e., during the G_1 period, and had synthesized DNA during germination.
In contrast, most of the anaphases with chromatid aberrations were
unlabeled—indicating that they had been in the split condition or in
the G_2 period when X-irradiated and had entered mitosis directly from
the G_2 period. In addition, the over-all ratio of chromosome to chromatid
aberrations was 1 : 2. Thus these three sets of results provide conclusive
evidence for the existence of G_1 and G_2 cell populations in wheat seeds.

The other reports in this category also demonstrate G_1 and G_2 cells
by autoradiography after germinating dry seeds in the continuous presence of TdR-H³. Bogdanov and Iordanskii (1964) estimate that around
10% of the cells in the dormant meristem of the embryonic radicle of
dry pea seeds are G_2 population cells. G_2 cells are also present in the
meristem of mature pea root segments (Torrey, 1961, and personal communication, 1964). The hypocotyl epidermis of lettuce seeds contains
G_2 cells (Feinbrun and Klein, 1962). The possibility of seed coat impermeability to TdR-H³ was excluded in this study by puncturing or
fracturing seeds prior to germination and obtaining the same results as
in the intact ones. Finally, G_2 population cells have been demonstrated
in shoot and in root tissues of germinating corn embryos (Stein and
Quastler, 1963).

C. Human Peripheral Leucocytes *in Vitro* (Procedure A)

The report on leucocytes by Gandini and Gartler (1964) is another
example of demonstrating G_1 and G_2 population cells according to Procedure A, Fig. 6. Two different substances, phytohemagglutinin (PHA)
obtained from *Phaseolus vulgaris* kidney beans, and a yeast extract
(YAF) prepared from *Saccharomyces cerevisiae,* were used to initiate
cell division in human mononuclear leucocytes *in vitro.* Leucocytes are
separated from freshly drawn human blood and cultured for 5–6 days in
a commercial tissue culture medium containing PHA, YAF, or both, and
TdR-H³, and mitoses are collected (arrested in metaphase) with colchicine during the last 8 hours of the experimental period. Mitotic counts
showed that there was an additive mitogenic effect when both PHA and
YAF were added to the same culture. Autoradiographic counts showed
that all metaphases in the PHA culture were labeled, whereas most of

the metaphases in the YAF culture were unlabeled, and that both appeared in the PHA + YAF culture.

The additive mitogenic effect with PHA and YAF suggests that there are two populations of mononuclear leucocytes with different sensitivities or requirements for cell division *in vitro*. The 100% labeling of metaphases in the PHA culture shows that PHA specifically activates G_1 population cells. The fact that most of the metaphases in the YAF culture were unlabeled shows that YAF activates G_2 population cells. The fact that there were many labeled metaphases (over 90%) in the YAF + PHA culture indicates that YAF in itself does not inhibit TdR-H^3 incorporation.

These experiments also demonstrate that G_2 population leucocyte cells can hold up in the G_2 period for as long as 6 days *in vitro*—and probably much longer since circulating leucocytes do not divide *in vivo* (Nowell, 1960, 1964). For comparison, the normal (or G_1 population) G_2 period in cultured leucocytes is around 4–6 hours (Bender and Prescott, 1962; Nowell, 1964). Thus, human peripheral mononuclear leucocytes are composed of a large population of G_1 cells and a small population of G_2 cells, and this can be demonstrated *in vitro* by selective activation with either PHA or YAF.

See Fig. 7 for a diagrammatic explanation of the experiments described above on tobacco pith tissue, germinating wheat seeds, and human leucocytes *in vitro*.

D. Mouse Kidney *in Vivo* (Procedures A and B)

We have recently used the experimental designs outlined in Procedures A and B to demonstrate G_1 and G_2 population cells in mouse kidney (Pederson and Gelfant, 1965). Cell division is initiated by unilateral nephrectomy according to the methods described by Argyris and Trimble (1964). Unilateral nephrectomy results in a marked increase in mitotic activity in the proximal tubules of the contralateral kidney— which reaches a peak around 48 hours after surgical removal of one kidney (Williams, 1961; McCreight and Sulkin, 1962; Argyris and Trimble, 1964).

In Procedure A, animals were unilaterally nephrectomized and exposed to TdR-H^3 for 48 hours—either by injecting TdR-H^3 (20 μc) every 6 hours or by supplying TdR-H^3 (20 μc per milliliter) in the drinking water. In Procedure B, animals were exposed to TdR-H^3 for 48 hours before being stimulated by unilateral nephrectomy. Colchicine was injected to collect mitoses during the last 6 hours of the experimental

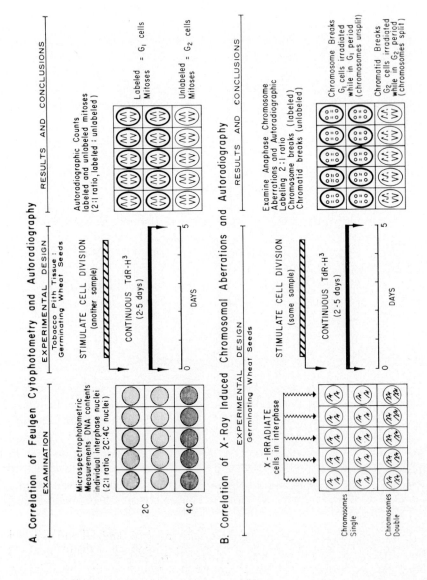

A. Correlation of Feulgen Cytophotometry and Autoradiography

EXAMINATION

Microspectrophotometric
Measurements DNA contents
individual interphase nuclei
(2:1 ratio, 2C:4C nuclei)

2C

4C

EXPERIMENTAL DESIGN

Tobacco Pith Tissue;
Germinating Wheat Seeds

STIMULATE CELL DIVISION
(another sample)

CONTINUOUS TdR-H³
(2-5 days)

0   DAYS   5

RESULTS AND CONCLUSIONS

Autoradiographic Counts
labeled and unlabeled mitoses
(2:1 ratio, labeled : unlabeled)

Labeled   = G₁ cells
Mitoses

Unlabeled   = G₂ cells
Mitoses

B. Correlation of X-Ray Induced Chromosomal Aberrations and Autoradiography

EXPERIMENTAL DESIGN

Germinating Wheat Seeds

X-IRRADIATE
cells in interphase

Chromosomes
Single

Chromosomes
Double

STIMULATE CELL DIVISION
(same sample)

CONTINUOUS TdR-H³
(2-5 days)

0   DAYS   5

RESULTS AND CONCLUSIONS

Examine Anaphase Chromosome
Aberrations and Autoradiographic
Labeling 2:1 ratio
Chromosome breaks (labeled)
Chromatid breaks (unlabeled)

Chromosome Breaks
G₁ cells irradiated
while in G₁ period
(chromosomes unsplit)

Chromatid Breaks
G₂ cells irradiated
while in G₂ period
(chromosomes split)

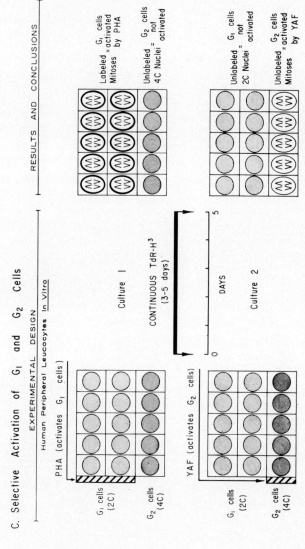

FIG. 7. Procedures used to verify and to demonstrate G_1 and G_2 cells. The ratios of G_1 and G_2 cells have been standardized for illustration purposes. (See text for further explanation; also refer to Procedure A, Fig. 6.)

periods, and autoradiographic counts were made of labeled and un-labeled mitoses. Around 6–8% of the mitoses collected in all experiments were unlabeled—showing that mouse kidney tubules contain G_1 and G_2 cell populations. (For further explanation see legends under Procedures A and B in Fig. 6.)

Procedure C has been used to demonstrate G_1 and G_2 cell populations in chicken esophagus and in rat kidney (see Fig. 6 for an illustrated explanation).

E. Chicken Esophagus *in Vivo* (Procedure C)

Inanition followed by feeding was used as a physiological tool to sup-press and then stimulate cell division in esophageal epithelium. (Cam-eron and Cleffmann, 1964). Newly hatched chicks were starved for 3 days, fed, given serial injections of TdR-H³, and killed at hourly intervals after each injection. Autoradiographic and mitotic counts were made to determine the DNA synthetic index (number of labeled interphase nuclei) and the mitotic index as a function of time after refeeding. Their results were essentially the same as those illustrated in Fig. 6. They also measured the S period of esophageal cells in another group of chicks and showed that the duration of S was not appreciably affected by starvation.

In summary, these results show that chicken esophagus contains two populations of cells which can be demonstrated by starvation and re-feeding. G_1 cells hold up in G_1, and G_2 cells hold up in the G_2 period during starvation (the S period is not affected). Refeeding stimulates G_2 population cells to enter mitosis immediately, and G_1 population cells to enter S after a delay of several hours. This is reflected by observing a prompt increase in the mitotic index—which occurs prior to the in-crease in the DNA synthetic index. If all esophageal cells had been sup-pressed in and stimulated from the G_1 period, there would have been no increase in the mitotic index until cells had synthesized DNA and passed through G_2. Such results were actually obtained in the duodenal epithe-lium (which is a rapidly dividing tissue) of starved-fed chicks. Thus, the existence of G_2 population cells may be related only to tissues having a slow-acting renewal rate (Gelfant, 1962).

F. Rat Kidney *in Vivo* (Procedure C)

Similar results were obtained by Stöcker *et al.* (1964) by producing compensatory hyperplasia of the rat kidney by ischemic shock. Cell di-vision was initiated in the right kidney by tying off the blood supply to the left kidney. Mitotic and DNA synthetic indices were recorded at hourly intervals for 8 hours after ischemia (see text above and Fig. 6 for how this was done). Once again the mitotic index rose rapidly during the first 6 hours, during which the DNA synthetic index remained level—before it too began to rise. And once again these data demonstrate the existence of G_1 and G_2 population cells—in this case, in a normally nondividing tissue, rat kidney.

It should be emphasized that the essential evidence for G_2 cells using Procedure C is the initial and immediate rise in the mitotic index and only during the first 3 or 4 hours after stimulation. After this, G_1 population cells which are also stimulated contribute to the rise in the mitotic index curve. For example, our results on mouse kidney indicate that G_1 cells are the predominant population and have a short G_2 period, so that over 90% of the mitoses collected by Procedure B were labeled G_1 cells, which probably entered mitosis during the last few hours of the 6-hour colchicine treatment.

IV. Additional Procedures for Detecting G_2 Cell Populations

The following is a brief description of some other procedures for activating and for detecting G_2 population cells.

a. Use of Heat Shock; Mouse Ear Epidermis in Vivo. It had previously been shown by Storey and Leblond (1951) that rat plantar epidermal cells were stimulated to enter mitosis by raising the temperature of the floor of the cage to 30°C. We have used this observation to activate and demonstrate G_1 and G_2 cell populations in mouse ear epidermis *in vivo*, according to Procedure B, Fig. 6. Mice are first exposed to the continuous presence of TdR-H³ for several days at an animal-room temperature of 22°–23°C. The cages are then moved to an environment of 35°C. After 1 hour of temperature equilibration, colchicine is injected and mitoses are collected for 5 hours at 35°C. This procedure results in a marked increase in epidermal mitotic activity of the ear and around 10% of the mitoses collected are unlabeled in autoradiographs. Thus heat shock initiates cell division in both G_1 and G_2 cells in mouse ear epidermis. This method can also be used to measure the ultimate duration of the G_2 population cells by administering TdR-H³ for many days or weeks prior to heat treatment.

b. Persistence of Heavily Labeled Nuclei after Pulse Administration of TdR-H³. This observation has been recorded in a number of experimental systems: in mouse body skin and ear epidermis *in vivo*, 10 days after a single injection of TdR-H³ (Gelfant, unpublished results, 1965); and in cultures of embryonic mouse pancreas, chick epidermis, and muscle cells, 5–8 days after a short exposure to TdR-H³ (Wessells, 1964a,b; H. Holtzer, personal communication, 1965). A pulse incorporation of TdR-H³ is usually dissipated by repeated cell divisions in the absence of label. This occurs within a relatively short period of time in a proliferating system, so that retention of a high level of labeling for many days may indicate that a cell was held up in the G_2 period and would therefore be considered a G_2 population cell. It should be emphasized that the persistence of heavy labeling in itself does not prove that these are G_2 cells. Additional information is required (i.e., quantitative grain counts or DNA measurements if the cell can be identified before autoradiography) to eliminate the possibility of a labeled daughter cell having held up in the G_1 period after the first division.

    c. Persistence of Unlabeled Nuclei after Continuous Exposure to TdR-H³. This is the reverse of the procedure described above and offers the advantage of easily combining autoradiography with Feulgen microspectrophotometry to verify results. In this case, a proliferating experimental tissue is exposed to the continuous presence of TdR-H³ for a period equal to 2 or 3 times the length of the total cell generation time—to make sure that all cycling cells have a chance to become labeled. The cells are then stained by the Feulgen reaction and autoradiographed. The presence of an unlabeled nucleus indicates that this cell has held up in either G_1 or G_2 throughout the period of TdR-H³ administration. The DNA contents of individual unlabeled nuclei are measured directly through the photographic emulsion by microspectrophotometry. In the absence of polyploidy, the appearance of 4C-DNA content unlabeled nuclei shows that these are G_2 population cells. This procedure is currently being used (De Cosse and Gelfant, 1966) to study cell populations in Ehrlich ascites tumor cells.

V. Methods for Detecting "Physiological Subpopulations" in Mouse Ear Epidermis *in Vitro*

    This section will illustrate the existence of further subgroups of epidermal cells which differ in their physiological requirements for mitosis *in vitro*. These "physiological subpopulations" belong to the major G_2

category and display their specific requirements during the G_2 period of the cycle.

A. The *in Vitro* Procedure

1. ANIMALS

Swiss albino (or C57 BL) male mice 3–5 months of age are used. Since adult male Swiss albino mice have a tendency to fight, our animals are separated at 7 weeks of age and kept in individually partitioned cages. Records of body weights and general fitness are kept and all unhealthy animals are eliminated. In addition to noting the general fitness of the animals, the ears of the mice are closely inspected for visible signs of injury or irritation, and only normal healthy ears are used. The experimental animals are uniformly killed at 10:00 hours by cervical dislocation. Their ears are detached with a pair of scissors and subsequently cultured *in vitro*. The specific methods for cutting the whole ear into smaller pieces prior to incubation *in vitro* are described in Figs. 8 and 9.

2. CULTURE TECHNIQUE

The technique for culturing mouse ear epidermis *in vitro* is derived from Bullough and Johnson (1951). The specific ionic composition of the culture medium is listed in Table IV. This standard medium is ordinarily maintained at a physiological pH by using a phosphate buffer (0.01 M Na_2HPO_4), pH 7.4. But in some cases a Tris buffer of the same molarity and pH is substituted to control for any phosphate ion effect on mitosis. Routine determinations of pH are made before and after each experiment. When glucose is used it is added to the standard medium in a final concentration of 0.002 M, which was found to be more effective than the 0.02 M concentration originally suggested by Bullough and Johnson (see Gelfant, 1959b). Colcemid dissolved in 0.04 ml standard medium is used in a concentration of 0.016 mg per flask.

The ears or pieces of ear, cut with a No. 11 detachable surgeon's blade, are placed in Warburg flasks containing 4 ml standard medium in the main vessel and 0.04 ml Colcemid in the side arm. If thymidine-H^3 is used it is added to the main vessel in a concentration of 5 μc/0.1 ml. The flasks, attached to the manometers, are placed in the constant temperature bath. After temperature equilibrium has been established, the joints and side-arm plugs are adjusted and tightened, and the whole assembly gently rocked at 38°C. After 1 hour of incubation Colcemid is tipped into the main vessel of the Warburg flask and incubation continued further for 4 hours.

3. Histological Technique and Determination of Epidermal Mitotic Activity

The ears or pieces of ears are fixed immediately upon removal from the Warburg flasks at the end of each experiment (Bouin's or acetic acid–alcohol). Since it was found that large pieces of ear or whole ears do not fix adequately, they are kept intact for incubation but are cut

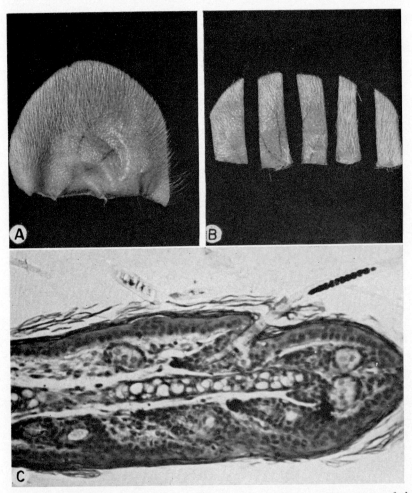

Fig. 8. Photographs of mouse ear epidermis as seen during various stages of the *in vitro* procedure. A: Whole intact ear after being removed from the animal. B: Five ear fragments cut from one ear. C: Longitudinal section of an ear fragment showing the complex histological composition—epidermis, dermis, hair follicles, sebaceous glands, and central cartilage.

into four longitudinal portions for fixation. Their original forms are re-constructed during the embedding procedure so that known serial sections across the whole ear can be made. (These special instructions are not required when using the regular sized ear fragments.) Longitudinal paraffin sections are cut at 7μ and stained with Ehrlich's acid hematoxylin and eosin solutions.

Each slide generally contains 30–60 sections, depending upon the size

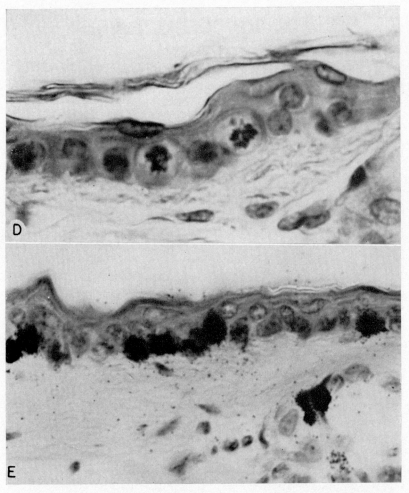

FIG. 8. (Cont'd.) Photographs of mouse ear epidermis as seen during various stages of the in vitro procedure. D: Mitotic figures in the basal layer of epidermis. Mitoses arrested by Colcemid in vitro. E: Autoradiographs; labeled interphase nuclei in the basal layer of epidermis of an ear fragment maintained in the presence of thymidine-H³ in vitro.

of the tissue. Epidermal mitotic activity is determined by counting the number of metaphase figures (arrested by Colcemid during the 4-hour incubation period) in unit lengths of 1 cm of epidermis. Counts are made on every other section, and five 1-cm counts are made per slide. When mitotic activity is studied in whole ear or in larger pieces of tissue, the longitudinal 7μ section is visually divided in half, and separate 1-cm counts are made of the tip and basal halves of the ear. In most cases, mitotic counts are made on ear fragments. These results are expressed as a basic unit count—which represents the average number of mitoses in twenty-five 1-cm unit counts obtained from five ear fragments. (See Fig. 8 for photographs of mouse ear epidermis during the various stages of the *in vitro* procedure.)

B. Cutting of the Ear as Mitotic Stimulant

We have alluded to the fact that cutting the ear into small pieces prior to incubation specifically induces epidermal cells to enter mitosis *in vitro* (see Fig. 8, Table I, and Section II,B). The diagrams and the results depicted in Fig. 9 provide evidence for this statement.

The usual method for studying mitosis in mouse ear epidermis *in vitro* makes use of fragments of ear approximately 2.5×5.0 mm in size, obtained from the ear as shown in diagram A (Fig. 9), and cultured in standard medium with glucose. Under these conditions, epidermal cells enter mitosis *in vitro* and mitotic activity occurs uniformly throughout the ear fragment. It had been assumed that the glucose in the culture medium and its subsequent conversion to energy within the cell initiated epidermal cells to enter mitoses *in vitro* (see Gelfant, 1960a, for a discussion of this question). The results in diagrams B to G disprove this assumption and show that cutting the ear prior to incubation is the initial stimulus required for epidermal mitosis *in vitro*. All tissues were cultured in the standard medium containing glucose.

When the whole ear is cultured intact (diagram B), the upper half is completely devoid of mitotic activity. Mitoses occur only in relation to the base of the ear where it was cut from the head. This suggestion that a cut edge is necessary for the development of mitosis is verified by the results in diagrams C and E, which demonstrate that epidermal mitoses occur only in relation to the cut edge and are distributed in the form of a mitosis gradient. Mitotic activity is highest wherever the ear has been cut, and falls off sharply with increasing distance from the cut edge.

Since facilitated penetration of nutrients through the cut edge might be responsible for these results, a paraffin sealing method was devised to distinguish the effects of penetration from those of cutting. The cut

edges of the ear are carefully sealed with melted paraffin to prevent the
entry of nutrients through the cut edges. The tissue is quickly dipped
into melted paraffin, and upon removal the paraffin hardens to produce
a narrow seal about 1 mm wide which covers the entire thickness of the
cut edge. The various tissues are cultured as usual. The paraffin is then

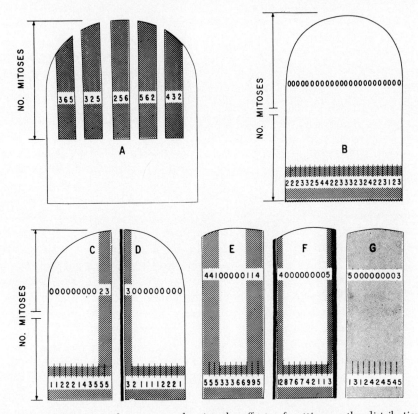

FIG. 9. Diagrams of mouse ear showing the effects of cutting on the distribution
and number of epidermal mitoses developing over a 4-hour period *in vitro*. Tissues
were cultured in standard medium containing 0.002 *M* glucose. The *shaded areas*
indicate the approximate location of mitoses; the *solid black line* shows where a cut
edge has been sealed with paraffin; the *stippled areas* indicate that the surfaces of
the ear have been sealed with paraffin—prior to incubation *in vitro*. A: The ear has
been cut into five ear fragments in the usual manner. B: The results of culturing the
whole intact ear. Mitoses occur only in relation to the base of the ear where it was
cut from the head. C: One half of an ear; the cut edge of the other half has been
sealed in D; E, F, and G: Large fragments of ear obtained by cutting the ear in
half and trimming the outer edges. E shows the distribution and number of mitoses
in such an ear fragment; both edges have been sealed in F; in G, the surfaces of
the ear fragment have been sealed with paraffin and the edges left free. (After
Gelfant, 1959a.)

easily stripped off at the end of each experiment before the tissues are fixed.

Penetration of nutrients is not responsible for mitosis, for when the cut edge is sealed (diagrams D and F) mitoses develop and in a specific relationship to the cut edge. Nor is penetration through the surface of the ear responsible for mitosis, for when the surfaces are sealed with paraffin (diagram G) mitoses again develop and are distributed in a specific gradient from the cut edge. In all cases, mitoses do not occur at all unless there is a cut edge, so that it seems warranted to conclude that cutting of the ear prior to its incubation is the essential and controlling stimulus for mitosis in mouse ear epidermis cultured *in vitro*.

C. Evidence for "Physiological Subpopulations"

These studies are made on mouse ear epidermis *in vitro* where the physiological environment can be controlled (Section V,A), where mitosis can be specifically initiated by cutting the ear (Section V,B), and where it is known that the epidermal cells which enter mitosis *in vitro* belong to the G_2 population (Section II,B). The question asked is: Do cells in which mitosis has already been initiated in the G_2 period require any additional physiological factor in order to actually move into mitosis? In these experiments, epidermal mitosis has already been initiated by cutting the ear into fragments prior to incubation *in vitro* (as shown in Figs. 8B and 9A). The evaluation of physiological factors involved in mitosis is made by specifically altering the culture medium to enhance the development of epidermal mitosis *in vitro*.

The experiments in Table IV show the effects of glucose and of sodium and potassium ion concentration on the development of epidermal mitosis *in vitro*. The addition of glucose to the standard medium enables epidermal cells to enter mitosis. It would appear that epidermal cells have an additional physiological requirement for mitosis—which is being satisfied by the presence of glucose in the medium. The suggestion that glucose serves as an energy requirement for mitosis has been emphasized in the past, but has been disputed by the present author (for an experimental analysis of this question, see Gelfant, 1960a). Experiments (part B, Table IV) show that the requirement for epidermal mitosis can be satisfied by merely altering the concentration of sodium ions in the standard medium—*in the absence of glucose*. A similar set of results is obtained (part C, Table IV) when the potassium concentration of the standard medium is reduced or KCl eliminated from the standard medium.

It would seem, on first impression, that the physiological requirement for mitosis in stimulated G_2 population epidermal cells can be satisfied either by adding glucose to the standard medium or by altering the Na^+ or the K^+ concentrations in the standard medium alone; second, that all three factors were satisfying the same epidermal cells—but by different

TABLE IV

EFFECTS OF GLUCOSE, Na^+, AND K^+ ON THE DEVELOPMENT OF
MITOSIS IN MOUSE EAR EPIDERMIS in Vitro[a]

A. Effects of glucose

Molar composition of standard medium (SM)		Number of mitoses	
			Standard medium
	Expt.	Alone	+ Glucose
NaCl, 0.123; KCl, 0.0048	1	0.1	5.3
CaCL$_2$, 0.0026; KH$_2$PO$_4$, 0.0036	2	0.1	6.8
MgSO$_4$, 0.001; NaHCO$_3$, 0.003	3	0.2	4.0
Na$_2$HPO$_4$ or Tris buffer, 0.01 (pH 7.4)			

B. Effects of Na^+ concentration

	Number of mitoses in expt.		
SM: NaCl (M)	1	2	3
0.123	0.1	0.2	0.6
0.092	4.0	2.2	2.4
0.082	2.7	1.9	3.3
SM + glucose	6.8	4.3	3.2

C. Effects of K^+ concentration

	Number of mitoses in expt.		
SM: KCl (M)	1	2	3
0.0048	0.1	0.4	0.1
0.0012	2.8	3.1	1.2
0.0	4.1	1.4	2.0
SM + glucose	5.8	4.6	2.4

[a] Five animals are used for each experiment. Each figure represents the average number of mitoses (arrested by Colcemid) per cm unit length of epidermis in 5 ear fragments incubated 4 hours at 38°C. (Adapted from Gelfant, 1963a.)

means. However, on the basis of the experiments shown in Table V, the present author developed an alternative explanation: namely, that those epidermal cells that respond to glucose are a distinct and separate population of cells from those that respond to sodium or potassium concentration changes; and that all three, glucose-, sodium-, and potassium-responding epidermal cells, are different and separate "physiological subpopulations."

The main observations indicating the existence of separate physiological subpopulations are based upon the fact that the effects of glucose and of sodium or potassium ion concentrations are *independent of one another,* and that their individual mitosis-activating effects are *additive.* Phosphate-buffered standard medium is used in the first experiment (Table V). The sodium population is activated by reducing the NaCl concentration from 0.123 M to 0.082 M. The potassium population is activated by omitting KCl from the standard medium (KCl, —). Dissociation of the sodium and the potassium populations is shown by the fact that each can be activated independently of the other. Second, when

TABLE V

DISSOCIATION AND ACCUMULATION OF THE SEPARATE GLUCOSE-, SODIUM ION-
AND POTASSIUM ION-RESPONDING PHYSIOLOGICAL SUBPOPULATIONS
IN MOUSE EAR EPIDERMIS *in Vitro*[a]

Standard medium (SM) (PO₄ or Tris buffer) (NaCl, 0.123 M; KCl, 0.0048 M)	Number of mitoses	
	SM (alone)	SM + glucose (0.002 M)
Experiment 1		
SM (PO₄)	4.4	11.3
SM (NaCl, 0.082 M)	10.7	20.0
SM (KCl,—)	7.2	14.9
SM (NaCl, 0.082 M; KCl,—)	18.6	32.1
Experiment 2		
SM (Tris)	2.6	12.0
SM (NaCl, 0.092 M)	6.2	18.3
SM (KCl,—)	8.8	17.4
SM (NaCl, 0.092 M; KCl,—)	13.7	24.7

[a] Five animals are used for each experiment. Each figure represents the average number of mitoses (arrested by Colcemid) per cm unit length of epidermis in 5 ear fragments incubated 4 hours at 38°C. (Adapted from Gelfant, 1963a.)

they are both activated at the same time (SM: NaCl, 0.082 M; KCl, —) their individual effects on mitosis are cumulative (10.7 + 7.2 = 18.6). The independently activated glucose population (number of mitoses, 11.3) specifically adds to each case of the sodium- and the potassium-activated populations. The last procedure in Experiment 1 shows that all three physiological subpopulations of epidermal cells are separate and independent of one another. The glucose population (11.3) adds to the combined number of epidermal mitoses activated by sodium and potassium alterations (18.6) to produce an accumulative figure of 32.1. The same kinds of result are obtained with Tris-buffered standard medium in Experiment 2. Particularly note the independent contributions of each population when the combined sodium and potassium effects (13.7) are added to the glucose effect (12.0) to produce an accumulation of 24.7 in the over-all number of epidermal cells now activated to enter mitosis.

One can also demonstrate these physiological subpopulations by selective inhibition. For example, G-strophanthin, a cation-transport inhibitor, selectively inhibits the sodium- and the potassium- but not the glucose-activated population. It can be shown that glucose-activated cells are also activated by other hexoses, and therefore represent a generalized sugar-responding epidermal cell population (Gelfant, 1963a).

D. Dissociation of G_1 and G_2 Cell Populations *in Vitro*

Finally, the *in vitro* procedure can be used to dissociate the two major G_1 and G_2 cell populations, as shown in Table VI. Adrenaline and its oxidation product adrenochrome are used for this purpose, and the ex-

TABLE VI

DISSOCIATION OF G_1 AND G_2 CELL POPULATIONS IN
MOUSE EAR EPIDERMIS *in Vitro*[a]

Standard medium (SM)[b]	Number of mitoses	Number of labeled interphases
SM	4.8	111
SM + glucose	11.8	91
SM + adrenaline	0.1	97
SM + adrenochrome	0.1	87
SM + glucose + adrenaline	0.1	87
SM + glucose + adrenochrome	0.1	105

[a] Five animals were used. Each figure represents the average number of mitoses (arrested by Colcemid), or the average number of labeled interphase nuclei per cm unit length of epidermis in 5 ear fragments incubated 4 hours at 38°C.

[b] Concentrations of additions are: glucose, 0.002 M; adrenaline, 10 µg; adrenochrome, 10 µg; and TdR-H³, 5 µc. (Adapted from Gelfant, 1963a.)

periments are carried out on cut ear fragments to stimulate both populations of cells. G_2 cells enter mitosis, are arrested by colchicine *in vitro*, and show up in mitotic counts. G_1 cells enter S *in vitro*, incorporate TdR-H³, and appear as labeled interphases in autoradiographs (see Table III, Section II,B).

Adrenaline is a powerful inhibitor of epidermal mitosis *in vitro* (Gelfant, 1960b). If adrenaline selectively inhibited mitosis (G_2 population) and not DNA synthesis (G_1 population), it would serve to experimentally dissociate between the expression of these two populations *in vitro*. The results of the experiment in Table VI are as predicted. Adrenaline and adrenochrome selectively inhibit epidermal mitosis, but do not inhibit DNA synthesis in the G_1 population *in vitro*.

One final experimental point is brought out in Table VI regarding the development of mitosis in the individual physiological G_2 subpopulations. The fact that adrenaline and adrenochrome equally suppress mitosis in both glucose-activated and nonglucose-activated cell populations indicates that, although the various physiological requirements for mitosis may be cell-specific (i.e., sugar, sodium, or potassium), they all have a common point of action on their specific cell types—as far as mitosis is concerned.

VI. General Conclusions and Over-all Concept of Discrete Cell Populations

Figure 10 depicts and summarizes the main conclusions developed in this report. Although mouse epidermis served as the primary example, the concepts depicted in Fig. 10 apply to other experimental systems (reviewed in Section III). To begin with, two inherent blocks in the cell cycle have been visualized: a G_1 block between the G_1 and S periods, and a G_2 block between G_2 and the period of mitosis. G_1 and G_2 cell populations are defined in relation to these two blocks.

The upper half of the first diagram in Fig. 10 illustrates the behavior of G_1 population cells and the lower part the behavior of G_2 population cells during the cell division cycle in a normal, unstimulated system. Both blocks are partially open for proliferating G_1 cells (i.e., in mouse epidermis). The G_1 block would be closed in nonproliferating G_1 cells (in dormant seeds or in human peripheral leucocytes which do not divide *in vivo*). There is no evidence on the G_1 block with regard to G_2 cells (this is indicated by an appropriate question mark in the diagram). However, there is evidence that G_2 cells are normally detained in the G_2 period of the cell cycle for more than 2 days in mouse epidermis and kidney, for as long as 16 days in tobacco pith tissue *in vitro*, for about 5 or 6 days in human peripheral leucocytes *in vitro*, and for at least 3 days in chicken esophagus *in vivo*. The G_2 block has therefore been closed in the diagram, showing a piling up of G_2 cells in the G_2 period of the cell cycle. The G_2 block would also be closed in nonproliferating G_2 cells (i.e., dormant seeds and human peripheral leucocytes *in vivo*). The evidence that these cells are detained in G_2—in readiness for mitosis— comes from the studies on mouse ear epidermis *in vitro*, mouse and rat kidney *in vivo*, and chicken esophagus *in vivo*, where G_2 cells can be specifically initiated to enter mitosis within an hour after an appropriate stimulus.

The lower diagram in Fig. 10 depicts the situation when an experimental system is stimulated. The main over-all effect of stimulation has to do with the opening of the G_1 and the G_2 cell-division blocks in both major cell populations. The primary effect of proliferative stimulation on G_1 cells is the opening of the G_1 block to the S period of the cell cycle. The secondary effect is the opening of the G_2 block to mitosis. G_1 cells stimulated to enter S (indicated by blackened nuclei in Fig. 10) pass through the S period and through their relatively short G_2 period, and are now also free to enter mitosis because of the opened G_2 block. If experiments are carried out by thymidine-H^3 autoradiography, stimulated G_1 cells appear as labeled mitoses in all experimental systems.

Now for the behavior of stimulated G_2 population cells as depicted in the lower diagram, Fig. 10. Since most of these cells have been held up in the G_2 period of the cycle, the primary effect of proliferative stimulation in this population is the opening of the G_2 block to mitosis. Evidence for the speed at which the G_2 block is opened depends upon the particular experimental system or stimulus used. Examples of a rapid release of the G_2 block are mouse epidermis in vivo after heat shock and in vitro after cutting the ear, mouse and rat kidney after unilateral nephrectomy or ischemic shock, and chicken esophagus after refeeding. If thymidine-H^3 autoradiography is used, G_2 cells appear as unlabeled mitoses, whether they come through rapidly or after a delay, and in all experimental systems.

Superimposed upon the behavior of stimulated G_2 population cells is the question of additional physiological requirements necessary for the actual movement of these cells into mitosis. Although the evidence on this question comes primarily from in vitro studies on mouse ear epidermis, there is strong indication that the same situation exists in cultured tobacco pith tissue—where in addition to a cutting stimulus, kinetin and indoleacetic acid are required for the development of pith cell mitosis in vitro (Das et al., 1956). Stimulated G_2 population mouse ear epidermal cells (in which the G_2 block has been opened by cutting the ear) still require some additional physiological factor in order to enter mitosis in vitro. This requirement is usually satisfied by adding glucose to the standard medium. It was shown, however, that this additional physiological requirement can also be satisfied by altering the sodium or potassium concentrations in the standard medium alone, and, moreover, that the glucose, sodium, and potassium culture-environment changes each activated a separate group of epidermal cells. Therefore, in Fig. 10, the present author postulates the existence of additional G_2 physiological subpopulations which have specific and different requirements for mitosis. No conclusions are drawn regarding the actual physiological significance of

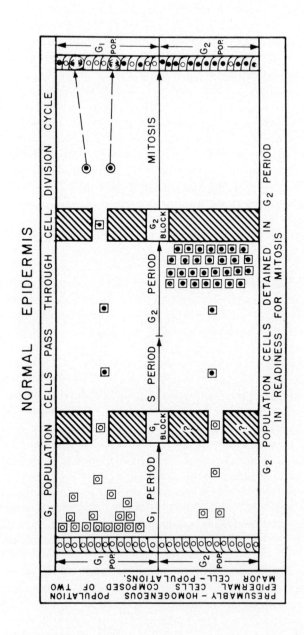

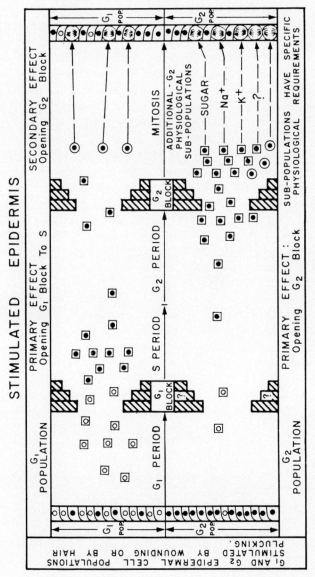

FIG. 10. Concept of discrete cell populations existing within a single layer of epidermal cells in the same tissue (after Gelfant, 1963a).

sugar, in terms of energetics, or of sodium or potassium, in terms of cation transport, and mitosis. These *in vitro* environmental factors have just been used as tools for activating and for discovering the existence of hidden physiological subpopulations.

What is being proposed here is a concept of heterogeneity: the existence within the same tissue of discrete cell populations having different patterns of cell division (i.e., G_1 and G_2 patterns) and even different requirements for mitosis (i.e., physiological subpopulations). G_1 and G_2 population cells exist within the same tissue in a wide variety of animal and plant tissues. Specific "physiological factor-requiring subpopulations" probably exist in mouse epidermis, in tobacco pith tissue, and in other experimental systems as yet unstudied.

It should be emphasized that the G_1 concept of cell division is not unusual. This pattern of behavior during the cell cycle and in response to injury or other stimuli is the generalized pattern of cell division for most animal and plant cells (see Gelfant, 1962, for references). The main contribution of the present report is to show that some cells (usually only a small proportion of the over-all population) have a different pattern of cell division, the G_2 pattern, and that within this G_2 category there may be still further subgroups of cells having different and specific physiological requirements for mitosis.

At this point, one cannot assess the significance of these various cell populations in terms of the over-all proliferative and functional capacities of any tissue in which they reside. The fact that G_2 cells exist in many animal and plant tissues establishes a generalization that requires investigation, and draws attention to the control processes which may be operating during the G_2 period of the cell cycle. It was the purpose of this chapter to describe the experimental procedures that can be used to study some of these problems.

REFERENCES

Argyris, T. S., and Trimble, M. E. (1964). *Anat. Record* **150**, 1.
Avanzi, S., Brunori, A., D'Amato, F , Nuti Ronchi, V., and Scarascia Mugnozza, G. T. (1963). *Caryologia* **16**, 553.
Bender, M. A., and Prescott, D. M. (1962). *Exptl. Cell Res.* **27**, 221.
Bogdanov, Y. F., and Iordanskii, A. B. (1964). *Zh. Obshch. Biol.* **25**, 357 (in Russian).
Bullough, W. S., and Johnson, M. (1951). *Exptl. Cell Res.* **2**, 445.
Bullough, W. S., and Laurence, E. B. (1957). *Brit. J. Exptl. Pathol.* **38**, 273.
Cameron, I. L., and Cleffmann, G. (1964). *J. Cell Biol.* **21**, 169.
Das, N. K., Patau, K., and Skoog, F. (1956). *Physiol. Plantarum* **9**, 640.
De Cosse, J., and Gelfant, S. (1966). *Proc. Am. Assoc. Cancer Res.* **7**, 17.
Feinbrun, N., and Klein, S. (1962). *Plant Cell Physiol.* (*Tokyo*) **3**, 407.
Gandini, E., and Gartler, S. M. (1964). *Nature* **203**, 898.
Gelfant, S. (1959a). *Exptl. Cell Res.* **16**, 527.

Gelfant, S. (1959b). *Exptl. Cell Res.* **18**, 494.

Gelfant, S. (1960a). *Ann. N.Y. Acad. Sci.* **90**, 536.

Gelfant, S. (1960b). *Exptl. Cell Res.* **21**, 603.

Gelfant, S. (1962). *Exptl. Cell Res.* **26**, 395.

Gelfant, S. (1963a). *Symp. Intern. Soc. Cell Biol.* **2**, 229.

Gelfant, S. (1963b). *Exptl. Cell Res.* **32**, 521.

Gelfant, S. (1963c). *Intern. Rev. Cytol.* **14**, 1.

Kopriwa, B. M., and Leblond, C. P. (1962). *J. Histochem. Cytochem.* **10**, 269.

McCreight, C. E., and Sulkin, N. M. (1962). *Am. J. Anat.* **110**, 199.

Nowell, P. C. (1960). *Cancer Res.* **20**, 462.

Nowell, P. C. (1964). *Exptl. Cell Res.* **33**, 445.

Patau, K., and Das, N. K. (1961). *Chromosoma* **11**, 553.

Pederson, T., and Gelfant, S. (1965). In preparation.

Prescott, D. M. (1964). *In* "Methods in Cell Physiology" (D. M. Prescott, ed.), Vol. 1, p. 365–370. Academic Press, New York.

Stein, O. L., and Quastler, H. (1963). *Am. J. Botany* **50**, 1006.

Stöcker, E., Cain, H., and Heine, W. D. (1964). *Naturwiss.* **51**, 195.

Storey, W. F., and Leblond, C. P. (1951). *Ann. N.Y. Acad. Sci.* **53**, 537.

Torrey, J. G. (1961). *Exptl. Cell Res.* **23**, 281.

Wessells, N K. (1964a). *In* "Differentiation and Development," p. 153. Little, Brown, Boston, Massachusetts.

Wessells, N. K. (1964b). *J. Cell Biol.* **20**, 415.

Williams, G. E. G. (1961). *Brit. J. Exptl. Pathol.* **42**, 386.

Chapter 14

Biochemical and Genetic Methods in the Study of Cellular Slime Mold Development[1]

MAURICE SUSSMAN

Department of Biology, Brandeis University, Waltham, Massachusetts

I. Introduction

The conceptual and methodological framework of developmental biology has been radically altered over the past 10 years. Recent monumental advances in microbial genetics and in the chemistry of macromolecular biosynthesis have been applied to problems with developmental import and have made it possible to answer questions previously unforeseen or unapproachable. These questions relate to the programmed

[1] Some of the work reported was performed with the assistance of a grant from the National Science Foundation (GB 1310).

appearance and disappearance of macromolecules that play crucial roles in developmental sequences and to the molecular and genetic bases of the regulatory programs themselves. As a consequence, a new technology has arisen which must be adapted to the peculiarities of each kind of biological material. The purpose of this chapter is to describe genetic, biochemical, and immunochemical techniques that have been successfully applied in our laboratory to the study of cellular slime mold development. It is intended that, by reading this compendium, an investigator previously unexposed to these organisms should be able to maintain and cultivate them and to perform several basic kinds of biochemical experiment without exceeding the background level of frustration. Earlier reviews dealing with the more classical aspects of cellular slime mold biology should also be consulted (Raper, 1951; M. Sussman, 1956). Two recent discussions of the physiology and biochemistry of these organisms (Wright, 1964; Gregg, 1964) are useful sources of methodological information.

II. Maintenance and Preservation of Stocks

A. Some Useful Species and Strains

The cellular slime molds are designated as an order, Acrasiales, within the phylum Myxomycophyta. Several genera and many species have been described. However, the bulk of the current investigation has been carried out with *Dictyostelium discoideum* (Fig. 1). This species has a particularly interesting morphogenetic sequence and constructs fruiting bodies rapidly and well under a wide variety of environmental conditions. Many mutant strains have been isolated which display a wealth of developmental aberrations.

Two related species, *D. mucoroides* and *D. purpureum,* have also been employed extensively. A member of another genus, *Polysphondylium pallidum,* has gained recent prominence as the first cellular slime mold to be cultivated axenically upon defined media (Hohl and Raper, 1963; M. Sussman, 1963). Unfortunately it constructs fruiting bodies more slowly than *D. discoideum* and more fastidiously. However, if morphogenetically suitable variants can be selected, *P. pallidum* may well become the species of choice.

B. Methods of Cultivation

Slime mold amoebae are routinely grown in two-membered culture in association with either *Escherichia coli* or *Aerobacter aerogenes.* The composition of the medium is not critical so long as the bacterial associate can grow rapidly and to high density without raising the level of toxicity above that which inhibits the slime mold fruiting process. Standard medium (SM) in our laboratory contains (in grams per liter) glucose, 10; bactopeptone, 10; yeast extract, 1; $MgSO_4$, 1; KH_2PO_4, 1.5; K_2HPO_4, 1; agar, 20. About 2×10^5 slime mold spores or amoebae and 4–6 drops of 24–48-hour bacterial broth culture are distributed over the agar surface with

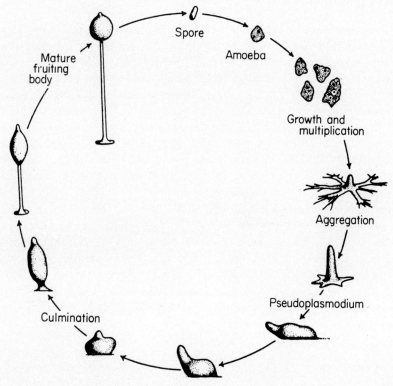

FIG. 1. A schematic diagram of the development of *Dictyostelium discoideum* (Ward, 1959).

a glass spreader. The bacteria initially produce a confluent lawn of cells. Subsequently the amoebae clear the plate of the bacteria and, after reaching the stationary phase, begin fruit construction. At 22°C and under these conditions, *D. discoideum* requires about 10 hours for spore germination and 30–35 hours to reach the stationary phase. The log phase generation time is about 3.5 hours and the total yield of amoebae is about 10^9 per plate (60 mg dry weight). The terminal stages of fruiting are attained after an additional 18–20 hours of incubation. Clonally isolated growth can be achieved by decreasing the inoculum size to less than

150–200 spores or amoebae. The progeny of each inoculated slime mold cell form a separate clear plaque within the bacterial lawn. Each plaque consists of a peripheral ring of vegetative amoebae and an inner zone in which those amoebae no longer in contact with their food supply aggregate and construct fruits. Where considerable cell yields are wanted, large Pyrex cake dishes covered with aluminum foil can be employed. The inocula should be scaled up proportionately to the area to be covered.

Two-membered cultivation can also be accomplished in a broth medium (M. Sussman, 1961) containing (in grams per liter): glucose, 5; bactopeptone, 5; yeast extract, 0.5; $MgSO_4$, 0.5; KH_2PO_4, 2.25; $K_2HPO_4 \cdot 12H_2O$, 1.5. About 5×10^5 slime mold spores or amoebae and 10 ml of A. aerogenes broth culture are inoculated in 100 ml of medium. The cultures are incubated with shaking at $22°C$. The generation time is 3.5 hours and the yield is about 1×10^{10} cells per liter (600 mg dry weight). Alternatively, washed E. coli (10^{10} cells per milliliter) in 0.05 M phosphate buffer pH 6.5 may be employed (Gerisch, 1959). The generation time and yield are comparable to the levels attained in the glucose-peptone broth.

Polysphondylium pallidum strain WS-320 can be grown axenically (Sussman, 1963) in a broth containing (in grams per liter): soybean lecithin, 0.2; lipid-free milk powder ("starlac"), 5; protose peptone, 10; and phosphate buffer, 0.05 M, pH 6.5. The generation time is 3.7 hours at $22°C$. The yield is 2×10^{10} cells per liter. Cultures as large as 5–7 liters have been grown in spinner flasks with good yields. Smaller cultures (500 ml) are incubated in Fernbach flasks with moderate shaking.

C. Preservation of Strains

Spores are easily lyophilized and can be retained indefinitely in this condition (Fennell, 1960). In the past the only strains which presented a problem were mutants that could not form spores. However, the standard methods of freezing in glycerol, previously developed for animal cells, have recently been found to be applicable to slime mold amoebae. Cells are suspended at very high density in cold 5% glycerol. Aliquots of 0.5–1 ml sealed in foil-wrapped ampules are frozen and retained in a liquid nitrogen refrigerator. For recovery, the cells are melted rapidly, diluted into cold water, centrifuged immediately, resuspended in water without glycerol, and plated. Thus far, cultures have been preserved for at least a year under these conditions.

D. Isolation of Mutant Strains

The following kinds of mutant strain have been isolated from *Dic-tyostelium* and *Polysphondylium* species. Some have arisen spontane-ously, others after ultraviolet treatment.

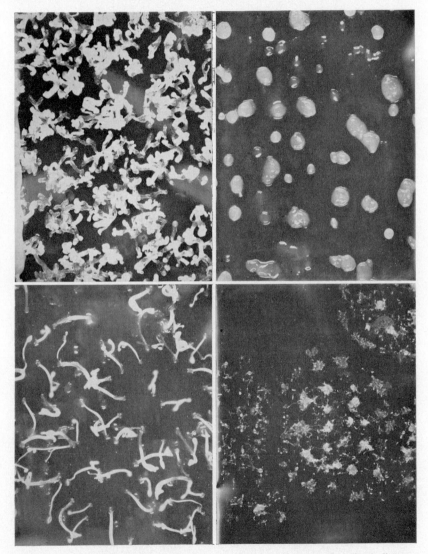

FIG. 2. Four mutant strains. *Upper right,* fruitless; *lower right,* "bushy"; *upper left,* "glassy"; *lower left,* "forked."

MAURICE SUSSMAN

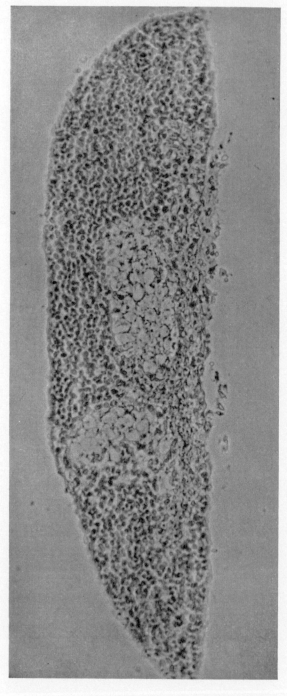

FIG. 3. A section taken through an aggregate of the mutant strain Fr-17 (Sonneborn *et al.*, 1963). Two sections of stalk and the surrounding spores can be observed.

Aggregateless. The amoebae grow normally but fail to aggregate after reaching the stationary growth phase. They retain their vegetative morphology and do not transform into stalk cells or spores (R. R. Sussman and Sussman, 1953; Rafaeli, 1962; Kahn, 1964).

Fruitless. Some fruitless strains develop no further than to construct amorphous or partially completed aggregates; others form tight, normal aggregates before development ceases (Fig. 2); still others can become organized into migrating slugs but do not construct fruiting bodies (R. R. Sussman and Sussman, 1953; Rafaeli, 1962). With the exception of one stock (see below) the fruitless strains thus far encountered do not (by themselves) produce stalk cells or spores. However, it should be noted that certain aggregateless and fruitless strains, when mixed with one another or with the wild type, can construct fruiting bodies synergistically (Kahn, 1964; M. Sussman, 1954; M. Sussman and Lee, 1955; Ennis and Sussman, 1958).

FR-17, a Temporally Deranged Mutant (Sonneborn *et al.*, 1963). This is a mutant of *D. discoideum* which forms flat, amorphous, papillated aggregates ultimately containing typical spores and coiled sections of stalk intermixed in a chaotic disarray (Fig. 3). This terminal stage of development is accomplished in about half the time required by the wild type to construct a mature fruiting body. The sequence of morphogenetic events and synthesis of specific end products detectable by biochemical and immunochemical assays are correspondingly accelerated.

Mutants that Construct Aberrant Fruit Bodies. As shown in Fig. 2, deviations in number and size of fruits ("fruity"), branching ("forked"), shape ("curly"), texture ("glassy"), and pigmentation ("white," "brown") are observed (R. R. Sussman and Sussman, 1953).

Drug-Resistant Strains. In the past many attempts have been made to select mutant strains resistant to drugs, temperature, ultraviolet, etc. Thus far only a few chloramphenicol-resistant mutants have been isolated (H. L. Ennis, unpublished results, 1961). In all other cases (using ultraviolet as the mutagenic agent), any resistant strains that could be isolated were unstable, reverting to the original level of sensitivity during a single passage in the absence of the lethal agent. Thus far, only ultraviolet has been employed as a mutagen. Base analogs, ethyl methyl ether, and nitrosoguanidine have not been investigated.

E. Maintenance of Genetic Constancy

As might be expected, sustained subculture of slime mold stocks without periodic clonal reisolation encourages the accumulation of variant phenotypes and can lead to experimental difficulties. Changes in ploidy offer an additional source of genetic variation including changes in morphogenetic capacity (R. R. Sussman and Sussman, 1961; M. Sussman and Sussman, 1963; M. Sussman, 1964). In *D. discoideum* three classes of strains inheriting different ploidal states can be recognized. These are stable haploid, stable diploid, and metastable. Each can be isolated from the others and reduction to the haplophase has been shown to be accompanied by segregation of genetic markers.

To avoid these sources of variation, it is the practice in our own laboratory to reclone actively carried stocks at 1–2 week intervals.

III. Preparation of Cells for Developmental Studies

Ideally a description of the biochemical events accompanying slime mold morphogenesis should be carried out with single aggregates, slugs, and fruiting bodies at the desired developmental stages. This approach has been exploited to examine changes in Q_{o_2}, dry weight, nitrogen content, reducing power, polysaccharides, sugars, certain enzyme activities, and antigenic reactivity (Gregg, 1950; Gregg and Bronsweig, 1956a,b; Gregg *et al.*, 1954; Takeuchi, 1963). Unfortunately, however, a single aggregate, slug, or fruit contains, at most, about 10^5 cells. The isolation and manipulation of such structures require delicacy and finesse and the small amount of material precludes the application of many analytical techniques whose limits of sensitivity lie above the needed range.

Alternatively, larger populations of amoebae (10^7–10^{10}) have been incubated on solid substrates in the absence of nutrients to initiate fruit construction and harvested at fixed time intervals or at desired developmental stages. Obviously, the results obtained with such samples depend for their pertinence upon the degree to which morphogenetic synchrony is established. The procedure generally followed (White and Sussman, 1961; Wright and Anderson, 1959) is to incubate growth plates until after the bacterial lawn has disappeared but before overt aggregation of the amoebae has begun, harvest the cells in cold water or salt solution (in grams per liter): NaCl, 0.6; KCl, 0.75; CaCl₂, 0.3 (Bonner, 1947), wash 3–4 times in the centrifuge (5 minutes at about $1000 \times g$), suspend the cells at 2–5 $\times$ 10^8 per milliliter, and spread 0.5-ml aliquots on nonnutrient agar plates (5-ml aliquots on Pyrex cake pans) containing streptomycin sulfate at 0.5 gm per liter. This system is somewhat limited by the fact that the cells cannot be spread perfectly evenly and resulting fluctuations in population density disturb morphogenetic synchrony to a small but significant degree. Furthermore agar, even if cleaned by repeated washings, is an ill-defined substrate and contains enough bound nutrients and inorganic ions to permit substantial growth by microbial contaminants. Both drawbacks are eliminated by the substitution of Millipore filters in place of the agar (M. Sussman and Lovgren, 1965). The washed cells are dispersed in 0.5-ml aliquots on 2-inch black Millipore filters resting on absorbent support pads, saturated with 2 ml salt solution (in milligrams per milliliter): KCl, 1.5; $MgCl_2$, 0.5; streptomycin sulfate, 0.5—phosphate buffer (0.05 M, pH 6) may be added if desired— and contained in 60-mm plastic Petri dishes. The dishes are then stored in a moist atmosphere. Under these conditions, morphogenetic synchrony is excellent and, in the case of *D. discoideum*, the entire sequence is completed in 24 hours, 8 hours sooner on Millipore than on agar. The cells are harvested by placing the Millipore on the lip of a centrifuge tube and squirting water or buffer at it with syringe or pipette. At any stage of development, the cells can be exposed to isotopically labeled compounds, enzyme poisons, etc., by adding the reagent to the fluid saturating a support pad and then shifting the Millipore to the new pad. Removal of the reagent can be accomplished by a second shift of the Millipore. Under these conditions, for example, the incorporation of uridine-H^3 or C^{14}-amino acids is constant from the time of the shift; the inhibition of protein synthesis becomes apparent immediately after addition of the antibiotic cycloheximide, and disappears within a few minutes after its removal.

Cell aggregation can also be carried out in liquid medium (Gerisch, 1960) in a manner similar to that originally employed for the cells of vertebrate embryos (Moscona and Moscona, 1952). Unfortunately, no further development occurs under these conditions but the system is useful for the analysis of aggregation per se.

A. Preparation of Extracts

All developmental stages including mature fruits can be satisfactorily homogenized by two passages through a French pressure cell. Microscopic examination reveals complete cell breakage and about 90% of the total protein remains in the supernatant after a 10,000 × g centrifugation. Assays of enzyme activity and serological reactivity in replicate samples display very good agreement. The pressure cell can be used on about 10 samples (3–5 ml) before rechilling becomes necessary. The Bramson sonifier (with micro tip) has also been used with excellent results in our laboratory. Samples as small as 3 ml can be employed and breakage occurs at low intensity (level 2) and short exposure (about 45 seconds). Mature fruits are more resistant than earlier developmental stages and require higher intensity. Although sonication is slightly less reproducible than the pressure cell, it is much more convenient and has become our method of choice. For experiments requiring gentler treatment (isolation of nuclear histones, polyribosomes, etc.), the Potter-Elvehjem grinder has been successfully employed. Total and immediate lysis can be achieved by treatment with 0.5% sodium dodecyl sulfate, useful for the isolation of ribonucleic acid (RNA) (R. R. Sussman, unpublished data, 1965), or 0.5% deoxycholate for isolation of polyribosome complexes (Phillips *et al.*, 1964).

In preparing extracts for enzymatic or serological assays, it is important to remember that the slime mold amoebae are meat eaters and as a consequence contain considerable amounts of proteolytic and nucleolytic enzymes. More important, these enzymes decrease greatly during morphogenesis, reflecting the over-all loss of cell protein (Gregg and Bronsweig, 1956b; White and Sussman, 1961). Thus the macromolecular composition of extracts from early developmental stages can be drastically modified under conditions of treatment and incubation which leave later extracts unaffected. In the past, dramatic increases in a variety of enzyme activities during development (Wright and Anderson, 1958, 1959) were later shown to be due to their differential stability (Wright, 1960). Fortunately, however, if assays are performed immediately after cell breakage and the reaction mixtures are incubated for reasonably short times (about 1 hour), this source of uncertainty is eliminated (Wright, 1960; M. Sussman and Osborn, 1964). The general instability of RNA during fractionation procedures is likewise a serious problem, but the presence of sodium dodecyl surface (polyvinyl sulfate for ribosomal separations), treatment with bentonite, and extraction at low temperatures obviate the difficulty here (R. R. Sussman, unpublished data, 1965).

B. The Problem of Bacterial Contamination

Dictyostelium discoideum grown on SM agar with *A. aerogenes*, harvested, washed by centrifugation, and resuspended at 2×10^8 amebae per milliliter is usually contaminated by about 10^5 per milliliter bacteria (by viable count). The bulk of these are ingested and destroyed by the amoebae within 1–2 hours after deposition on washed agar or Millipore filters. Nevertheless equivocal results may be obtained if one is measuring a metabolic capacity common to both organisms (RNA synthesis, pool turnover, etc.) over long periods of incubation. This problem can be avoided by harvesting (centrifuging and resuspending) the amoebae in SM broth containing streptomycin (0.5 gm per liter) and incubating them (2×10^7 amoebae per milliliter) for 2 hours at 22°C with moderate shaking (Inselburg, 1965). Under these conditions the bacterial count is reduced to less than 10^3 per milliliter and their contribution to the observed metabolic alterations becomes negligible. Two kinds of control have been employed in this laboratory to rule out suspected bacterial contributions. Incubation of the amoeboid population at 37°C for 30–60 minutes kills the amoebae. Subsequent incubation at 22°C in the presence of labeled nucleosides or amino acids and examination of incorporation into the trichloroacetic acid- (TCA) insoluble fraction can then serve as a measure of contaminating bacteria. In studies of RNA metabolism the RNA has been purified from cells pulsed with $P^{32}O_4$ at various stages and then chromotographed on methylated albumin-Kieselguhr (MAK) columns (Inselburg, 1965). The elution pattern of bacterial ribosomal RNA is significantly different from the slime mold counterpart (which resembles animal cells in this respect), and the distribution of isotope could thereby serve as a sensitive criterion for the presence of contamination.

IV. Isolation of Nucleic Acids

Dictyostelium discoideum deoxyribonucleic acid (DNA) is especially interesting in view of its extremely low guanine + cytosine content (about 20%). It has been isolated (Schildkraut *et al.*, 1962) and employed as a template for *E. coli* RNA polymerase (Hurwitz *et al.*, 1963). The standard phenol extraction–ethanol precipitation procedure was used for isolation. However, because of the small quantity of DNA per cell and the presence of a low level of contaminating DNA from the bacterial associate, the procedure included preliminary fractionation in a CsCl gradient (Schildkraut *et al.*, 1962). RNA has also been isolated by standard phenol extraction–ethanol precipitation procedures and further purified in sucrose gradients and on MAK columes (R. R. Sussman, unpublished data, 1965; Inselburg, 1965). No particular technical problems were encountered excepting those mentioned (i.e., the presence of nucleases and the danger of contamination with RNA from the bacterial associate).

V. Analyses of Cell Constituents

The following is an annotated bibliography listing sources of useful information on analytical methods that have recently been employed in the study of slime mold development.

Serology. A considerable number of antigenic determinants have been examined by quantitative complement fixation (Sonneborn *et al.*, 1964, 1965), agglutination assays (Sonneborn *et al.*, 1965; Gregg and Triggsted, 1958; Gregg, 1956), agar double diffusion (Sonneborn *et al.*, 1964, 1965; Gregg and Triggsted, 1958; Gregg, 1960), immunophoresis (Gregg, 1961), and application of fluorescent labeling techniques (Takeuchi, 1963).

Major cell components. Both static and developmental studies have been made of standard cell fractions, including dry weight (Gregg and Bronsweig, 1956a; White and Sussman, 1961), protein (Gregg *et al.*, 1954; White and Sussman, 1961), polysaccharides (Gezelius and Ranby, 1957; Mühlethaler, 1956; White and Sussman, 1963a,b), disaccharides (Clegg and Filosa, 1961), simple sugars (Gregg and Bronsweig, 1956b; White and Sussman, 1961), lipids (Davidoff and Korn, 1963a), steroids (Heftmann *et al.*, 1959), RNA (White and Sussman, 1961), and amino acid pool composition and replenishment turnover (Wright and Anderson, 1960a,b; Krivanek and Krivanek, 1959).

Intermediary metabolism. These include a survey of respiratory pathways (Wright, 1963; Liddel and Wright, 1961), glucose catabolism (Liddel and Wright, 1961; Wright and Bloom, 1961), glutamate oxidation (Wright and Bard, 1963; Brühmüller and Wright, 1963), histidine metabolism (Krichevsky and Love, 1964), and the biogenesis of lipids (Davidoff and Korn, 1963b), and steroids (Johnson *et al.*, 1962).

Enzyme studies. The developmental kinetics of a variety of enzymes have been examined, including succinic dehydrogenase and cytochrome oxidase (Takeuchi, 1960), esterases and acid phosphatases (Solomon *et al.*, 1964), alkaline phosphatases (Krivanek, 1956; Gezelius and Wright, 1965), glucose-6-phosphate dehydrogenase and isocitric dehydrogenase (Wright, 1960), UDPG cellulose transferase (Wright, 1965), and UDP-Gal polysaccharide transferase (M. Sussman, 1965; M. Sussman and Sussman, 1965; M. Sussman and Osborn, 1964).

REFERENCES

Bonner, J. T. (1947). *J. Exptl. Zool.* **106**, 1.
Brühmüller, M., and Wright, B. E. (1963). *Biochim. Biophys. Acta* **71**, 50.
Clegg, V. S., and Filosa, M. F. (1961). *Nature* **192**, 1077.
Davidoff, F., and Korn, E. D. (1963a). *J. Biol. Chem.* **238**, 3199.
Davidoff, F., and Korn, E. D. (1963b). *J. Biol. Chem.* **238**, 3210.
Ennis, H. L., and Sussman, M. (1958). *J. Gen. Microbiol.* **18**, 433.
Fennell, D. I. (1960). *Botan. Rev.* **26**, 79.
Gerisch, G. (1959). *Naturwiss.* **46**, 654.
Gerisch, G. (1960). *Arch. Entwicklungsmech. Organ.* **152**, 632.
Gezelius, K., and Ranby, G. G. (1957). *Exptl. Cell Res.* **12**, 265.
Gezelius, K., and Wright, B. E. (1965). *J. Gen. Microbiol.* **38**, 309.

Gregg, J. H. (1950). *J. Exptl. Zool.* **114**, 173.
Gregg, J. H. (1956). *J. Gen. Physiol.* **39**, 813.
Gregg, J. H. (1960). *Biol. Bull.* **118**, 70.
Gregg, J. H. (1961). *Develop. Biol.* **3**, 757.
Gregg, J. H. (1964). *Physiol. Rev.* **44**, 631.
Gregg, J. H., and Bronsweig, R. D. (1956a). *J. Cellular Comp. Physiol.* **47**, 483.
Gregg, J. H., and Bronsweig, R. D. (1956b). *J. Cellular Comp. Physiol.* **48**, 293.
Gregg, J. H., and Triggsted, C. W. (1958). *Exptl. Cell Res.* **15**, 358.
Gregg, J. H., Hackney, A., and Krivanek, R. C. (1954). *J. Cellular Comp. Physiol.* **107**, 226.
Heftmann, E., Wright, B. E., and Liddel, G. U. (1959). *J. Am. Chem. Soc.* **81**, 6525.
Hohl, H., and Raper, K. B. (1963). *J. Bacteriol.* **85**, 199.
Hurwitz, J., Evans, A., Babinet, A., and Skalka, A. S. (1963). *Cold Spring Harbor Symp. Quant. Biol.* **28**, 59.
Inselburg, J. (1965). In manuscript.
Johnson, D. F., Wright, B. E., and Heftmann, E. (1962). *Arch. Biochem. Biophys.* **97**, 232.
Kahn, A. J. (1964). *Develop. Biol.* **9**, 1.
Krichevsky, M. I., and Love, L. I. (1964). *J. Gen. Microbiol.* **34**, 483.
Krivanek, J. O. (1956). *J. Exptl. Zool.* **133**, 459.
Krivanek, J. O., and Krivanek, R. C. (1959). *Biol. Bull.* **116**, 265.
Liddel, G. U., and Wright, B. E. (1961). *Develop. Biol.* **3**, 265.
Moscona, A., and Moscona, M. (1952). *J. Anat. (London)* **86**, 287.
Mühlethaler, K. (1956). *Am. J. Botany* **43**, 673.
Phillips, W. D., Rich, A., and Sussman, R. R. (1964). *Biochim. Biophys. Acta* **80**, 508.
Rafaeli, D. E. (1962). *Bull. Torrey Botan. Club* **89**, 312.
Raper, K. B. (1951). *Quart. Rev. Biol.* **26**, 169.
Schildkraut, C. L., Mandel, M., Levisohn, S., Smith, J. E., Sonneborn, D. R., and Marmur, J. (1962). *Nature* **196**, 795.
Solomon, E. P., Johnson, E. M., and Gregg, J. H. (1964). *Develop. Biol.* **9**, 314.
Sonneborn, D. R., Sussman, M., and Levine, L. (1964). *J. Bacteriol.* **87**, 1321.
Sonneborn, D. R., Levine, L., and Sussman, M. (1965). *J. Bacteriol.* **89**, 1092.
Sonneborn, D. R., White, G. J., and Sussman, M. (1963). *Develop. Biol.* **7**, 79.
Sussman, M. (1954). *J. Gen. Microbiol.* **10**, 110.
Sussman, M. (1956). *Ann. Rev. Microbiol.* **10**, 21.
Sussman, M. (1961). *J. Gen. Microbiol.* **25**, 375.
Sussman, M. (1963). *Science* **193**, 338.
Sussman, M. (1964). *Nature* **201**, 216.
Sussman, M. (1965). *Biochem. Biophys. Res. Commun.* **18**, 763.
Sussman, M., and Lee, F. (1955). *Proc. Natl. Acad. Sci. U.S.* **41**, 70.
Sussman, M., and Lovgren, N. (1965). *Exptl. Cell Res.* **38**, 97.
Sussman, M., and Osborn, M. J. (1964). *Proc. Natl. Acad. Sci. U.S.* **52**, 81.
Sussman, M., and Sussman, R. R. (1963). *J. Gen. Microbiol.* **30**, 349.
Sussman, M., and Sussman, R. R. (1965). *Biochim. Biophys. Acta* **108**, 463.
Sussman, R. R., and Sussman, M. (1953). *Ann. N.Y. Acad. Sci.* **86**, 949.
Sussman, R. R., and Sussman, M. (1961). *J. Gen. Microbiol.* **28**, 417.
Takeuchi, I. (1960). *Develop. Biol.* **2**, 343.
Takeuchi, I. (1963). *Develop. Biol.* **8**, 1.
Ward, J. M. (1959). *Proc. 4th Intern. Congr. Biochem., Vienna, 1958* Vol. 6, pp. 1-26. Pergamon Press, Oxford.

White, G. J., and Sussman, M. (1961). *Biochim. Biophys. Acta* **53**, 285.
White, G. J., and Sussman, M. (1963a). *Biochim. Biophys. Acta* **74**, 173.
White, G. J., and Sussman, M. (1963b). *Biochim. Biophys. Acta* **74**, 179.
Wright, B. E. (1960). *Proc. Natl. Acad. Sci. U.S.* **46**, 798.
Wright, B. E. (1963). *Ann. N.Y. Acad. Sci.* **102**, 740.
Wright, B. E. (1964). *In* "Biochemistry and Physiology of Protozoa" (S. H. Hutner, ed.), Vol. 3, p. 270. Academic Press, New York.
Wright, B. E. (1965). In manuscript.
Wright, B. E., and Anderson, M. L. (1958). *In* "Chemical Basis of Development" (W. D. McElroy and B. Glass, eds.), p. 108. Johns Hopkins Press, Baltimore, Maryland.
Wright, B. E., and Anderson, M. L. (1959). *Biochim. Biophys. Acta* **31**, 310.
Wright, B. E., and Anderson, M. L. (1960a). *Biochim. Biophys. Acta* **43**, 62.
Wright, B. E., and Anderson, M. L. (1960b). *Biochim. Biophys. Acta* **43**, 67.
Wright, B. E., and Bard, S. (1963). *Biochim. Biophys. Acta* **71**, 45.
Wright, B. E., and Bloom, B. (1961). *Biochim. Biophys. Acta* **48**, 342.

Author Index

Numbers in italics indicate the pages on which the complete references are listed.

411

416 AUTHOR INDEX

Padilla, G. M., 219, 227
Painter, R. B., 329, 330, 356
Palade, G. E., 268, 309
Pantelouris, E. M., 18, 19, 36, 42, 60
Pappas, G. D., 307, 309
Patau, K., 371, 391, 394, 395
Pavan, C., 42, 59
Pearson, H. E., 319, 320
Pease, D. C., 287, 309
Pedersen, K. J., 280, 309
Pederson, T., 375, 395
Pekarek, J., 202, 214
Pelc, S. R., 231, 240, 249, 253, 255, 309, 326, 356
Pelling, C., 86, 90, 92
Perdue, S. W., 312, 321
Pereira, J. P. M., 348, 349, 353, 356
Perez-Silva, P., 66, 91
Perrota, C. A., 332, 356
Perry, R. P., 331, 356
Peters, H., 278, 309
Pfeffer, W., 144, 145, 162, 214
Pfendt, E., 101, 111
Phillips, W. D., 406, 409
Picheral, B., 23, 27, 35, 36
Pilgrim, C., 331, 332, 333, 345, 347, 356
Pirson, A., 154, 197, 214
Plaut, W., 255, 309
Porter, K. R., 3, 36
Potter, R. L., 329, 356
Potter, V. R., 329, 356
Prát, S., 160, 197, 207, 214, 215
Prescott, D. M., 114, 130, 256, 270, 309, 330, 356, 363, 375, 394, 395
Price, C. A., 223, 227
Pringsheim, E. G., 217, 227
Pringsheim, N., 145, 215
Pringsheim, O., 217, 227
Prud'homme van Reine, W. J., Jr., 196, 197, 215
Puck, T. T., 94, 111

Quastler, H., 299, 309, 329, 330, 331, 333, 334, 340, 341, 345, 356, 357, 374, 395

Rafaeli, D. E., 403, 409
Ramsay, I. A., 168, 215
Ranby, G. G., 408, 408
Rangan, S. R. S., 321

Raper, K. B., 398, 409
Rasmussen, R. E., 330, 356
Rauber, A., 1, 36
Rdzok, E. J., 275, 280, 310
Rege, D. V., 224, 227
Reichard, P., 329, 357
Repp, G., 166, 215
Resühr, B., 181, 182, 215
Reuter, L., 197, 215
Revel, J. P., 230, 252, 253, 257, 286, 296, 301, 309
Rey, V., 75, 76, 92
Reynolds, E. S., 243, 245, 253, 305, 309
Rich, A., 406, 409
Richardson, K. C., 279, 281, 309, 312, 321
Ris, H., 38, 54, 56, 60
Rittenberg, S. C., 223, 228
Robbins, E., 101, 104, 111, 318, 321
Robbins, W. J., 223, 228
Robert, M., 86, 92
Robinson, C. F., 279, 309
Rogers, M. E., 42, 60
Rolshoven, E., 351, 357
Romijn, C., 197, 215
Rosen, S. I., 318, 319, 321
Ross, G. I. M., 223, 227
Rostand, J., 1, 36
Rouiller, C., 117, 130
Rubini, J. R., 329, 357
Rückert, J., 37, 60
Runge, J., 258, 276, 309
Ryter, A., 268, 309, 312, 316, 320

Salpeter, M. M., 230, 231, 232, 233, 235, 236, 237, 239, 240, 242, 243, 245, 246, 249, 251, 252, 253
Sambuichi, H., 12, 36
Samuel, D. M., 312, 321
Sandell, E. B., 221, 228
Sankewitsch, E., 199, 213
Sapranauskas, P., 317, 320
Scarascia Mugnozza, G. T., 373, 394
Scarth, G. W., 182, 195, 215
Schaefer, G., 154, 195, 197, 198, 199, 200, 201, 202, 203, 214, 215
Schiff, J. A., 223, 227
Schildkraut, C. L., 127, 130, 407, 409
Schimper, A. F. W., 206, 215
Schmidt, E. L., 72, 92

Schmidt, H., 199, *215*
Schneider, E., 209, *215*
Schoenberg, M. D., 278, *309*
Schor, N. A., 67, 75, 76, 92
Schultz, J., 1, *36*
Schwalbach, G., 280, *310*
Sechaud, J., 312, 316, *320*
Seidel, F., 25, *36*
Sekiyama, S., 278, *310*
Sheffield, H. G., 318, *321*
Shelokov, A., 101, *111*
Sherman, F. G., 329, 340, 341, 345, *356*
Shumway, W., 6, 7, 15, *36*
Signoret, J., 19, 20, 22, 23, 24, 27, 28, 34, *36*
Silk, M. H., 257, 298, 299, 301, *310*
Simmel, E. B., 330, 331, *355*
Sirlin, J. L., 67, 75, 76, 85, *91*, 92
Sisken, J. E., 346, *357*
Sitte, P., 144, *215*
Skalka, A. S., 407, *409*
Sketon, F. R., 283, *309*
Skoog, F., 391, *394*
Slatyer, R. O., 147, 148, *215*
Slizynski, B. M., 90, 92
Smale, N. B., 278, 281, 282, 283, *309*
Smith, J. E., 407, *409*
Smith, L. D., 28, *36*
Smith, S., 18, 32, *34*
Smuckler, E. A., 282, *310*
Sobotka, H., 218, 219, *227*
Solomon, E. P., 408, *409*
Somers, C. E., 114, *130*
Sonneborn, D. R., 402, 403, 407, 408, *409*
Spanner, D. C., 147, *215*
Sparvoli, E., 318, *321*
Spemann, H., 1, 25, *36*
Spence, I. M., 257, 298, 299, 301, *310*
Spendlove, R., 101, *111*
Sprinz, H., 351, *355*
Spurway, H., 43, *58*
Stadelmann, E., 144, 145, 146, 151, 152, 155, 161, 179, 180, 181, 185, 189, 190, 191, 192, 193, 194, 195, 197, 198, 204, 205, 206, 209, *215*
Staub, M., 73, *91*
Stebbins, M. E., 223, *228*
Steel, G. G., 355, *355*
Stein, O. L., 374, *395*

Steinberg, M., 6, 14, *36*
Stern, H., 329, *356*
Stevens, A. R., 266, *310*
Stevens, C. E., 324, *356*
Stewart, D. R., 181, 182, *214*
Stocker, O., 199, *215*
Stocking, C. R., 170, *213*
Stöcker, E., 378, *395*
Stone, G. E., 132, *142*, 263, *310*, 312, *321*, 328, *357*
Stone, R. S., 277, *308*
Storey, W. F., 346, *357*, 379, *395*
Storm, J., 218, 221, *227*
Stroeva, O. G., 12, *36*
Subtelny, S., 12, 24, 27, 28, 30, *36*
Sulkin, N. M., 375, *395*
Sussman, M., 398, 400, 402, 403, 404, 405, 406, 408, *409*
Sussman, R. R., 403, 404, 406, *409*
Sutton, J. S., 318, *321*
Suzuki, T., 278, *310*

Takada, H., 203, *214*
Takamine, N., 196, 197, 198, *215*
Takeuchi, I., 404, 408, *409*
Tamiya, H., 151, *215*
Tartar, V., 2, *36*
Taylor, J. H., 256, *310*
Taylor, S. A., 147, 148, *215*
Thoenes, W. Z., 279, 281, *310*
Thöni, H., 168, *215*
Thrasher, J. D., 330, 332, 333, 345, 346, 348, 349, 350, *356, 357*
Tolmach, L. J., 94, 96, 103, *111*
Tomlin, S. G., 54, *60*
Tonna, E. A., 337, *357*
Torrey, J. G., 374, *395*
Triggsted, C. W., 408, *409*
Trimble, M. E., 375, *394*
Trump, B. F., 282, 284, *309, 310*
Twitty, V. C., 6, *36*
Tzitsikas, H., 274, 275, 280, *310*

Unger, F., 151, *215*
Ursprung, A., 144, 145, 155, 156, 161, 163, 167, 168, 169, 173, 174, 175, 177, 195, *215, 216*
Ursprung, H., 12, 31, *35, 36*

Vallee, B. L., 223, *227*

Subject Index